RECENT ADVANCES IN
ANIMAL NUTRITION — 1989

Recent Advances in Animal Nutrition

1989

W. Haresign, PhD

D. J. A. Cole, PhD
University of Nottingham School of Agriculture

BUTTERWORTHS
London   Boston   Singapore   Sydney   Toronto   Wellington

First published 1989

© **The several contributors named in the list of contents, 1989**

British Library Cataloguing in Publication Data

Recent advances in animal nutrition.—(Studies in
  the agricultural and food sciences).—1989
  ·1. Animal nutrition
  I. Series
  636.08′52   SF95

ISBN 0-408-041498

Photoset by Butterworths Litho Preparation Department
Printed and bound in Great Britain by Courier International Ltd., Tiptree, Essex

PREFACE

This volume is based on the Twenty-third University of Nottingham Feed Manufacturers Conference and considers a range of issues important to the feed industry.

The first section addresses various aspects of legislation relevant to the feed compounder and producers of livestock products. There is a plethora of legislative changes with which the feed industry must comply. These include changes and rationalization in the detailed labelling of animal feeds, new regulations relating to maximum aflatoxin levels and pesticide residues in feeds, and the compulsory registration of feed mills designed to control the sale and distribution of medicated feed. All of these issues are clearly discussed in the first chapter. The presence of drug residues in animal products is a topic of considerable concern to consumers, and this is considered in the second chapter. Information is provided on the way in which the licensing system for veterinary medicinal products is designed to ensure consumer safety. Results from statutory monitoring for residues in animal products illustrates the responsible attitude of the industry in observing the correct withdrawal periods. The third chapter in this section highlights the opportunities for the UK feed compounding industry from the removal of trade barriers and the creation of a single European market in 1992.

The first chapter in the section on poultry and rabbit nutrition addresses the issue of vitamin requirements of chickens and turkeys. The working party considered information available before and since the 1974 ARC Review, but reassessed it on the basis of dose-response relationships to provide new guidelines on requirements, and their translation into suggested dietary allowances for the different classes of poultry. A second chapter on poultry nutrition indicates why pelleting of the feed given to meat birds usually improves performance, but emphasizes the need to maintain pellet quality and to supply a diet with adequate protein to ensure that any additional productive energy will be directed towards lean rather than fat deposition. The final chapter in this section highlights the special features of digestion in the rabbit and seeks to provide guidelines on diet composition designed to maintain a high level of performance while minimizing digestive upsets.

Silage represents a major feed input in many systems of ruminant production, but rationing systems in the past have had to contend with errors in predicting its nutritive value. As illustrated in the first chapter on ruminant nutrition, the advent of the NIR technique offers the promise of more accurate estimates of nutritive value and hence more precise rationing systems. A further chapter in this section indicates that silage additives can improve animal performance, especially when fermentation characteristics are less than ideal, but that wilting of silage often reduces digestibility and thus animal performance. The formulation of compounds designed to complement high intakes of silage aim to minimize substitution rate yet encourage efficient microbial utilization of nitrogen in the rumen, and a chapter is devoted to a consideration of how this might be achieved. In recent years there has been increased interest in systems of early lamb production in which the lambs are early weaned and finished indoors on all-concentrate diets. It is fitting that the final chapter in the section on ruminant nutrition should concentrate on various aspects of the nutrition of intensively reared lambs.

The final section is devoted to non-ruminant nutrition. The first chapter tackles the difficult area, often overlooked in the past, of electrolyte balance and its importance in animal nutrition. A further chapter describes the use of a factorial procedure to estimate nutrient partitioning in the sow and suckling piglet during lactation, and then uses this information to build a robust computer model to predict the outcome of different feeding strategies on sow productivity. Information on the design and interpretation of experiments to determine amino acid requirements of pigs and poultry is covered in another chapter, together with information on how to translate requirements into allowances. The final chapter considers the aetiology of diarrhoea in pigs and pre-ruminant calves, and indicates that the problem is complex and can have a number of different causes, but suggests that a sound management strategy for preventing diarrhoea is to formulate weaner diets with highly digestible nutrients, preferably with a composition as close as possible to that of whole milk.

The organizers and the University of Nottingham are grateful to BP Nutrition (UK) Ltd for their support in the organization of this conference.

W. Haresign
D. J. A. Cole

CONTENTS

Preface   v

I   **Legislation**   **1**

1   PRESENT AND FUTURE LEGISLATION AND ITS IMPLICATION
   FOR THE FEED COMPOUNDER   3
   D.R. Williams, *BOCM Silcock Ltd, Selby, Yorks YO8 7DT, UK*

2   RESIDUES OF VETERINARY DRUGS IN ANIMAL PRODUCTS   13
   A.R. Peters, *Hoechst UK Ltd, Walton Manor, Milton Keynes, UK*

3   IMPACT OF EEC LEGISLATION AND REMOVAL OF TRADE
   BARRIERS ON THE UK ANIMAL FEED INDUSTRY   27
   B. Rutherford, c/o *UKASTA, 3 Whitehall Court, London, UK*

II   **Poultry and Rabbit Nutrition**   **33**

4   VITAMIN REQUIREMENTS AND ALLOWANCES FOR POULTRY   35
   C.C. Whitehead, *AFRC Institute for Grassland and Animal Production,
   Poultry Department, Roslin, Midlothian EH25 9PS* and J.I.
   Portsmouth, *Peter Hand (GB) Ltd, Stanmore, Middlesex HA7 4AR,
   UK*

5   EFFECT OF PELLET QUALITY ON THE PERFORMANCE OF
   MEAT BIRDS   87
   E.T. Moran, Jr, *Poultry Science Department and Alabama Agricultural
   Experiment Station, Auburn University, AL 36849–5416, USA*

6   NUTRITION OF RABBITS   109
   G. Santomá, J.C. de Blas, R. Carabaño and M.J. Fraga, *Departamento
   de Producción Animal, Universidad Politécnica de Madrid, Spain*

III Ruminant Nutrition   **139**

  7 PREDICTING THE NUTRITIVE VALUE OF SILAGE   **141**
      G.D. Barber and N.W. Offer, *West of Scotland College, Auchincruive,*
      Ayr and D.I. Givens, *ADAS Feed Evaluation Unit, Alcester Road,*
      Stratford-upon-Avon, UK

  8 EFFECT OF SILAGE ADDITIVES AND WILTING ON ANIMAL
      PERFORMANCE   **159**
      F.J. Gordon, *Agricultural Research Institute of Northern Ireland,*
      Hillsborough, Co. Down, UK

  9 OPTIMIZING COMPOUND FEED USE IN DAIRY COWS WITH
      HIGH INTAKES OF SILAGE   **175**
      D.G. Chamberlain, P.A. Martin and S. Robertson, *The Hannah*
      Research Institute, Ayr KA6 5HL, UK

 10 NUTRITION OF INTENSIVELY REARED LAMBS   **195**
      R. Jones, *NAC Sheep Unit, Stoneleigh, Warks,* R. Knight, *BP Nutrition*
      (UK) Ltd, Wincham, Cheshire and A. White, *Avondale Veterinary*
      Group, Warwick, UK

IV Non-Ruminant Nutrition   **209**

 11 THE PHYSIOLOGICAL BASIS OF ELECTROLYTES IN ANIMAL
      NUTRITION   **211**
      J.F. Patience, *Prairie Swine Centre, University of Saskatchewan,*
      Saskatoon, Saskatchewan, Canada

 12 PREDICTING NUTRIENT RESPONSES OF THE LACTATING SOW   **229**
      B.P. Mullan, W.H. Close, *AFRC Institute for Grassland and Animal*
      Production, Pig Department, Church Lane, Shinfield, Reading RG2
      9AQ and D.J.A. Cole, *University of Nottingham School of Agriculture,*
      Sutton Bonington, Loughborough, Leics LE12 5RD, UK

 13 AMINO ACID NUTRITION OF PIGS AND POULTRY   **245**
      D.H. Baker, *Department of Animal Sciences, University of Illinois,*
      Urbana, Illinois, USA

 14 AETIOLOGY OF DIARRHOEA IN PIGS AND PRE-RUMINANT
      CALVES   **261**
      J.W. Sissons, *AFRC Institute for Grassland and Animal Production,*
      Shinfield, Reading, Berks, UK

      LIST OF PARTICIPANTS   **283**

      INDEX   **295**

I

Legislation

PRESENT AND FUTURE LEGISLATION AND ITS IMPLICATION FOR THE FEED COMPOUNDER

D. R. WILLIAMS
BOCM Silcock Ltd, Selby, Yorks YO8 7DT, UK

New legislation covering the manufacture, sale and distribution of feeds was introduced in the UK and EEC in 1988 and more will be agreed in early 1989 for implementation in 1989/90. This chapter describes the most important aspects which require feed manufacturers to introduce new practices in the mills and additional warranties and information on feed labels.

Areas which are addressed are the marketing of feeds, medicated feeds, feed additives, and health and safety at work regulations.

Most of the legislation which is specific to feeds originates in the EEC, and the 1992 target date for the removal of trade barriers is encouraging the legislators to introduce regulations for implementing harmonized feed laws by that date. The clear implications of this are that feed manufacturers will be faced with the costs of complying, and the commitment of management time to ensure compliance.

Marketing of feeds

The EEC have proposed further substantial changes for harmonizing feed labels. When the original 'marketing' directive was adopted in 1979 (79/373) it contained several derogations permitting member states to exercise options for declaring some the constituents of feeds on labels. It was considered that this has given rise to barriers in the free movement of feeds in the Community. The new proposals, listed below, effectively cancel the derogations. The Commission has tried to balance a number of requirements: the need to provide the livestock producer with meaningful information which is accurate; the need for the compounder to provide objective information whilst safeguarding formulation expertise arising from costly research and development investments; and also the need, in the public interest, to ensure that declarations can be adequately policed. The following are the principal changes which are likely to be adopted in 1989.

DECLARATION OF INGREDIENTS

By October 1988 no clear decision had been made, but it is probable that compulsory listing of ingredients by category will be agreed. The European feed trade interests have considered possibilities and many, including UKASTA, favour the listing of categories, in decreasing order of inclusion in the feed. The following is a suggested list being put forward for consideration by FEFAC:

(1) Oil meals and other protein products of vegetable origin.
(2) Protein of animal origin.
(3) Cereals and other carbohydrate products and by-products.
(4) Oils and fats of animal and vegetable origins.
(5) Dried products (fruit, leaves, etc.).
(6) Minerals.
(7) Premixes.

It is clearly in the interests of compounders to keep the list of categories as short as possible whilst still offering information on the types of new materials used. The size of labels may have to be increased to accommodate the new declaration (and others) and it will be necessary to ensure the correct order of inclusion. Enforcing the declaration will pose problems and may lead the authorities to inspect formulations rather than try to apply optical microscopy to determine the accuracy of the ingredient statement.

STORAGE LIFE

Two forms of declaration are proposed namely:

(1) 'use before' followed by the date for highly perishable feeds, or
(2) 'best before' followed by the date for other feeds.

DATE OF MANUFACTURE

It is likely that the actual date of manufacture will be required although UKASTA and other European Trade Associations have advised of the practical problems involved. Month of manufacture is more sensible!

Compounders already declare a warranty period for vitamins A, D and E and additional declaration of dates of manufacture and shelf-life seem to be superfluous.

MOISTURE

The moisture content will have to be stated on compound feeds if it exceeds 14%. This requirement will probably lead to all compounds containing significant quantities of cereals and cereal by-products carrying a 14% declaration simply to protect the compounder. Cereals and beans are bought at up to 16% moisture! Attempts by UKASTA to have the limit raised were not supported by the Commission. An upper tolerance of 1% is anticipated.

ASH INSOLUBLE IN HYDROCHLORIC ACID

This new control and declaration is particularly useless and takes a characteristically complex form as follows:

'The level of ash insoluble in hydrochloric acid shall not exceed 3.3% of the dry matter in the case of compound feeds composed mainly of rice by-products and 2.2% of dry matter in other cases.

However, that level may be exceeded in the case of:
- compound feeds containing authorized mineral binding agents,
- mineral compound feeds, and
- compound feeds containing more than 50% of sugar beet chips or sugar beet pulp, provided that the level is declared as a percentage of the feed as such'.

LYSINE AND METHIONINE

Lysine will be a compulsory declaration of all pig feeds and methionine for all poultry feeds. The implications for the compounder arise mainly from the inadequate tolerances of only 15%. There is no official method of analysis and policing of these declarations will be a problem until realistic tolerances are agreed on the basis of collaborative tests on samples by an officially agreed method.

ENERGY VALUE

The declaration of ME will be optional and not compulsory as at present in the UK. Declarations of energy for pig and ruminant feeds using national official methods will also be optional.

There are several amendments in the new proposals but they apply mainly to small volume categories of feed.

AFLATOXIN LEVELS

The new controls on aflatoxin came into force in December 1988 and apply to six raw materials and their by-products:
groundnut, cottonseed, maize, palm kernel, babassu and copra and from 3 December 1988, an offence will be committed if:

(1) a shipper/merchant sells, or has in his possession with a view to sale, for use as an ingredient, any of the listed raw materials if the level of aflatoxin B_1 is above 0.2 mg/kg;
(2) a shipper/merchant sells, or has in his possession with a view to sale, for use as an ingredient, any of the listed raw materials if the level of aflatoxin B_1 is above 0.05 mg/kg unless the material is for use by a recognized manufacturer and it is accompanied by a document stating:
    (a) that the material is for use by a recognized manufacturer;
    (b) that the material must not be fed unprocessed to livestock;
    (c) the amount of aflatoxin in the material.

One of the conditions of registration is that a manufacturer must have facilities, equipment (including that for analytical testing), and skills necessary to ensure:

(1) the totally even distribution (homogeneous) of the aflatoxin B_1 throughout the premixture or finished feed;
(2) that the quantities of contaminated ingredients used are such that the maximum permitted level of aflatoxin B_1 in the feed is not exceeded.

The Ministry of Agriculture, Fisheries and Food confirms that it is not necessary for a compounder to have in-house laboratory facilities. Access to outside consultancy laboratories will be sufficient. They also point out that any manufacturer wishing to check on the aflatoxin level could use one of the rapid test kits, which are now on the market, to give a rough guide.

It is not necessary for a shipper/merchant to apply for membership of the national list if he is selling any of the six raw materials covered by the legislation but not mixing them into a compound feed.

MAFF advise that the person who either imports for sale from a third country or is the first to place one of the six raw materials on the market in the UK, is responsible for checking the aflatoxin B_1 content. Anyone importing from another member state of the Community should request the appropriate information from his suppliers, should the aflatoxin B_1 level fall between 0.05 and 0.2 mg/kg. Any consignment of raw materials must be accompanied by a document stating:

(1) that the raw material is for a manufacturer on the MAFF list;
(2) that the raw material may not be fed unprocessed to livestock;
(3) the aflatoxin B_1 level in the raw material.

It is clearly in the interests of receivers of these raw materials further down the distribution chain, within the UK, to check the aflatoxin B_1 level.

As with the current legislation, the compounder would commit an offence if the level of aflatoxin B_1 in the finished feed exceeded the maximum permitted level. In addition, a compounder would commit an offence if he sold a listed raw material for use as a feed ingredient, and the aflatoxin B_1 content exceeded the maximum permitted level (i.e. the offence is commited by sale not by possession for own use).

The onus would be on the trader (merchant or compounder) to make sure that the material was not above 0.2 mg/kg. There should be no material on sale in the UK that contains levels of aflatoxin B_1 above 0.2 mg/kg.

PESTICIDE LEVELS

The maximum pesticide residue limits which apply to raw materials, straights and feeds are shown in Table 1.1. These come into force from December 1990.

Analyses of several hundred samples of raw materials and feeds from UK compounders are nearing completion. MAFF organized the survey and the analyses were carried out at the Laboratory of the Government Chemist.

The results to date show that the feeds comply with the new requirement.

UK compounders do not have the same opportunities for cross border trade in feeds as most other member states, so it may be argued that a harmonized feed label is not so advantageous to the UK, and incurs additional costs. On the other hand, the continuous review of labelling is necessary to keep pace with nutritional and public health advances.

Medicated feeds

This section of the chapter is applicable to the UK and describes the next steps in the completion of legislation in the form of regulations under the Medicines Act to control the sale and distribution of medicated feeds.

Table 1.1   MAXIMUM PERMITTED LEVELS OF PESTICIDES IN FEEDS

Substances	Feeding stuffs	Maximum content in mg/kg of feeding stuffs referred to a moisture content of 12%
Aldrin ⎫ singly or combined expressed as Dieldrin ⎭ dieldrin	All feeding stuffs except fats	0.01 0.2
Camphechlor (Toxaphene)	All feeding stuffs	0.1
Chlordane (sum of *cis* and *trans* isomers and of oxychlordane	All feeding stuffs except fats	0.02 0.05
DDT (sum of DDT, TDE and DDE isomers, expressed as DDT)	All feeding stuffs except fats	0.05 0.5
Endosulphan (sum of alpha and beta isomers and of endosulphan sulphate, expressed as endosulphan)	All feeding stuffs except maize oilseeds complete feeding stuffs for fish	0.1 0.2 0.5 0.005
Endrin (sum of endrin and delta, keto endrin, expressed as endrin)	All feeding stuffs except fats	0.01 0.05
Heptachlor (sum of heptachlor and of heptachlor epoxide, expressed as heptachlor)	All feeding stuffs except fats	0.01 0.2
Hexachlorobenzene (HCB)	All feeding stuffs except fats	0.01 0.2
Hexachlorocylohexane (HCH) alpha isomer	All feeding stuffs except fats	0.02 0.2
beta isomer	Straight feeding stuffs except fats Compound feeding stuffs except compound feeding stuffs for dairy cattle	0.01 0.1 0.01 0.005
gamma isomer	All feeding stuffs except fats	0.2 2.0

This new legislation follows on from the Medicines (Medicated Animal Feeding Stuffs) Regulation 1988, which came into force in July 1988, and containing the following provisions:

(1) A medicinal product may not be incorporated into an animal feedingstuff unless it is incorporated in accordance with a Product Licence, an Animal Test Certificate or a Veterinary Written Direction.
(2) With effect from 1 July, 1988 any person incorporating a medicinal product at a rate below 2 kilograms per tonne must register the premises where the incorporation is to take place in Part A of the Register to be kept by the Royal Pharmaceutical Society of Great Britain/Department of Agriculture for Northern Ireland (RPSGB/DANI). (In any case, with effect from 1 July, 1989, he must be registered with either Part A or Part B.) A fee of £150 for Register A and £50 for Register B is payable annually.

(3) A person not registered in Part A of the Register may not incorporate into an animal feedingstuff a medicinal product for which there is no Product Licence or Animal Test Certificate.

(4) Manufacturers must comply with the relevant Codes of Practice, A or B, which cover quality assurance, personnel and training, documentation, premises and equipment. Register B will be appropriate for on-farm mixers and a few of the smaller compounders.

(5) If a person operating mobile mixing equipment chooses to register, then this will be in respect of the premises where the equipment is normally kept.

(6) The Regulations prohibit a person from selling or supplying any animal feedingstuff in which a medicinal product (not being a prescription-only medicine) has been incorporated unless the medicinal product was incorporated in accordance with a Product Licence, an Animal Test Certificate or a Veterinary Written Direction.

(7) The Regulations generally prohibit a person from selling or supplying any animal feedingstuff in which a prescription-only medicine has been incorporated or from importing any such animal feedingstuff except in accordance with a Veterinary Written Direction.

REGISTRATION

Distinction is made between Register A manufacturers who also have under their control animals either as their sole or part of their business activities and those who do not. The keeping or maintenance of animals for research or educational purposes is not regarded as being a sole or part of the business activity.

VETERINARY WRITTEN DIRECTION

These Regulations also introduce a revised standard form of Veterinary Written Direction, the essential features of which are:

Vets may authorize the use of medicinal products for which there are no licences for use in feeds (i.e. vets' specials) and also for their incorporation by unregistered manufacturers in emergencies. (A veterinary special is 'any medicinal product, whether licensed or not, other than a licensed medicinal feed additive'. It does not include a licensed medicinal feed additive used outside the terms of its product licence, e.g. Romensin in sheep feed).

A certificate of analysis is required on the final medicated feed showing the contents of all medicinal products in the feed if a medicinal product has been incorporated for which there is no Product Licence or Animal Test Certificate for its use in animal feeds (i.e. a certificate of analysis is required for a veterinary special). Individual compounders have to decide on the guaranteed level they give on a feed label for any unlicensed product in the feed. Also, they have to decide who should pay for the analysis of the feed containing a veterinary special. Individual compounders have to resolve this issue with their farmer customers and veterinary surgeons. There is a legal requirement to ensure that the veterinary special drug levels comply with the permitted limits of variation and that the period of potency guarantee has therefore to be considered.

Analysis is required before sale and supply except in an emergency authorized on the Veterinary Written Direction by the vet. The new Regulations also introduced new emergency procedures. If the veterinary surgeon decides to invoke these procedures then the appropriate part of the new section has to be completed in triplicate by the veterinary surgeon. If the emergency section is completed the manufacturer must send a copy of the form to the RPSGB or DANI.

The new controls on the manufacture, sale and distribution of medicated feeds have two principal aims: to ensure high standards of safe manufacture; and to ensure that medicinal products, e.g. medicated feeds, are uniquely identified from source through incorporation in feeds, their sale and distribution. Since a large production of medicated feeds, especially intermediate medicated feeds (e.g. protein concentrates and supplements) are distributed through agricultural merchants, legislation was required to extend the current registration scheme for merchants who sell medicinal products, to those who sell only intermediate medicated feeds. At present, there are no restrictions on the latter. These merchants will be required to register with the RPSGB (or DHSS in Northern Ireland) and comply with a simple code of practice.

The Code will require documentation of receipt and sales of intermediate medicated feeds and good storage. This will enable the authorities to satisfy EEC requirements to trace the origin and destination of medicated feeds and completes the only remaining gap in the chain.

This new legislation will be published early in 1989 and apply from July 1989.

Feed additives

A number of important developments in the control of additives are in progress and the EEC are preparing Directives for agreement in 1989.

The main categories under consideration are probiotics, prophylactics and growth promoters, beta-agonists, and products of biotechnological processes.

Of immediate concern is the proposal to transfer prophylactic medicinal products such as coccidiostats and growth promoters from the additives directive (70/524) to the medicinal products directive (81/851). Unless special exemptions were made to maintain their current status, they would become prescription only medicinal products. This would have the effect of making all medicated feeds as defined in the UK available only on veterinary written direction (prescription).

All European feed trade associations, and FEFAC are seeking to maintain these products in the additives directive but the outcome is uncertain.

Beta-agonists are to be added as a new category to the additives directive (70/524) but may of course be transferred to the medicinal products along with growth promoters.

There are now more than 30 probiotics available on the UK market and even more throughout the EEC. Legislative control is imminent and two options are being considered, namely that they should be a new category in the additives directive or become subject to a probiotics directive. In either case provisions will be made for each probiotic to be assessed for safety, quality and efficacy and for appropriate labelling requirements to be made. Scientific dossiers will be required and once the controls are in operation, only listed products could be marketed.

A major change in licensing procedures for medicinal products is under discussion which will have two fundamentally important aspects. The first is the proposal to institute Community Marketing Authorization for Veterinary Medicinal Products (centralized licensing) and the second is to require a new category of assessment for growth promoters namely their 'social and economic impact' in addition to safety, quality and efficiency. Environmental impact is also newly specified but this aspect is usually covered by other legislation. The proposed requirement for 'social and economic impact' is being strenuously opposed since it introduces a subjective political aspect into the licensing procedure. The issue is in sharp focus with the controversy surrounding the injection of dairy cows with bovine somatotrophin (BST) and the current licence application being considered by the Commission expert group. The implications to feed compounders are far reaching in that public opinion may be prejudiced against important livestock products and the feeding systems and nutritional aspects could be influenced.

Health and safety legislation

New UK regulations on the Control of Substances Hazardous to Health (COSHH) come into force on the 1 October, 1989. They were laid before Parliament on the 12 October, 1988. They are meant to protect the health of people exposed to substances hazardous to their health from work activities. They apply to all places of work including feed mills (and offices and laboratories). These regulations are the most far-reaching health and safety legislation since the introduction of the Health & Safety at Work Act in 1974.

A substance hazardous to health means any substance in preparation which creates a hazard to health of any person and includes substances already classified as very toxic, toxic, harmful, corrosive or irritant; a microorganism which creates a hazard is also included as is dust of any kind when present at a substantial concentration in air (limits already exist e.g. $5 \, mg/m^3$, 8-h time weighted average of respirable dust). The legislation was prepared in consultation with industry and the trade unions, and the main requirements may be summarized as follows.

ASSESSMENT OF RISKS

The employer will have a statutory duty to assess the risk to which his employees and others might be exposed through inhalation of or contact with hazardous substances at the work place. The assessment will entail a systematic review to ensure that the employer knows from available data the types of substances his employees are liable to encounter, the health effects of those substances, the places where the substances are likely to be and the ways and extent to which his workers and others on, and in the vicinity of, his premises could be exposed. Details will have to be available to employees and their representatives.

CONTROL OF EXPOSURE

The employer will have to ensure that exposure of employees and others to substances identified as hazardous is prevented or controlled below set exposure limits (where they exist) to the extent required to prevent adverse health effects.

CONTROL MEASURES

If the employer's assessment were to identify a potential risk of exposure, he will be required to institute control measures commensurate with that risk. Control will be achieved through systems of work or, if necessary, engineering controls. If they are not reasonably practicable, personal protective equipment would be allowed.

MAINTENANCE

The employer will be obliged to ensure that monitoring is carried out to the extent it will be required to confirm that exposure was being adequately controlled. Records will need to be kept of monitoring procedures and their results.

HEALTH SURVEILLANCE

This will be obligatory when an employee or other person has been exposed to one of the few specified substances, or to a number of other less well-defined substances (like carcinogens) where adverse health effects are likely to result and where there are valid techniques for detecting those effects.

RECORDS

The employer will be advised to keep and retain monitoring and health records for up to 30 years.

Conclusions

The steady appearance of more and more legislation affecting the manufacture, sale and distribution of feed is the inevitable result of a number of factors – consumer demands, EEC harmonization, developments in feed technology and nutrition, environmental and public health consideration. It is vital that feed compounders' interests are safeguarded during the formative stages of legislation at national and European levels. UKASTA and FEFAC are the focal points for representing these interests.

2

RESIDUES OF VETERINARY DRUGS IN ANIMAL PRODUCTS

A. R. PETERS
Hoechst UK Ltd, Walton Manor, Milton Keynes, UK

Introduction

A standard dictionary definition of the term 'residue' is 'that which remains behind'. In the context of veterinary medicine, this normally means the portion of an administered substance including any metabolite which remains in the animal's carcass. In Council Directive 86/469 EEC 'residue' is defined as 'residue of substance having a pharmacological action and of conversion products thereof and other substances transmitted to meat and which are likely to be dangerous to human health'. These various definitions immediately lead to some confusion since the use of pharmacological agents in animals is almost certain to result in minute residual quantities of those substances remaining in the carcass. This means that it is vital to differentiate between safe and unsafe residual concentrations rather than being able to insist on zero residues (Council of Europe, 1986; Hoffman, 1987; Saunders, 1988).

The compounds

Approximately 2000 products are currently licensed for use as veterinary medicines in the UK. Of course many of these are identical generics sold under different names by different companies and many are not used in food-producing animals. Nevertheless a large number of compounds is used in farm animals. These products fall into a number of categories (see Table 2.1), the major ones being:

(1) Antimicrobials – used to treat and prevent bacterial infections and including several classes of compound.
(2) Anthelmintics – drugs which are active against helminth parasites, i.e. roundworms, tapeworms and flukes.
(3) Ectoparasiticides (pesticides) – used to kill external parasites, such as sheep scab and warble fly.
(4) Antiprotozoals – e.g. coccidiostats.
(5) Hormones – used for therapeutic purposes mainly in fertility control or for growth promotion and include a wide variety of chemical classes.

Table 2.1   MAJOR CLASSES OF COMPOUNDS USED IN FARM ANIMALS

Class	Subclass	Examples
Antimicrobial	β-lactams	Penicillin
	Aminoglycosides	Streptomycin
	Tetracylines	Oxytetracycline
	Macrolides	Tylosin
	Sulphonamides	Sulphadimidine
Anthelmintics	Benzimidazoles	Albendazole
	Imidathiazoles	Levamisole
	Pyrimidines	Morantel
	Avermectins	Ivermectin
Pesticides	Organochlorines	Lindane (BHC)
	Organophosphates	Diazinon
	Avermectins	Ivermectin
Antiprotozoals (coccidiostats)	Ionophores	Monensin
	Sulphonamides	Sulphaquinoxaline
	Quinolones	Clopidol
Hormones	Steroids	Oestradiol
		Trenbolone
	Prostaglandins	Cloprostenol
	Gonadotrophins	Human chorionic gonadotrophin
	Releasing-factor peptides	(HCG)
		Gonadotrophin releasing
		hormone (GnRH)

(6) Other categories including vaccines, anaesthetics, vitamins and numerous miscellaneous compounds not considered important here from the public health point of view.

Pharmacokinetic principles

The fate of veterinary drugs within the animal body is highly variable. For example, some compounds are metabolized or eliminated from the body very quickly, e.g. prostaglandins. However other products, e.g. some antibiotics, are much more persistent. Persistence in the body is mainly dependent on the following factors:

(1) the chemical nature of the drug substance,
(2) the route of administration, e.g. oral, intravenous, subcutaneous,
(3) the formulation, i.e. rapid or slow release,
(4) the distribution in and binding to different body tissues,
(5) the metabolic fate, e.g. is the drug excreted unchanged or is it converted into other metabolites?

The elimination or metabolism of any drug normally follows an exponential decay curve. This means that the initial rate of elimination is high but that the rate decreases as the concentration falls. Eventually very low concentrations may be reached when further decrease is extremely slow and consequently the residues may not be eliminated completely. The fact that many drugs are converted to other metabolites is a complicating factor since the metabolites have to be identified and their potential toxicity and tissue concentrations assessed separately.

Licensing of veterinary products

Veterinary medicinal products are licensed in the UK by the Ministers of Agriculture and of Health under the Medicines Act 1968. Companies wishing to market such products are required to demonstrate their safety, quality and efficacy by carrying out appropriate tests. These are the sole requirements. The system is administered by officials at MAFF with technical input from the Medicines Unit, Central Veterinary Laboratory and the Department of Health. The Veterinary Products Committee is a body of independent scientists set up under section 4 of the Medicines Act to

(1) give advice to the licensing authority on safety, quality and efficacy, and
(2) promote the collection and investigation of information on adverse reactions to veterinary products.

Under the terms of the Medicines Act a medicinal product is defined as any substance or article administered to a human or animal for a medicinal purpose. A medicinal purpose includes:

(1) treating or preventing disease,
(2) diagnosing disease or a physiological condition,
(3) contraception,
(4) anaesthesia,
(5) any interference with a physiological function.

It will be seen immediately from this list that the Medicines Act covers the use of substances for very different purposes and this is where some difficulty, at least in understanding, has arisen in recent years. For example, the licensing criteria for therapeutic products are exactly the same as for products like growth promoters, there are no differences.

The adequacy of the animal medicines licensing system has recently been reviewed at the request of the Licensing Authority (Cunliffe, 1988). Confidence was expressed in the current system and the other main conclusions were that:

(1) licensing should remain a ministerial responsibility,
(2) there should be closer integration of the licensing system as one directorate,
(3) the licensing system should be financed by the pharmaceutical companies applying for product licences.

It is apparent that most of the findings are currently being implemented.

Procedures developed in the EEC are outlined in Table 2.2 and have differed in some respects from those in the UK. Two Council directives 81/851 and 81/852 lay down procedures for the control of veterinary medicines within the Community. Directive 81/851 also provides for the establishment of the Committee on Veterinary Medicinal Products (CVMP) comprising representatives from each member state to advise on the implementation of the Directive. The aim of these directives has been the approximation of laws of the member states relating to veterinary products and their testing. However there has been no central licensing procedure as such. Under the more recent directive 87/22, it is a requirement for all products produced by new biotechnological methods to be referred to the CVMP for its opinion, before licensing in any member state. Bovine somatotrophin (BST) is one of the first products to be considered under this new regimen. Additionally new proposals have been drafted by the European Commission for the complete

Table 2.2 SUMMARY OF MAJOR EEC LEGISLATION ON VETERINARY MEDICINES
AND RESIDUES

Number	Title
70/524	Concerning additives in feedingstuffs
81/851	On the approximation of the laws of the member states relating to veterinary medicinal products
81/852	On the approximation of the laws of the member states relating to analytical, pharmacotoxicological and clinical standards and protocols in respect of the testing of veterinary medicinal products
87/22	On the approximation of national measures relating to the placing on the market of high-technology products, particularly those derived from biotechnology
81/602	Concerning the use of certain substances having a hormonal action and of any substance having a thyrostatic action
85/649	Prohibiting the use in livestock of certain substances having a hormonal action
86/469	Concerning the examination of animals and fresh meat for the presence of residues
86/363	On the fixing of maximum levels for pesticide residues in and on foodstuffs of animal origin
87/410[a]	Laying down the methods to be used for detecting residues of substances having a hormonal action and of substances having a thyrostatic action

[a] Denotes a Commission Decision not a Council Directive.

centralization of veterinary product registration for the whole of the EEC and based in Brussels (Anon, 1988a), removing the requirement for separate national systems. In addition to the present requirements for safety, quality and efficacy it has been proposed that assessment of social and economic effects may also be required at least for some products.

Currently, medicinal feed additives are specifically excluded from the terms of the veterinary medicine directives described above and instead are dealt with under directive 70/524, by a different directorate. As such they are considered largely as nutritional supplements rather than as medicinal products.

Assessment of the safety of veterinary medicines

Under the terms of the UK Medicines Act, safety includes:

(1) safety to the target animal,
(2) safety to the consumer,
(3) safety to any operator, e.g. in manufacture, use or processing,
(4) safety to the environment.

Within this chapter safety is confined to the consumer and therefore the other aspects will not be considered further.

Of central importance to consumer safety is the establishment of acceptable residue tolerances (maximum residue levels; MRLs) of the drug and any metabolites. In order to do this, an assessment of the drug's toxicological profile is necessary. Generally the effects which are of major concern are acute toxicity and any chronic effects including genotoxicity, carcinogenicity, mutagenicity, terato-genicity, ability to affect reproductive function and immunopathological effects, e.g. allergies (Council of Europe, 1986). A battery of toxicological tests is carried out in a variety of species to build up a toxicological profile of the drug as laid down in Directive 81/852 EEC.

From these studies, the maximum dose which produces no toxic effect is determined as a 'no effect level' (NEL). The lowest NEL value determined from this series of studies is then used to calculate 'acceptable daily intake' and 'maximum residue level' values (see Table 2.3). First the NEL is divided by a safety factor which allows for extrapolation from laboratory animals to humans. A factor of 1000 is used unless toxicity data from lifespan studies exist and show no evidence of carcinogenicity in which case a factor of 100 is used. Thus the safety-corrected NEL is 100–1000 times lower than that determined in the toxicity studies. The acceptable daily intake (ADI) is then calculated by multiplying the corrected NEL by the 'standard' human body weight of 70 kg. Then assuming a daily intake of 500 g meat or offal, this can be converted into a maximum residue level (MRL) or tolerance (see Table 2.3).

Residue studies are normally required on each target species including data on muscle, fat, liver, kidney and site of injection, milk and eggs if appropriate. The tolerance or MRL is then set in relation to the tissue with the highest residue concentration. A worked calculation of an ADI and tolerance is given in Table 2.3. The time taken for all tissue concentrations of the residue to fall below the tolerance is then set as the withdrawal or withholding period of the drug. International agreement on tolerances are gradually being achieved under the auspices of such bodies as the World Health Organization, Food and Agriculture Organization and the Council of Europe.

Table 2.3 CALCULATION OF AN 'ACCEPTABLE DAILY INTAKE' AND RESIDUE TOLERANCE USING ZERANOL (RALGRO) AS AN EXAMPLE

'No (hormonal) effect level' of zeranol in experimental animals (Lamming *et al.*, 1987)	=	25 µg/kg
A safety factor of 1/1000 is applied to the NEL to allow extrapolation to human	=	25/1000 µg/kg
Therefore corrected NEL	=	25 ng/kg
Assuming a standard human body weight of 70 kg, the corrected NEL is equivalent to an 'acceptable daily intake' (ADI) of 25 × 70	=	1750 ng
	=	1.75 µg
It is assumed that 0.5 kg of meat or offal is eaten daily		
Assuming the 'ADI' of 1.75 µg is present in a daily intake of 0.5 kg of meat or offal,		
The concentration would be 1.75/0.5	=	3.5 µg/kg
Therefore the tolerance should be		3.5 µg/kg
Maximum concentration of zeranol found in any tissue of Ralgro-treated bulls (Peters *et al.*, 1988)	=	478 ng/kg or 0.48 µg/kg
Therefore margin of safety even in this 'worst case'	=	3.5/0.48
	=	7.3 times

Although several countries including the UK use tolerances to determine withdrawal periods this approach is not universal. Other countries, e.g. Denmark and Italy, insist on 'zero residues' using the most sensitive assay available. This poses the possibility that as analytical techniques are improved and become more sensitive, so residues become more readily detectable, and therefore longer and longer withdrawal periods are required. This is clearly an unacceptable situation scientifically and indeed Saunders (1988) has recently proposed that concentrations of a drug or metabolite that are present below a tolerance calculated on toxicological grounds, as described above, should be defined as 'zero residues'.

Legislation on residues

It is perhaps fair to say that the problem of public confidence over residues has arisen partly because the withdrawal period itself is not legally enforced in this country. It is a requirement of Directive 81/851 that product literature should bear a stated withdrawal period. However its observance is not statutory. In the UK, ministers attempted to fill this loophole in 1985 by publishing a Code of Practice for the Safe Use of Veterinary Medicines on Farms (MAFF, 1985). The code was distributed to all livestock producers and gives advice on storage and administration of veterinary medicines with particular reference to record keeping and observance of withdrawal periods.

Concern in Europe over the consequences of use and abuse of steroid hormones and related products as growth promoters probably led to the legislation on veterinary residues as a whole. The 'Residues Directive' 86/469 lays down procedures for the collection of material, laboratory analysis and other measures. This is being implemented in the UK at the present time under a series of statutory instruments (1988 Nos. 848 and 849). These essentially give powers to officials to take samples from abattoirs and farms, to analyse them for various substances and to take action where positives are found, including the prosecution of offenders. Producers can be fined up to £2000 for an offence under this legislation.

Residue monitoring under Directive 86/469 began in the UK in January 1988 for steroid-type hormones only. The statistical sample is based on 0.15% of the slaughter population and currently includes about 4000 cattle. Samples are taken at random from abattoirs (bile) and farms (blood or faeces). If positives are found then the intensity of sampling is increased to 0.25% of the slaughter population, mainly in the geographical area(s) where the initial violation was detected, including thorough investigation of the farm of origin.

Officials have powers to prosecute and to restrict animals on a premises. However, at present there are no powers in Great Britain to detain or condemn violative carcasses in an abattoir. MAFF also has powers to take additional samples from animals or carcasses in circumstances where there is suspicion of use of illegal substances, e.g. the presence of implants in a carcass or atypical conformational changes. It is apparent that a number of recent violations in this country have been identified in this way.

Statutory monitoring for antimicrobial residues will begin in 1989 comprising approximately 32 000 samples from sheep, pigs and cattle, being about 0.10% of slaughtered animals. Most sampling will be carried out in abattoirs and will consist mainly of kidney samples. In addition to these, 1200 pig carcasses will be sampled for the presence of sulphonamides and 300 carcasses per species for chloramphenicol.

Sampling will also begin in 1989 for certain pesticide residues including both organochlorine and organophosphorus compounds. This is to be carried out under Directive 86/469 and Directive 86/363 on the fixing of maximum levels for pesticide residues in and on foodstuffs of animal origin (see Table 2.2). Analysis for arsenic and the heavy metals (lead, cadmium and mercury) will also begin in 1989. Arsenical compounds are used as pesticides but also to treat gut infections and as growth promoters in pigs and poultry. Heavy metal residues may be present as environmental contaminants but are included in this legislation. Monitoring for anthelmintic residues is unlikely to begin before 1990 as the analytical methodology is not yet fully developed.

Under the terms of The Animal and Fresh Meat (Examination for Residues) Regulations 1988 (Statutory Instrument No. 848), UK producers are now required to keep records of all veterinary medicines administered to their animals. It is also possible that legal enforcement of withdrawal periods may be introduced at some time in the future.

There is an obligation on member states to introduce monitoring procedures for veterinary residues in poultry meat and farmed fish, and a proposal from the Commission is due to be put to the Council of Ministers in January 1989. It is likely that regulations will be extended eventually to game birds and rabbits.

Methods of analysis

The criteria for methods of analysis to be used in monitoring for hormones and thyrostatics under Directive 86/469 are set out in Commission Decision 87/410 (Table 2.2). This document includes criteria for both screening and for reference methods although the latter are not yet fully agreed.

The steroid hormones and the xenobiotic zeranol are screened for using well-documented radioimmunoassay (RIA) procedures (Heitzman, 1986a; N. F. Cunningham, personal communication). These generally have a sensitivity of 0.1–1.0 parts per 10^9 (μg/kg). Confirmation of positives is by repeat RIA preceded by a high-performance liquid chromatography (HPLC) clean-up step, but the final reference method is gas chromatography–mass spectrometry (GCMS). Detection of banned xenobiotic compounds, such as trenbolone, zeranol, stilboestrol or hexoestrol, is comparatively straightforward. However, determination of the source of natural steroids, i.e. whether they are endogenous or exogenous, remains problematical and there is no firm policy on this aspect, although the EEC Scientific Veterinary Committee is currently considering the issue.

The Four Plate Test (FPT) is used as the standard screening method for antimicrobials (Pennycott, 1987). This utilizes four nutrient-agar plates each seeded with *Bacillus subtilis* or *Micrococcus luteus* under different conditions. Samples of tissue (kidney) are incubated on the plates and the presence of inhibitory substances is identified by zones of inhibition of growth around the colonies of bacteria. The sensitivity of the FPT varies, depending on the antimicrobial compound. It is very sensitive to the β-lactams, aminoglycosides, macrolides and tetracyclines but not to chloramphenicol and sulphonamides (Heitzman, 1986b) and, of course, cannot distinguish between compounds. Samples positive on the FPT are then tested using high voltage electrophoresis (HVE) for supporting evidence. In the UK at least, thin-layer chromatography (TLC) is used for sulphonamides and an enzyme-linked immunosorbent assay (ELISA) card test is used for chloramphenicol (Pennycott, 1987).

There is little information on anthelmintic residues or on methods to measure them. Most data are based on manufacturers' own radiometric studies (Heitzman, 1986a). Gas chromatography is used to detect organochlorine and organophosphorus pesticides and atomic absorption spectrometry is used for arsenic.

It will be apparent from this section that with the exception of steroid hormones and some antimicrobials, analytical methods for veterinary residues are rather poorly developed.

Residue monitoring in the UK to date

Residue monitoring has been carried out in the UK since 1980 (Lindsay, 1983). This was initially administered by the MAFF Food Science Division as the National Meat Monitoring Programme until 1983 when it was taken over by the State Veterinary Service (SVS) as the National Surveillance Scheme. Sampling was based on a statistical design originating from a USA monitoring programme (see Cordle, 1988). It was shown that if a residue problem exists in 1% of animals and if 300 samples are taken, there is a 95% chance of detecting it. To increase the assurance rate to 99% for example, would necessitate an unjustifiable increase in the number of samples required. In the UK, survey monitoring was initially confined to cattle, sheep and pigs but in 1985 it was extended to poultry and farmed trout.

Results of the survey up to 1986 have recently been published (MAFF, 1987). The proportion of quadruped samples containing low concentrations of stilbene residues declined dramatically from 8.2% in 1981 to 0.4% in 1985 corresponding to the withdrawal of stilbene products from the market in 1982. Illegal use was suspected in six adult cattle, one calf and five pigs and successful prosecutions were brought in two cases. No cases of stilbene residues were found in poultry meat. Additionally no stilbenes were found in veal imported from the Netherlands. Very low concentrations of trenbolone and zeranol were found in a proportion of cattle as would be expected with their routine use prior to 1986.

The MAFF survey also analysed small numbers of samples for the presence of levamisole and benzimidazoles using HPLC, but no samples contained these anthelmintics at a limit of determination of 0.1 and 0.05 μg/kg respectively.

Results of antimicrobial testing for the period 1980–83 are shown in Table 2.4 and indicate that positive results were obtained in about 1% of samples. Antimicrobial activity was detected more frequently in kidney than in muscle, consistent with known distribution of some antibiotics (Nouws and Ziv, 1977; 1978). However, definite identification of specific antimicrobials in positive samples was not carried out.

Table 2.4 SURVEY OF ANTIMICROBIAL ACTIVITY IN MEAT AND KIDNEY SAMPLES FROM VARIOUS SPECIES SLAUGHTERED IN GREAT BRITAIN 1980–1983 (MAFF, 1987)

Species	No. of samples tested	No. of samples positive
Cattle (adult)	815	10
Calves	891	16
Pigs	853	11
Sheep	823	4

Table 2.5 SURVEY OF SULPHONAMIDE RESIDUES IN PIG KIDNEYS (MAFF, 1987)

Period	No. of samples	No. positive	No. > 0.1 mg/kg[a]	Percent
1980–83	163	34	20	12.3
1985	50	14	7	14.0

[a] Maximum tolerance level recommended by Veterinary Products Committee

Kidney samples from 75 cattle, 75 calves, 75 sheep and 75 pigs were analysed for chloramphenicol by a GC method sensitive to 10 μg/kg but no residues were found.

Fifty-six samples of farmed trout were analysed for the presence of tetracyclines. Seven contained between 8 and 40 μg/kg oxytetracycline and one contained 30 μg/kg tetracycline only. No trout samples were found to contain furazolidone at a limit of detection of 20 μg/kg.

Perhaps the only disturbing feature of the results of the survey (MAFF, 1987) is the consistent finding of sulphadimidine residues in a significant proportion of pig kidneys. The results are shown in Table 2.5 and indicate that between 12 and 15% of pig kidneys contain 0.1 μg/kg sulphadimine or more. It should be noted that the Licensing Authority has recommended a maximum residue level for sulphonamides of 0.1 μg/kg.

The sulphonamide issue

Sulphonamides generally, and sulphadimidine (sulphamezathine) in particular, are widely used in the pig industry throughout the world, usually as feed additives, to treat and control diseases of the gastro-intestinal and respiratory tracts. The problem of residues in pig meat is not new nor is it confined to the UK. For example it has been a problem in the USA for some years (Bevill, 1986) and has also recently been recognized as a problem in milk in the USA (Anon, 1988b). Of particular concern are recent reports that sulphadimidine caused an increase in incidence of thyroid and liver adenomas in experimental mice (see Young, 1988). However, these preliminary data are yet to be confirmed and are being treated with caution in the UK since there has been no previous evidence of any toxicological problem with these compounds. The occurrence of sulphadimidine residues has been attributed to:

(1) poor husbandry and on-farm feed-mixing practice,
(2) poor commercial feed-mill practice,
(3) cross-contamination of feeds following electrostatic adherence of the drug to the mixer,
(4) recycling of faeces and urine by coprophagic pigs.

In the UK a number of measures have been taken to reduce the problem. MAFF has embarked on a campaign of farmer-education about the problem, emphasizing the importance of withdrawal periods. This involves visits by SVS staff together with literature distribution (e.g. MAFF, 1986). One of the main problems, however, is that most sulphonamides are rather old products with 'product licences of right', approved before current stringent requirements came into force.

Consequently there are few data on their individual pharmacokinetics and elimination on which to set a withdrawal period. Therefore 'standard withdrawal periods' have been introduced as an interim measure to apply in such cases to individual species rather than relating to the product used. The standard withdrawal periods are, cattle and sheep 28 days, pigs 10 days and poultry 7 days. There is an EEC requirement to review all licences of right using modern criteria by 1991 so that only full product licences will then be permitted.

Additionally MAFF has taken a number of measures designed to improve the control of all medicated feed. These include:

(1) registration of feed mills (see Williams, 1989),
(2) introduction of the standard 'veterinary written direction' (VWD) to replace the prescription,
(3) improved labelling of medicated feeds and premixes,
(4) 'VWD licences' for previously unlicensed drugs,
(5) a proposal to ban the use of many unlicensed drugs.

Residues in milk

Whilst current EEC regulations do not cover drug residues in milk, there are proposals for such measures. However, in the UK, milk has been tested routinely by the Milk Marketing Boards for antibiotic residues since 1965. The most recent survey published for England and Wales (Booth and Harding, 1986) reported that in 1984–85, 99.6% of the 2×10^6 samples tested were negative at a sensitivity of 0.02 IU/ml penicillin or equivalent. The most frequent reason for test positives was failure to withhold milk for the full withdrawal period after intramammary treatment for mastitis. The terms of sale of milk to the MMB require amongst other things that 'milk shall not include . . . that from a cow which has had . . . antibiotics unless a . . . sufficient withdrawal period has elapsed between treatment and production'. Financial penalties are imposed on offending producers and this appears to ensure a well-regulated system.

Anabolic steroids

The major residue issue in recent times has been that of the anabolic steroids, sufficiently important to warrant a Select Committee Report (House of Lords, 1985) and resulting in the ban on stilbene derivatives (Directive 81/602) and the natural steroids and trenbolone and zeranol (Directive 85/649). Following a preliminary report on the safety of natural steroids (Lamming, 1984), the long-awaited report on trenbolone and zeranol was published recently (Lamming *et al.*, 1987). The authors concluded that the five compounds testosterone, progesterone, oestradiol-17β, trenbolone and zeranol are not harmful to human health, but they were careful to emphasize that this applied only following 'accepted husbandry practice', i.e. correct use.

However the Lamming Report also clarifies some concerns over the dangers of misuse of hormones. As discussed above, a major concern has been the possible non-observance of withdrawal periods. According to the Report all the products are safe even with a zero withdrawal period and indeed trenbolone has been licensed accordingly in the USA. Similarly, repeat implantation of steroids would present no residue problem. High residue concentrations could possibly arise following injection of steroids directly into an edible part of the carcass, but this would require an injectable material not an implant, and legal availability of hormone implants would possibly remove the incentive for misuse of injectable materials. There have certainly been reports of widespread illegal use of steroids in EEC countries since the hormone ban (Anon, 1988c; d).

The EEC hormone ban has also resulted in unnecessary restrictions on the use of steroids for therapeutic purposes. Of equal concern is that part of the 'Residues Directive' (86/469) which requires monitoring for the presence of natural steroids when scientific opinion has indicated that the source of natural hormones, i.e. endogenous or exogenous, cannot always be pragmatically determined.

The development of new performance enhancing drugs

The steroid hormone debate may have set a precedent in determining how future developments are viewed. There is already controversy over bovine growth hormone (somatotrophin; BST) and it is still some way from the marketplace. There are few published data on BST residues in milk since these are confidential between the pharmaceutical companies concerned and the Licensing Authority. However, the data that are available indicate that milk concentrations of BST are not significantly elevated in BST-treated cows (Table 2.6). Equally there is good evidence that BST is rapidly broken down into inactive peptides and amino acids before absorption from the gut.

Perhaps of greater significance is the effect of BST treatment on the concentration of insulin-like growth factor 1 (IGF1) in milk. Somatotrophin treatment stimulates the secretion of IGF1 from the liver, and plasma and milk concentrations increase as a result (Prosser, Fleet and Corps, 1988). Since bovine IGF1 is identical to human IGF1, it will be important to establish that raised IGF1 concentrates in milk from BST-treated cows have no effect on human gut epithelium before digestion.

Another class of compounds which is beginning to cause concern in relation to meat residues are the β-adrenergic agonists. Whilst at least one of these, clenbuterol, is already available in veterinary medicine for therapeutic use, a number of analogues are under commercial development as carcass repartitioning agents (see Peters, 1989). However, there have already been several reports of use of clenbuterol as a growth promoter in calves (Anon, 1988e; f) when it is not licensed for this purpose. The non-availability of licensed anabolic steroid products may have contributed to this situation.

Table 2.6 BOVINE SOMATOTROPHIN (BST) CONCENTRATIONS IN MILK (BULK SAMPLES)

Concentration in milk	Commercial farms (untreated)	Trial farms[a]
Number of samples	125	56
<0.3 ng/ml	112 (89.6%)	52 (92.9%)
0.3–0.8 ng/ml	13 (10.4%)	4 (7.1%)

[a] Approximately half of the cows on the trial farms were treated with BST (Source Monsanto plc).

Conclusion

The subject of veterinary residues has become an issue of public interest over recent years and there has been much sensational media coverage. This high profile is unfortunate since irrational fears may be generated with little valid reason. It is considered by some observers that such concerns are an inevitable reflection of an affluent society with substantial surpluses of food. Nevertheless it is right that consumers continue to be vigilant in order to ensure that adequate safeguards are maintained.

Finally it is likely that the issue of residues in food will continue to be contentious for some time, not necessarily as a scientifically important public health issue but as an emotive one. With a background of costly food surpluses the development of new techniques to improve the efficiency of animal production, such as β-agonists, somatotrophins and related products, will no doubt fuel the controversy.

References

ANON. (1988a). *Animal Pharm,* **162**, 2–4
ANON. (1988b). *Animal Pharm,* **156**, 16
ANON. (1988c). *Animal Pharm,* **161**, 2
ANON. (1988d). *Animal Pharm,* **162**, 6
ANON. (1988e). *Animal Pharm,* **159**, 1
ANON. (1988f). *Animal Pharm,* **157**, 3
BEVILL, R.F. (1986). In *Practices in Veterinary Public Health and Preventive Medicine in the United States,* pp. 291–298. Ed. Woods, G. T. Iowa State University Press, Ames, IA
BOOTH, J.M. and HARDING, F. (1986). *Veterinary Record,* **119**, 565–569
CORDLE, M.K. (1988). *Journal of Animal Science,* **66**, 413–433
COUNCIL OF EUROPE (1986). *Residues of Veterinary Drugs in Food of Animal Origin. Partial Agreement in the Social and Public Health Field.* Strasborg
CUNLIFFE, P.W. (1988). *Review of Animal Medicines Licensing.* HMSO, London
HEITZMAN, R.J. (1986a). In *Control and Manipulation of Animal Growth*, pp. 315–330. Ed. Buttery, P.J., Haynes, N.B. and Lindsay, D.B. Butterworths, London
HEITZMAN, R.J. (1986b). In *Drug Residues in Animals*, pp. 205–219. Ed. Rico, A.G. Academic Press, London
HOFFMANN, B. (1987). In *Beta-Agonists and their Effects on Animal Growth and Carcass Quality*, pp. 1–12. Ed. Hanrahan, J.P. Elsevier, London
HOUSE OF LORDS (1985). *Select Committee on the European Communities*, 3rd report. HMSO, London
LAMMING, G.E. (1984). *Commission of the European Communities (Agriculture),* Report Eur 8913, pp. 4–25
LAMMING, G.E., BALLARINI, G., BAULIEU, E.E., BROOKES, P., ELIAS, P.S., FERRANDO, R., GALLI, C.L., HEITZMAN, R.J., HOFFMANN, B., KARG, H., MAYER, H. H. D., MICHEL, G., POULSEN, E., RICO, A., VAN LEEUWEN, F.X.R. and WHITE, D.S. (1987). *Veterinary Record,* **121**, 389–392
LINDSAY, D.G. (1983). *Veterinary Record,* **112**, 469–471
MAFF (1985). *Code of Practice for the Safe Use of Veterinary Medicines on Farms.* HMSO, London

MAFF (1986). *ADAS special publication. 'Just Pigs'*, November 1986
MAFF (1987). *Anabolic, Anthelmintic and Antimicrobial Agents*, Food surveillance paper no. 22. HMSO, London
NOUWS, J.F.M. and ZIV, G. (1977). *Tijdschrift voor Diergeneeskunde*, **102**, 1187–1196
NOUWS, J.F.M. and ZIV, G. (1978). *Tijdschrift voor Diergeneeskunde*, **103**, 435–444
PENNYCOTT, T.W. (1987). *State Veterinary Journal*, **41**, 153–159
PETERS, A.R. (1989). *Veterinary Record*, **124** (in press)
PETERS, A.R., SOUTHGATE, J.R., AUGHEY, E. and DIXON, S.N. (1988). *Animal Production*, **47**, 215–221
PROSSER, C.G., FLEET, I.R. and CORPS, A.N. (1989). *Journal of Dairy Research*, **56**, 17–26
SAUNDERS, J.C. (1988). *Veterinary Record*, **123**, 464–467
WILLIAMS, D.R. (1989). In *Recent Advances in Animal Nutrition – 1989*, pp. 3–11 Ed. W. Haresign and D.J. Cole. Butterworths, London
YOUNG, F.E. (1988). *Federal Register*, **53**, 15886–15890

3

IMPACT OF EEC LEGISLATION AND REMOVAL OF TRADE BARRIERS ON THE UK ANIMAL FEED INDUSTRY

B. RUTHERFORD
c/o UKASTA, 3 Whitehall Court, London, UK

Introduction

The objective of creating a single 'common' market in the European Economic Community goes back to the Treaty of Rome. The Treaty clearly envisaged from the outset the creation of a single integrated internal market, free of restrictions on the movement of goods; the abolition of obstacles to the free movement of persons, services and capital; the institution of a system ensuring that competition in the common market is not distorted; the approximation of laws as required for the proper functioning of the common market; and the approximation of indirect taxation in the interest of the common market.

Progress was slow during the years that the six member states grew to 12, and momentum was lost particularly through the onset of recession and partly through lack of confidence and vision. Pressure to bring about the single market gathered pace in the early 1980s. In December 1982, in Copenhagen, the European Summit Council instructed the Council to decide on the priority measures proposed by the Commission to reinforce the internal market. In June 1984 at Fontainebleau they requested a study to bring about 'the abolition of all police and customs formalities for people crossing intracommunity frontiers'. There was a further push in Dublin 1984. In Brussels, in March 1985, the Council laid emphasis on 'Action to achieve a single large market by 1992 thereby creating a more favourable environment for stimulating enterprise, competition and trade'. It called upon the Commission to draw up a detailed programme with a specific timetable before its next meeting.

During 1985, the EEC heads of government committed themselves to completing the single market progressively by the 31 December, 1992 and this commitment was included in the package of Treaty Reforms known as the Single European Act (SEA) which came into operation on 1 July, 1987.

The first paragraph of Article 1 says 'the European Communities and European Political Cooperation shall have as their objective to contribute together to making concrete progress towards European unity'. Article 13 is the commitment; 'The Community shall adopt measures with the aim of progressively establishing the internal market over a period expiring on 31 December, 1992'.

The Commission's response to the European Council's request, that they draw up a detailed programme with a specific timetable to achieve a single large market by 1992, was speedy and it is all included in COM (85) 310 Final. It is a White Paper from the Commission to the European Council entitled 'Completing the Internal Market' and it is dated Brussels 14 June, 1985 and runs to 94 pages. It sets out the essential and logical consequences of accepting that commitment, together with an action programme for achieving the objective. This objective of completing the internal market has three aspects:

(1) The welding together of the 12 individual markets of the member states into one single market of 320 million people.
(2) Ensuring that this single market is also an expanding market – not static but growing.
(3) To this end, ensuring that the market is flexible so that resources, both of people and materials, and of capital and investment, flow into the areas of greatest economic advantage.

The year 1992 is primarily about the first aspect of welding the 12 individual markets into one single market and the first 60 pages of the White Paper discuss the changes that are necessary, and the last 34 pages set out the detailed timetable for completing the internal market by 1992 giving subject, date of Commission's proposal and expected date for adoption by Council. Many of these proposals are familiar, some have already been adopted and some still require discussion.

The timetable is in three parts:

PART 1   THE REMOVAL OF PHYSICAL BARRIERS

(a) The control of goods, culminating in the full abolition of all remaining import formalities and controls on goods between member states.
(b) Veterinary and phytosanitary controls covering such matters as maximum levels of pesticide residues, trade in embryos, veterinary inspection problems.
(c) Control of individuals, including the directive on the approximation of drugs legislation.

PART 2   THE REMOVAL OF TECHNICAL BARRIERS

(a) Free movement of goods.
(b) Approximation of laws, mainly motor vehicles and agricultural machines.
(c) Food law – additives, labelling, sampling, analysis and so on.
(d) Chemical products such as dangerous substances and liquid fertilizers.
(e) Construction and construction products.
(f) Other items such as textiles, flammability and protective devices.
(g) Public procurement.
(h) Free movement for labour and the professions.
(i) Common market for services covering banks, insurance, transport, TV and radio. Capital movements. The creation of suitable conditions for industrial cooperation including company law. Intellectual and industrial property.

PART 3   THE REMOVAL OF FISCAL BARRIERS

The first part is mainly concerned with VAT harmonization, then follows excise duties, alcoholic drinks, cigarettes, mineral oils ending with proposals concerning the gradual abolition or reduction of excises not covered by the common system and giving rise to border formalities.

No apologies are made for the long introduction to this chapter because it is necessary to know the background to appreciate and anticipate the changes that have and will take place. Already more than 150 individual measures have been approved since the 1992 Commitment was agreed in 1985, and current priorities include capital movements, standards under the EEC's new approach, public purchasing and freedom of establishment for the professions. At this stage it is necessary to be more specific and discuss some of the opportunities and challenges to the UK animal feed industry.

Will 1992 lead to greater competition in the import and export of compound feed? With some reservations, the answer is likely to be that it will not affect the animal feed industry too much. There are two main reasons for this view and they both depend on the economics of moving stuff, i.e. raw materials and finished product.

Consider first the raw materials. Broadly speaking, imported by-products have to be economic against UK grain and the fact is that a large proportion are bought trans-shipped from Rotterdam at a delivered UK mill premium of up to £15/tonne over CIF Rotterdam. Similarly, the UK is deficient in soya and an importer of other proteins, and in general the ex-mill UK price is at a similar premium to imported by-products, that is some £10 to £15/tonne. It is estimated that imported by-products and protein, some 30% of the raw materials used in compounds, lead to a higher price of between £3 and £5/tonne. If one bears in mind that each tonne of compound exported will bear the export cost of some £10–£15/tonne to get it to the continent, all in all it seems likely that the present situation of virtually nil exports to other EEC countries will continue.

Of course this view is reinforced by the way the UK animal feed industry had developed over the past 20 years. The larger companies have endeavoured to build their mills some 30–40 miles apart to take advantage of receiving local homegrown wheat and barley and for distributing compounds. It is particularly difficult to trade over 50 miles, and that is likely to be the maximum distance contemplated whatever the size of the feed company.

As already discussed, 1992 is unlikely to affect the animal feed industry too much, with little change for normal products. However, there should be opportunities for higher added value and specialized products like pig creep, calf milk replacers and fish food, where transport costs are of less importance as a percentage of delivered price. It is suggested there is the added possibility of making significant nutritional advances in the formulation of such rations which would make them more competitive over greater distances.

The cynics will ask and perhaps justifiably, whether there will be local health problems associated with free trade. There should not be barriers and it is to remove current ones and ensure free movement in the future that trade associations and federations have spent so much time discussing the legislation in Brussels – not least the feed label to get a workable, commercial and policeable solution. The White Paper says, under Removal of Technical Barriers, 'The general principle should be approved that, if a product is lawfully manufactured and marketed in one member state, there is no reason why it should not be sold freely throughout the Community', and goes on to say that 'Legislative harmonization will in future be restricted to laying down essential health and safety requirements, which will be obligatory in all member states. Conformity with this will entitle a product to free movement'. Much of the 1992 feed legislation programme is, or is almost, complete.

The National Office of Animal Health (NOAH) points out the urgent need to get the licensing of animal medicines on to a common standard throughout the Community. At present it is extremely complex, and the aim is to get all member states in step with compatible systems and equal, fair standards based on the scientific criteria of safety, quality and efficacy and nothing else. The UK wish and are working to preserve intact the UK distribution system for animal medicines which ensures farmers have ready access to Pharmaceutical Merchants List (PML) products at competitive prices while providing veterinary control over prescription only medicines (POM) products which require professional expertise.

Similarly, the UK Animal Health Distributors intend to ensure that 1992 does not destroy the access of some 800 rural companies to their M£42 share of the UK animal health product market. This could come about by a directive regulating distribution itself or by a common European classification of animal health products; harmonization on a German model rather than a UK one could end distribution as we know it in the UK. The commission's current proposals imply that if there is licensing of veterinary medicinal products this could be achieved by allowing distribution to remain appropriate to each member state.

It is appropriate at this stage of the discussion on the import and export of compound feed to consider the implications of the cross-channel tunnel, for that may have as significant an effect on matters after 1992. It is unlikely to alter the points made but it does suggest that exporting and importing may become easier and certainly more simple, though the cost of using the tunnel shouldn't be significantly different from using ships. However, one can envisage an active trade in raw milk through the tunnel, and the question is whether this can be made to the export advantage of UK farmers, or whether the competition will be such that imports flow into the UK.

It will certainly be easier for companies in the animal feed industry and any other industry for that matter, to buy companies in different countries, i.e. across national Community borders. Currently there are restrictions, especially in the agricultural field and this should ease. Paragraph 143 of the White Paper says that 'If it is to satisfy the needs of a genuine internal market, the Community cannot concentrate simply on the arrangements for creating subsidiaries or branches in order to make it easier for enterprises to set up in other member states. Enterprises must also be able to engage in cross-border mergers within the Community . . . with the straightforward acquisition of a shareholding.' However, it may well prove to be easier to buy smaller companies and more difficult for large companies to take over large companies.

A report of the Agricultural Co-op Training Council indicated that, with the advent of 1992, the Dutch co-ops are expecting to expand their trade to other European countries. With little opportunity for major internal mergers they are working on three optional strategies – joint trade agreements with co-ops or private organizations in other countries, federalization of European co-operatives or mergers between European co-operatives. The report went on to say that the Dutch are working hard towards encouraging and developing a single European statute for co-operatives by 1992–94. With over 50% of Dutch compounds being produced by co-operatives compared with some 10% by co-ops in the UK there are obvious opportunities, challenges or threats to the UK market.

Research should be aided with greater harmonization, yet the greatest threat in this area is the so-called 'open declaration' regulation for the marketing of compound feeds which are currerntly being fought. Open declaration is likely to bring animal feed research and trial work on raw material use to a speedy end, particularly if the results, often generated at enormous expense, have to be shared with all via the label.

One area where the free trade arrangements of 1992 are likely to have a dramatic effect is between Northern Ireland, with a compound production of about 1M tonnes, and the Republic of Ireland with twice that level of production. It will be interesting to see how raw material and finished product flows change with this 1992 harmonization. Hopefully one topic of discussion will disappear and that is the value and volume of smuggling across that border.

The policy to remove border controls within the Community and to maintain external protection by a common customs tariff will continue. January 1988 saw the introduction of the single customs form to replace more than 190 such documents. Speaking in Brussels to a Trade Meeting, Mr Pierru (Head of Division in DG XXI responsible for customs) said that 'The aim is to secure a Community customs procedure with open links between Community custom computers and the computers of trading firms. An interface will be created to enable information to pass from computer to computer. He also pointed out that the Single European Act permits decisions on the removal of technical barriers to be taken by qualified majority in this as well as other areas. This of course must change MAFF and the UK Trades' approach to Commission proposals. Instead of asking 'how can we block this proposal?' it may well become 'what must we change in this proposal in order to be able to live with it?'

It is disappointing that, to date, minimal progress has been made over harmonization of sampling for customs purposes. The UK would like this to be based on GAFTA terms so that it would know exactly where it stood commercially.

It is extremely difficult to understand how a free market can exist after 1992 without the elimination of quotas for milk and for sugar. Such a change would surely be a challenge and opportunity for the UK animal feed business.

The Commission's proposal for VAT is to have the harmonized bands of 4–9% and 11–20%. The UK – under a Labour government – with all member states agreed that zero rates were transitory only. However the present government under Mrs Thatcher has said that harmonization is not necessary for VAT since market forces will move them into competitive positions. The UK may well not agree to harmonization into the proposed bands, but one compromise would be to have the lower band, not at 4–9% but zero to say 6%! The Feed Industry certainly wishes to remain zero rated as far as VAT is concerned.

Monetary Compensatory Amounts (MCAs) have a fundamental effect on cereals and on all raw material prices, due to economic price relativities. Technically the Commission has proposed a progressive dismantling of the agrimonetary system of green currencies and MCAs, and the plan is to dismantle them by 1992. This is also bound up with implications of whether the UK should join the exchange rate mechanism of the European Monetary System (EMS). It is a subject that the experts could debate for a long time; it seems that the UK will probably still wish to stay out of the EMS at least in the short term, although when Spain joins the EMS will be reorganized, and this may provide an opportunity for the UK to reconsider its position. Certainly, this whole area of finance may be one where harmonization is less than complete by 1992, but there are still 4 years to go, and even a week is a long time in politics.

In conclusion, attempts have been made to give the history of 'Completing the internal market' and it has been possible to do little more than touch on a dozen areas of major importance to the animal feed industry. With 4 years still to go there is a lot of work left to be done. Significant changes are, and will, take place. The harmonization should lower the costs of doing business in the Community. It is obviously crucial that all in the animal feed business and ancillary industries are prepared for the opportunities opening up and the increase in merger and acquisition activity that will result. Many of the changes are of a highly technical nature, and it will be up to members of the UK industry to master them. It is to be hoped that the UK Trade Associations in conjunction with the Federations in Brussels, can do this in a manner which will maintain and improve the position of the UK in the EEC agribusiness.

Acknowledgements

I would like to thank the following who have aided me or whose papers I have made use of: FEFAC, UKASTA, GAFTA, HGCA, MAFF, NOAH, AHDA, ASI, CELCAA, Unilever, BOCMS, MMB, University of Newcastle, Midland Bank, AMC Review and DTI.

II

Poultry and Rabbit Nutrition

VITAMIN REQUIREMENTS AND ALLOWANCES FOR POULTRY

C. C. WHITEHEAD
AFRC Institute for Grassland and Animal Production, Poultry Department, Roslin, Midlothian EH25 9PS, UK

and

J. I. PORTSMOUTH
Peter Hand (GB) Ltd, Stanmore, Middlesex HA7 4AR, UK

Introduction

The subject of vitamin requirements and allowances for poultry is large and complex. The metabolic roles of the vitamins are as participants in fundamental biochemical pathways. As such, the need for any individual vitamin can be markedly influenced by many factors, such as the level of performance of a bird, its health and the balance of nutrients, including other vitamins. There are also different criteria by which the need for a vitamin can be judged, e.g. performance, health, carry over to progeny, metabolic function. Thus, many considerations have to be taken into account when assessing vitamin needs, and the concept of a value that represents the requirement, the minimum amount of the nutrient needed by the bird, has many limitations.

In the subsequent sections, current information on vitamin needs and responses are reviewed and interpreted. Some of the original data have been reassessed when it was thought that the authors had used an inappropriate method of data analysis, e.g. significance of differences between treatments rather than dose-response analysis. Two major practical problems have been apparent when assessing individual vitamins.

The first problem arises over the wide differences that can exist between different experimental estimates of a vitamin requirement. Variability may be explained by the interaction factors alluded to above, but often these interactions are not sufficiently quantified to justify a precise statement of requirement. The value chosen can therefore be somewhat subjective, but any uncertainty can be taken into account in the setting of practical allowances, as discussed in the final section.

The other main problem lies in the amount and relevance of data. For some classes of poultry, data are very few. For instance, in the case of turkey breeders, experimental information is not available on the requirements for several vitamins. For other vitamins and classes of poultry, requirements have been studied many years ago using strains of birds whose performance was poor by modern standards.

Experimental diets or conditions may have even prevented birds achieving those limited performance potentials. The relevance of requirements established under these conditions to modern high performance strains can therefore be questioned. Quantitative relationships between bird performance and nutrient requirement have been established for amino acids, but this approach has not been extended to vitamins except tentatively in the case of broiler biotin requirements (Whitehead, 1986). Requirements in this chapter are therefore discussed in relation to existing data, and the values presented can be viewed as a further, but not final, step in the continuing process of understanding and quantifying vitamin nutrition in poultry.

Metabolism and requirements

VITAMIN A

Biochemical functions

Vitamin A activity is shown by members of the retinoid class of compounds. Direct activity is shown *in vivo* by three such compounds, namely retinol, retinaldehyde and retinoic acid. Retinol is the most important compound, showing the full range of vitamin A activity. Both retinaldehyde and retinoic acid are derived from retinol within the bird and take part in vital functions. However, retinoic acid cannot support all functions, particularly reproduction. Retinaldehyde in the diet on the other hand can support all vitamin A-dependent functions, but it must first be converted in the gut to retinol before it can be absorbed in the normal way as an ester of palmitic acid.

Members of a wider class of compound, the carotenoids, show provitamin A activity since they can be converted into retinol. β-carotene is the most widely occurring provitamin A and though it is itself only poorly absorbed by poultry, it can be degraded in the gut to give retinol. Theoretically cleavage in the middle of one molecule of β-carotene could give two molecules of retinol, but it is possible that degradation starts at one end of the molecule since vitamin A activity of β-carotene is seldom observed to be greater than 50% on a weight basis, and may be as little as 10% at high dietary concentrations. Carotenoids used as yolk pigments usually do not give rise to retinol because of the position of a hydroxyl substituent. An exception however is apocarotene, often used as a yellow pigmenter, which can be degraded to retinol and may show up to 40% vitamin A activity.

Vitamin A is involved in a range of physiological functions. The best understood mechanism is in the visual process where retinaldehyde combines with the protein opsin to form a photosensitive pigment rhodopsin in the retina. Vitamin A also fulfils an active role in metabolic processes in epithelial and other tissues including adrenal function, and in spermatogenesis.

Deficiency

Vitamin A can be stored in the liver and when birds are fed a deficient diet, signs of the deficiency may take time to develop, depending upon the size of the store. In chicks, a deficiency causes a deterioration of the mucous membranes and epithelial linings brought about by a failure of glycoprotein synthesis. Hyperkeratosis occurs, involving the synthesis of keratin of higher than normal molecular weight. Hyperkeratosis of the cornea and a cloudiness of the eyes are characteristic of xerophthalmia and cause blindness. White deposits build up in the mouth, oesophagus, crop and respiratory tract. Kidney degeneration leads to a build up of uric acid and visceral gout. Abnormal bone development occurs: bones in the skull may be enlarged, vertebrae distorted and leg bones spongy with a build up of cartilage in the epiphyses.

Gross signs of deficiency include decreases in feed intake and growth. Feathers are distorted and birds show signs of weakness and incoordination. Nerve degeneration can result in ataxia. There is a general pallor, especially of combs and wattles and the birds die in very poor bodily condition.

Similar abnormalities, together with decreases in egg production and hatchability, occur in adult birds. Hens given retinoic acid as the sole source of vitamin A show none of the lesions described above and lay eggs (that fail to hatch) at a normal rate but become blind. In deficient cockerels, depressed sperm count and motility diminish fertility.

Considerable attention has been paid to the interaction between vitamin A and disease. Deficient birds show a greater susceptibility to a range of diseases and this had led to suggestions that vitamin A is involved in the production of antibodies and the build up of disease resistance.

Requirements

Vitamin A activity is often expressed in International Units; 1 IU corresponds to the activity of 0.300 µg retinol or 0.344 µg retinyl acetate.

Growth and reproduction have been taken most frequently as experimental criteria of vitamin A requirement. Liver concentrations have also beeen used, but may not give such a meaningful indication of requirement. The growth requirement of the chick seems to be about 1400 IU/kg. Data of Donovan (1965), reinterpreted by Fisher (1974) suggested that the broiler requirement was 1300 IU/kg, with the requirement of Leghorn chicks somewhat less than this. Data of Singh and Donovan (1973) indicated the requirement was not higher than 1500 IU/kg.

Results from studies on turkey growth requirements are inconsistent. A requirement of 1760 IU/kg was estimated by Stoewsand and Scott (1961). However Couch, Creger and Cavez (1971) found a response to up to 4000 IU/kg in growth to 12 weeks. More recently Prinz *et al.* (1979) concluded on the basis of large-scale field trials that 2000 IU/kg were adequate for growing and fattening turkeys. This value is taken as the requirement, though it is recognized that more may be needed on some occasions for reasons discussed later.

In adult hens, data of Reid *et al.* (1965) suggest a requirement of 3300 IU/kg and those of Reddy, Panda and Rao (1977) 3600 IU/kg. In turkeys a level of 2600 IU/kg was reported to be needed for maximum egg production and hatchability (Stoewsand and Scott, 1961) but data of Jensen (1965) suggested that 3500 IU/kg was needed.

Factors affecting the requirement

Dietary fat can enhance the absorption of retinol but high dietary protein levels can decrease liver storage and accelerate the onset of a deficiency.

Vitamin A seems to be involved in adrenal function and plays an important role in the metabolic response to stress. It also shows many interactions with disease, especially parasitic diseases. Coccidiosis is more severe in deficient birds and one of its effects is to further decrease vitamin A status by hindering intestinal absorption. Higher requirements have been demonstrated in infected birds: for instance Panda, Combs and De Volt (1964) found that 2400 IU/kg gave better growth than 790 IU/kg. In *Capillaria* infection, 10 800 IU/kg gave a better performance than 3600 IU/kg (Bauernfeind and De Ritter, 1959).

The interaction between vitamin A and other diseases is less clear. Lowered reserves of vitamin A have been observed in field outbreaks of Newcastle disease and other infections, but this may have been due to decreased feed intake, rather than a specific effect. A different insight into the relationship between the vitamin and disease is provided by the work of Tengerdy and Nockels (1975) and Tengerdy and Brown (1977) which has shown beneficial responses in haemagglutinin titres and mortality in chicks infected with *Escherichia coli* and fed high levels of vitamin A (30 000 and 60 000 IU/kg) or vitamin E. High levels of these vitamins seem to potentiate the immune system.

It is thus apparent that practical needs for vitamin A are considerably higher than the minimum requirement levels when prophylactic and therapeutic roles are considered. This leads to the feeding of relatively high levels of vitamin A in practice, as is discussed later.

Effects of excessive vitamin A

Feeding very high levels of vitamin A can cause toxicosis. The subacute toxic threshold for chicks has been estimated to be 64 mg retinol/kg body weight/day. This is very high, but nutritionists have been more concerned by reports that smaller excesses of vitamin A have sometimes been associated with impaired performance. For instance depressed growth has been reported in broilers fed 12 000 IU/kg (Jensen *et al.*, 1981; Jensen *et al.*, 1983). However, other researchers have reported no adverse effects from feeding up to 48 000 IU/kg (Marusich *et al.*, 1983). Interactions between vitamin A and other fat-soluble vitamins or diseases probably explain these inconsistent observations. High levels of vitamin A can depress absorption of other fat-soluble nutrients, including the vitamins. Interrelationships between vitamins A and D have been confirmed by Veltman, Jensen and Rowland (1983) who observed that the adverse effects of large oral doses of vitamin A, such as depressed growth and bone ash content, could be alleviated by additional vitamin D. Similar effects have been demonstrated in relation to status of vitamins E and K. In order to minimize the effects of such interactions, it is therefore important to maintain a proper balance between all the fat-soluble vitamins in practical diets.

VITAMIN D

Biochemical functions

The most active form of this vitamin in poultry is vitamin D_3 (cholecalciferol); compounds with lesser activity are vitamins D_2 (ergocalciferol), D_4, D_5, D_6 and D_7. Under natural conditions vitamin D can be formed by the action of sunlight on a naturally-occurring skin lipid, 7-dehydrocholesterol in the case of vitamin D_3. However, the lack of exposure of intensively-housed chickens to sunlight means that the full requirement for vitamin D must be supplied in the diet.

The metabolic function of vitamin D closely resembles that of a steroid hormone. Cholecalciferol is not biologically active itself; it must undergo a number of modifications. It is first hydroxylated in the liver to form 25-OH-D_3. This is then transported to the kidney where it is further hydroxylated to give a series of metabolites such as 1,25-$(OH)_2$-D_3, 24,25-$(OH)_2$-D_3 and 25,26-$(OH)_2$-D_3. Further modifications give a complex cascade of other metabolites. These compounds between them give rise to the range of biological responses attributable to vitamin D, though there is much that remains to be understood regarding the roles and interactions of these individual compounds.

The vitamin D metabolites regulate calcium metabolism and maintain, along with the peptide hormones calcitonin and parathyroid hormone, calcium and phosphorus homeostasis. 1,25-$(OH)_2$-D_3 regulates absorption of both calcium and phosphorus in the intestine by its influence on transport proteins. This metabolite also influences bone mineralization and, along with 24,25-$(OD)_2$-D_3, can also stimulate calcium mobilization from bones. Apart from the synthesis of metabolites, the kidney is also an important organ for vitamin D metabolism in relation to the control of reabsorption of calcium and phosphate from the glomerular filtrate.

Deficiency

Rickets is the classic sign of vitamin D deficiency in young animals. In chicks this takes the form of decreased bone calcification, especially at the epiphyses, with the result that the bones become soft and rubbery. All bones can be affected: the legs become bent and the spine and sternum may be crooked, with the rib-ends beaded. Growth is retarded and feathering is poor.

In hens, osteomalacia can be induced by feeding a vitamin D deficient diet. Bones become soft and may break easily and the hens adopt a characteristic squatting position. As the deficiency progresses egg shells become thinner, then soft shelled and production falls. Hatchability is also markedly decreased and embryonic abnormalities include deformed bones and oedema. The effects of a deficiency on a hen can be more severe when moderately inadequate amounts of vitamin D are fed; this allows the hen to continue egg production with depletion of body calcium reserves. When there is an extreme lack of vitamin D, egg production terminates rapidly, sparing the hen from other effects of calcium depletion.

Requirements

The discovery of the D_3 metabolites had led to investigations on whether their use in addition to or in place of cholecalciferol can enhance poultry performance. Several studies have suggested that, for normal egg production and shell quality, cholecalciferol can be replaced by either 1-OH-D_3, 25-OH-D_3, or 1,25-$(OH)_2$-D_3 (Abdulrahim, Patel and McGinnis, 1979; Soares *et al.*, 1979; Hamilton, 1980) though there was no evidence that these characteristics were improved by use of the metabolites. In order to support normal hatchability 1,25-$(OH)_2$-D_3 must be fed in conjunction with 24,25-$(OH)_2$-D_3 (Henry and Norman, 1978; Norman, Leathers and Bishop, 1983). One of the problems associated with feeding the metabolites is that they have a low toxic threshold in relation to the biologically effective dose. Thus stimulatory effects on shell calcification have been observed from feeding the correct amount of a metabolite. However, the situation is complicated by the synergistic effect of feeding combinations of metabolites and by the existence of synthetic derivatives (e.g. fluorinated compounds) that have greater biological effects than the natural metabolites (Rambeck, 1988). The situation regarding the feeding of the metabolites is thus highly complex. Much research is in progress, but as yet there does not appear to be any clear-cut strategy or advantage for their use. The requirements for vitamin D are therefore discussed in terms of cholecalciferol.

Requirements for vitamin D are often expressed in terms of international chick units (1 ICU = 0.025 µg cholecalciferol). Growth is a relatively insensitive indicator of vitamin D requirements and, since larger amounts of the vitamin are often required for optimum bone mineralization, bone ash content is generally taken as the most satisfactory criterion.

Estimates of the vitamin D requirement of growing chicks usually fall in the range 200–400 ICU/kg. The data of Waldroup, Ammerman and Harms (1963) suggested a requirement for broilers of 400 ICU/kg but the results of a later experiment (Waldroup *et al.*, 1965) were consistent with a requirement of 200 ICU/kg, provided dietary calcium was not less than 10 g/kg. A level of 200 ICU/kg gave maximum blood calcium levels in Leghorns (Parsons and Combs, 1979) but another study concluded that 500 ICU/kg were needed for optimum bone mineralization in this strain (Valinietse and Bauman, 1982). McNaughton, Day and Dilworth (1977) reported that no significant increase in bone mineralization occurred above 200 ICU/kg but a dose-response analysis of the data suggested that about 400 ICU/kg were required for maximum bone ash. On the basis of these data, the minimum requirement is therefore estimated to be 400 ICU/kg.

The requirement of the growing turkey for vitamin D was previously estimated to be 900 ICU/kg but recent data (Stevens, Blair and Riddell, 1983; Cantor, Musser and Bacon, 1980) have suggested that bone mineralization responds to higher levels and that 1200 ICU/kg may be a more suitable requirement value.

Egg shell characteristics and hatchability are the most sensitive criteria of requirements in adults. Results from several recent studies suggest that 400 ICU/kg are needed by Leghorn hens to satisfy these criteria (Abdulrahim, Patel and McGinnis, 1979; Hamilton, 1980; Shen, Summers and Leeson, 1981). There is little recent information on the needs of other strains.

In breeding turkeys a dietary level of 900 ICU/kg has been found to give maximum hatchability and poult production (Stevens *et al.*, 1984). However a related study (Stevens, Blair and Salmon, 1984) showed that a higher level (2700 ICU/kg) resulted in more active vitamin D metabolism and better bone ash in the young poults.

Metabolic relationships exist between vitamin D and several other nutrients. Because of the role of the vitamin in calcium and phosphorus homeostasis, requirements are increased if dietary levels of these minerals are deficient or imbalanced. Thus in chicks, data of Waldroup *et al.* (1965) suggested that whereas the vitamin D requirement was 200 ICU/kg with 10 g Ca and 7 g P/kg, it rose to about 800 ICU/kg with 5 g Ca and 7 g P/kg and to about 1700 ICU/kg with 5 or 10 g Ca and 5 g P/kg. However, calcium retention depresseed by high dietary fat levels cannot be restored by additional vitamin D (Whitehead, Dewar and Downie, 1972).

Excessive dietary levels of vitamin A can depress vitamin D absorption and induce signs of rickets. This effect can be overcome by additional vitamin D. Ascorbic acid also influences vitamin D status, by playing a role in the renal hydroxylation of the metabolites.

Rachitogenic substances that can markedly increase the vitamin D requirement have been reported in raw soyabean meal, isolated soya protein and rye.

Hypervitaminosis D has been demonstrated in humans and animals but poultry have a wide tolerance. The lowest level that has been reported to have a slightly adverse effect on chicks is 40 000 ICU/kg

VITAMIN E

Biochemical functions

Vitamin E activity is shown by two series of naturally occurring compounds based on a two-ring hydroquinone structure, chroman-6-ol. Highest activity is shown by the tocopherols, where the chromanol contains a saturated 16-carbon chain and methyl substituents. The corresponding tocotrienols, in which the side chain contains three double bonds, have lower biological activity. The relative vitamin activities of these compounds are given in Table 4.1. In nutritional practice it is assumed that only α-tocopherol makes an effective contribution as the vitamin; 1 international unit of vitamin E is equivalent to 1 mg *dl*-α-tocopherol acetate.

The metabolic role of vitamin E is as an antioxidant. It is a very effective scavenger of free radicals that can initiate chain reactions in the peroxidation of lipids. The structure of vitamin E enables it to form part of mitochondrial, microsomal, plasma and other membranes and it can thus protect from degradation the polyunsaturated fatty acids present in membrane phospholipids. Some synthetic antioxidants can also fulfil this role but they lack the structural specificity so that their solubility in membranes is much less than of vitamin E. For instance, butylated hydroxyanisole is 25 times less efficiently bound to microsomal membranes than vitamin E and can only depress, but not prevent, peroxidation.

Table 4.1   BIOLOGICAL ACTIVITIES OF
DIFFERENT VITAMIN E COMPOUNDS

Compound	*Activity (relative to dl-α-tocopherol acetate = 100)*
d-α-tocopherol acetate	136
d-α-tocopherol	80
dl-α-tocopherol	59
d-β-tocopherol	45
d-α-tocotrienol	13
d-γ-tocopherol	13
d-β-tocotrienol	4

Leth and Søndergaard (1977)

The other natural antioxidant system is based upon selenium as a constituent of glutathione peroxidase. This enzyme is present in plasma and in the cytosol within the cell and removes hydroperoxides formed in the first step of fatty acid oxidation. The two systems are thus complementary, but neither can be a complete substitute for the other. For instance glutathione reductase is not so effective in protecting membranes in the brain as vitamin E. Likewise vitamin E cannot replace selenium in its role in other reactions involving products of hydroperoxides.

Deficiency

Vitamin E deficiency causes several distinct pathological abnormalities, mostly associated with defective membrane structures.

Encephalomalacia, or 'crazy chick disease', affects chicks 2–6 weeks of age. The head becomes twisted either backwards or forwards and uncoordinated muscular spasms affect the legs. Prostration and death ensue. The main lesions are in the cerebellum which is soft, swollen and oedematous often with small surface haemorrhages. Selenium is not effective in preventing this condition.

Abnormal permeability of capillary walls in subcutaneous tissues leads to oedema in a condition known as exudative diathesis. Leakages of blood fluid through the capillaries and from minor haemorrhages in muscles gives the oedematous fluid a bluish-green colour that can be seen through the skin. The condition can occur at any age but is most prevalent in young growing chickens and turkeys. It is responsive to dietary additions of vitamin E or selenium.

A type of muscular dystrophy associated with vitamin E deficiency occurs in chickens, turkeys and chicks. The condition is most noticeable in breast and thigh muscles which develop pale streaks of dystrophic degenerated fibres of high lipid content. Degeneration of the gizzard and heart muscles has also been seen in affected turkeys. The condition responds to vitamin E or selenium.

Other abnormalities in vitamin E deficient birds include increased susceptibility of erythrocytes to haemolysis, testicular degeneration in males and depressed egg production and hatchability in females. Inadequate carry over to the egg can lead to poor viability of progeny, with increased susceptility to encephalomalacia and exudative diathesis.

Requirements

Normal health is the main criterion for vitamin E requirement. The need for vitamin E is highly dependent upon two other dietary factors, selenium and polyunsaturated fatty acids (PUFA). The requirement for vitamin E falls as the selenium content of the diet increases and it is experimentally difficult to demonstrate a precise requirement. It is suggested that poultry diets should contain a minimum of 10 mg vitamin E/kg.

Increasing the PUFA content of a diet increases the metabolic need for antioxidants. When vitamin E levels are inadequate, increasing dietary PUFA can depress egg production and hatchability in hens and increase the susceptibility of chicks to the vitamin E deficiency conditions. Quantitative relationships between PUFA levels and vitamin E needs have been established on the basis of haemolysis of rat erythrocytes. It has been calculated that an additional 1 mg vitamin E was required by chicks per 1 g of linoleic acid (Weiser and Salkeld, 1977).

Vitamin A also influences vitamin E status and additional levels of vitamin E need to be fed to maintain tissue concentrations at high vitamin A intakes, as is discussed later.

VITAMIN K

Biochemical functions

Compounds with vitamin K activity are based on the structure of menadione (2-methyl-1,4-naphthoquinone). The lipid soluble vitamin K_1 or phylloquinone series occurs naturally in plants and contains a phytyl substituent of variable size at the 3-position. The menaquinones or vitamin K_2 series are also lipid soluble and are synthesized by microorganisms. Their structures are similar to those of the phylloquinones save that the side chain is composed of 4, 5 or 6 polymerized isoprene units.

Phylloquinone and menaquinone both exert direct biological activity and can be incorporated into the egg if fed directly. Menadione, or vitamin K_3, acts as a provitamin and must first be converted in the liver into menaquinone before it can exert a biological effect or be incorporated into the egg. However, menadione can be manufactured more readily than the other forms and is the usual basis for synthetic supplements. Menadione itself is lipid soluble but is unstable and can be an irritant to the skin and respiratory tract. Moreover it has a lower biological activity on a molar basis than phylloquinone. There are various water soluble derivatives of menadione containing a bisulphite substituent at the 4-position that have higher biological activity than menadione and are about as active as phylloquinone. The relative potencies of phylloquinone and menadione sodium bisulphite (MSB) have been found to be similar and 2.5 times greater than that of menadione on a molar basis. The range of these derivatives also includes sodium bisulphite complex (MSBC), menadione dimethylpyrimidol bisulphite (MPB) and menadione nicotinamide bisulphite (MNB). These derivatives are all more stable than menadione and are widely used in practical nutrition.

There is conflicting evidence on the relative activities of the various menadione derivatives. Griminger (1965) has suggested that MPB is more effective than MSBC by a factor of 1.4 in lowering prothrombin time in the presence of an inhibitor. However, Weiser and Tagwerker (1981) have reported similar activities for MSB and MPB using a different test. Charles (1977) compared MSB and MSBC from several manufacturing sources and found an approximately similar range of activities for the two compounds. However, there was considerable variation in activity between different samples of each product. Chemical studies showed structural differences between samples, leading to the conclusion that vitamin K derivatives from different processes might contain isomers which alter the overall biological activity of the product. The quality of the individual product is therefore an important factor in determining the overall effectiveness of a vitamin K source.

In animal metabolism, vitamin K is involved in the structural modification of several pre-formed proteins. It is required for the carboxylation of certain glutamate residues in peptides and proteins to give γ-carboxyglutamate units. These units will strongly bind calcium ions and it seems that vitamin K-dependent proteins play an important part in all major areas of calcium metabolism. For instance the proteins are found in bones (osteocalcin), shell gland and egg shell, several other organs including the kidney and in pathological calcifications such as renal calculi and calcified tissues. However, it is in relation to blood clotting that the metabolic role of vitamin K is best understood. The vitamin is needed for the synthesis of several proteins, including prothrombin, that play essential roles in this process. When tissue damage occurs a variety of factors combine to convert prothrombin into thrombin. Thrombin catalyses the conversion of fibrinogen into fibrin which forms into strands, trapping blood cells and forming a clot.

Deficiency

Vitamin K deficiency results in lowered synthesis of proteins containing γ-carboxyglutamate. The consequences of this are most readily observed in the blood clotting process which becomes retarded. This effect can be quantified in terms of the 'prothrombin time' which can be increased from a normal 17–20 s up to periods of several minutes and means that prolonged bleeding can arise from relatively minor injuries or bruises. A mild deficiency can be manifest by small haemorrhagic blemishes in the skin and other tissues and organs throughout the body. With more severe deficiency, generalized haemorrhaging can occur in muscles and other tissues and in extreme cases the bird can bleed to death. Birds are pale and anaemic both from loss of blood and hyperplasia of bone marrow. Eggs from deficient hens have depressed hatchability and signs of haemorrhaging may be apparent in the embryos. Carry over of the vitamin to the hatching chick is important and an inadequacy can lead to haemorrhagic conditions in very young chicks. In bones, lack of vitamin K results in lowered concentrations of some proteins and impairment of the mineral phase of bone development.

Requirements

Quantitative aspects of bird performance are not greatly influenced by vitamin K status. Instead, greater importance is assumed by qualitative criteria of requirement, such as good blood clotting ability and resultant freedom from haemorrhages and other blemishes on the skin or subcutaneously in live or processed birds. In this regard, measurement of prothrombin time is usually taken as a quantitative basis for the assessment of vitamin K requirement or the comparison of the relative activities of different forms.

Estimates of the vitamin K requirements of young growing birds are very variable. Edens *et al.* (1970) found that the response to vitamin K was similar in two lines of chicks selected for fast or slow prothrombin times; maximum response was achieved with a dietary concentration of about 0.2 mg menadione/kg. However, earlier studies by Nelson and Norris (1960; 1961a) concluded that the requirement of the chick was about 0.5 mg phylloquinone/kg. Corresponding requirements were 0.35 mg MSB/kg (1 mg phylloquinone equimolar with 0.73 mg MSB) or 0.48 mg menadione/kg (menadione only 40% active). These values are the preferred estimates of the requirement in chicks. The turkey poult has been estimated to need 1.1 mg MSB or 1.5 mg phylloquinone/kg for normal prothrombin time (Griminger, 1957).

Information on the requirements of adult birds is limited. Fry *et al.* (1968) reported that 0.2–0.4 mg MSBC/kg was required by breeding hens but the preferred estimate of requirement is taken as 1 mg phylloquinone/kg on the basis of the findings of Griminger (1964). Experimental evidence for the requirement of the breeding turkey is lacking.

Menaquinone can be synthesized by intestinal microorganisms but reingestion of faeces rather than direct absorption is likely to be the main way the bird could benefit from this. A number of drugs can interact with vitamin K. Sulphonamides can induce signs of deficiency, probably by inhibition of vitamin K metabolism in tissues rather than by inhibition of intestinal synthesis. The addition of 1–2 g sulphaquinoxaline/kg diet can increase the phylloquinone requirement 4- to 7-fold (Griminger and Donis, 1960; Nelson and Norris, 1961b). However, the relative abilities of the different forms of vitamin K to counteract the effects of the drug seem to differ, so that requirements for other forms may be even higher. Thus, Nelson and Norris (1961b) estimated that the activities of MSB and menadione were 70% and 40% respectively relative to phylloquinone in countering the effects of 1 g sulphaquinone/kg diet. Frost, Perdue and Spruth (1956) suggested that menadione was only 10–17% as effective as MSB.

Anticoagulants such as dicoumarol and warfarin act by inhibiting directly vitamin K metabolism in the coagulation process. Thus, the requirement for phylloquinone is approximately doubled by the presence of 100 mg warfarin/kg diet. Vitamin K_3 compounds are ineffective in countering these anticoagulants, presumably because their conversion to metabolically active forms is also inhibited.

Other growth promoters such as tetracycline antibiotics and arsanilic acid can also inhibit vitamin K but the effects are relatively small and can be overcome by moderate increases in vitamin K intake. Likewise high levels of vitamin A also necessitate a small increase in dietary vitamin K.

Disease and stress also influence vitamin K status in the bird. Infections involving haemorrhaging are particularly serious. Thus mortality from coccidiosis is enhanced by a lack of vitamin K. The implications of this relationship for vitamin K requirement depend upon the particular derivative being fed. Thus, the requirement for phylloquinone is little changed (Harms, Waldroup and Cox, 1962) but, perhaps as a result of destruction or impaired absorption, requirements for menadione derivatives are higher, perhaps by factors of 2 and 5 for MSBC and menadione respectively. Increased prothrombin times have been reported to accompany a range of other diseases and it is likely that stress of whatever origin can increase the need for vitamin K. This may account for the occasional observations of field outbreaks of haemorrhagic syndromes responsive to vitamin K occurring with diets not thought to be initially deficient in the vitamin.

THIAMIN

Biochemical functions

Thiamin, also known as vitamin B_1 or aneurine, is a cationic combination of pyridine and a thiazole ring. It is isolated most frequently as the chloride hydrochloride. The biological activity is greatly diminished by even small changes in molecular structure; the only derivatives with substantial activity are those containing ethyl or *n*-propyl moieties in place of the usual methyl substituent on the pyridine ring. The metabolically active forms of thiamin are the phosphorylated derivatives. Thiamin pyrophosphate (TPP) is the most common but mono- and triphosphates also occur.

The best understood role of thiamin is as a cofactor for enzymes involved in several areas of carbohydrate metabolism. The reactions involve decarboxylation of various substrates and contribute to the use of carbohydrates as energy sources. Thus in glycolysis, pyruvate is converted to acetyl coA and CO_2 by the thiamin-dependent enzyme pyruvate dehydrogenase. In the citric acid cycle, the conversion of α-ketoglutarate to succinyl coA and CO_2 is also thiamin-dependent. TPP is a coenzyme for transketolase, the enzyme in the pentose pathway that converts hexoses to pentoses in the synthesis of nucleotides and nucleic acids. It is also a cofactor in the decarboxylation of glyoxylic acid and has been associated with the activities of lactate dehydrogenase and glycerophosphate dehydrogenase. Thiamin also plays a role in the operation of the nervous system. Thiamin is liberated by nerve action and can stimulate the synthesis of acetyl choline. This neurological role of thiamin appears to be independent of its action as an enzyme cofactor, since TPP is without effect, but the precise mechanism is not understood.

Deficiency

A moderate deficiency of thiamin in young birds results in abnormalities in carbohydrate metabolism. Impairment of glycolysis causes elevations in blood pyruvate and lactate levels. Inhibition of the citric acid cycle causes a build up of α-ketoglutarate. These and other metabolic blockages mean that the metabolizable energy derived from carbohydrates is decreased in thiamin deficiency. More severe deficiency results in polyneuritis in both chicks and poults. The first signs can occur as early as 3 or 4 days after hatching and death can ensue within 7–10 days in an extreme deficiency. The initial signs are weakness and nervousness. Then spasms occur, the head becomes retracted and the legs extend so that the birds topple over backwards. Lesions have been observed in the duodenum and pancreas but not, at the microscopic level, in brain or nervous tissue. In adult birds thiamin deficiency can result in decreased food intake, egg production and hatchability.

Requirements

Though the activity of transketolase can give specific information on thiamin status, production responses have so far generally been used to determine thiamin requirements.

Requirements for thiamin can be affected by environmental temperature. The requirement to prevent polyneuritis has been found to be three times as high at 32°C than at 21°C. It is probable that abnormally low temperatures can also increase the requirement. Since thiamin is needed for the utilization of carbohydrate, the requirement for it is higher in diets of high carbohydrate content. Results of Peacock (1970) suggest a probable requirement of 1.0 mg/kg for a diet of 12.6 MJ metabolizable energy (ME) and 30 g fat/kg, decreasing to 0.75 mg/kg for a diet containing 230 g fat/kg. Other studies by Thornton and Shutze (1960) have suggested that an adequate dietary level of broiler chicks, extrapolated to a dietary content of 12.6 MJ ME/kg, was 1.5 mg/kg and this value is taken as the preferred estimate of requirement. For growing turkeys data of Robenalt (1960) and Sullivan, Heil and Armintrout (1967) suggest that 2 mg/kg is required for maximum growth rate and viability. A dietary level of 0.7 mg/kg has been found to give maximum hatchability in breeding hens (Polin, Wynosky and Porter, 1962a, b; 1963).

A wide range of other factors can influence thiamin status of birds or feeds. Coccidia compete with the host for thiamin in the intestine and can depress status. Amprolium, an anticoccidial drug, blocks thiamin metabolism and can induce a deficiency in the bird if dietary thiamin levels are inadequate. Furazolidone, if used in excess, can also impair thiamin status. Other anti-thiamin compounds can modify the thiamin molecule and inactivate it. These thiaminases occur in a range of plants, fish, shellfish and crustacea and can inactivate thiamin during feed processing. However, these factors are not thought to have a major effect on thiamin nutrition under practical conditions usually encountered in the UK.

RIBOFLAVIN

Biochemical functions

Riboflavin, or vitamin B_2, is a ribityl derivative of dimethylisoalloxazine. In crystalline form it is stable to heat and resistant to oxidation but can decompose in alkaline solution or on exposure to light. In the body it exists as free riboflavin usually only in urine and in the retina. In other tissues it is present as riboflavin monophosphate or flavin adenine dinucleotide. As such it is a cofactor for enzymes involved in a wide range of oxidation reactions generating energy from lipid, carbohydrate and amino acid sources. Thus acyl coA dehydrogenases participate in the oxidation of fatty acid coA esters of varying chain length, glucose oxidase is specifically involved in the oxidation of glucose and there are various amino acid oxidases. Succinate dehydrogenase and NADH dehydrogenase also take part in central oxidative pathways. Comparatively high concentrations of riboflavin are therefore present in metabolically active tissues such as muscles, liver and kidney.

Deficiency

Young birds fed a deficient diet have a depressed food intake and growth rate. They may suffer from diarrhoea due to inflammation of the mucous membranes of the digestive tract. The skin becomes rough, with a mild scaly dermatitis. Characteristic leg abnormalities, involving a curled toe paralysis occur in which the toes become compressed together and curl inwards. The bird resorts to walking on the side of its feet and on its hocks. Ultimately the birds lie prostrate, sometimes with legs extended in different directions. This leg abnormality is brought about by degeneration of the myelin sheaths of peripheral nerves, with the sciatic nerve showing a particularly marked enlargement. Young turkeys can show a severe dermatitis of mouth and legs and eyes may become closed by the crusting over of a sticky exudate. Reproduction is impaired in adult birds. Rate of egg production is depressed and there may be high embryonic mortality, especially during the first or second week of incubation. Deficient embryos are stunted and oedematous. They may also show a 'clubbed down' in which the feathers are coiled and fail to rupture the surrounding sheath. Hatching chicks may also show this condition. Clubbed down is not an inevitable sequel to riboflavin deficiency, however, since it also requires the presence of a specific gene related to down pigmentation. Thus, deficient embryos lacking this particular gene do not show clubbed down.

Requirements

Requirements of growing birds have been assessed using growth rate, freedom from toe lesions or the activity in erythrocytes of glutathione reductase as criteria. Requirements of both chickens and turkeys seem to be in the range of 4–5 mg/kg. Thus Jeroch (1970) has reported a requirement of 3.5 mg/kg for young broilers and Yoshida, Hoshii and Morimoto (1966) a value of 5 mg/kg (converted to a diet of 13 MJ ME/kg). More recently Ruiz and Harms (1988a) have suggested requirements of 4 mg/kg for optimum growth but about 5 mg/kg for complete freedom from toe lesions. The last value is taken as the preferred requirement for broilers. Available evidence suggests that 4 mg/kg is adequate for turkeys (Jeroch, Prinz and Hennig, 1978; Lee, 1982; Ruiz and Harms, 1986a).

Riboflavin requirements for egg production are met by 2–2.5 mg/kg. However, requirements for hatchability are higher. Tinte and Austic (1974) suggested that 2.75 mg/kg was marginal and the preferred estimate is taken as 4 mg/kg on the basis of data of Jeroch (1971). Data on the requirements of breeding turkeys are very sparse and a requirement value of 4 mg/kg is suggested.

NICOTINIC ACID

Biochemical functions

Nicotinic acid or niacin (3-pyridinecarboxylic acid) and its amine derivative are biologically active in animals. Nicotinic acid itself occurs widely in plants but in animals it is converted into nicotinamide and this compound is thus the source of vitamin activity in feedstuffs of animal origin.

In animal metabolism nicotinamide is the functional group for two coenzymes, nicotinamide adenine dinucleotide (NAD) and nicotinamide adenine dinucleotide phosphate (NADP). These enzymes are involved in the metabolic transfer of hydrogen as receptors or, in the reduced forms, as donors. They are essential in processes involving the production and use of metabolic energy, such as the citric acid cycle, and play fundamental roles in amino acid, carbohydrate and fatty acid metabolism. Nicotinic acid when supplied at pharmacological levels can have an antilipolytic effect.

Deficiency

Nicotinic acid deficiency has a depressing effect on the activities of NAD and NADP and their dependent metabolic pathways. In humans the deficiency condition is pellagra, characterized by dermatitis, dementia and diarrhoea. In poultry the first signs are general in nature, e.g. depressions in growth and viability of young birds or egg production and hatchability in adults. Dermatitis, enlargement of the hock joint, bending of the leg bones and slippage of the gastrocnemius tendon are more specific lesions occurring in growing chicks and poults. 'Black tongue' has also been reported to be a specific effect in chicks.

Relationship with tryptophan

The main nutritional interaction of this vitamin involves tryptophan. NAD is one of the degradation products of tryptophan and if the tryptophan intake is high, breakdown of the excess can result in a sparing of the dietary requirement for nicotinic acid. This sparing is likely to be minimal with dietary tryptophan levels near the requirement when most of the vitamin is used for growth or production and little degradation takes place. Nicotinic acid cannot spare the requirement for tryptophan since the reaction is not reversible.

The quantitative relationship between tryptophan and nicotinic acid is dependent upon (a) the amount of excess tryptophan, and (b) competition between different breakdown pathways. Degradation of tryptophan can proceed by two routes, only one of which gives nicotinic acid. Genetic variation in the balance between these two pathways can lead to strain differences in nicotinic acid requirements. Nutritional factors can also influence the conversion of tryptophan to nicotinic acid. For instance, riboflavin and pyridoxine are involved in different steps and a deficiency of either can inhibit the process. High levels of dietary fat, especially saturated fat, can also suppress the interconversion. Variation in these factors is the likely explanation for the wide range in the estimates of the efficiency of the conversion of L-tryptophan to nicotinic acid in the chicken. Estimates range from 40 to 45:1 on a weight basis (Baker, Allen and Kleiss, 1973), through 50 to 70:1 (Yen, Jensen and Baker, 1977) up to 187:1 (Manoukas, Ringrose and Teerie, 1968). In practice, diets seldom contain excessive amounts of tryptophan and requirements are discussed on the assumption that there is minimal contribution from tryptophan.

Requirements

The vitamin is available commercially as nicotinic acid or nicotinamide and comparisons of the relative biological activities of these compounds is of interest. Since the vitamin is biologically active in the animal as the amine, it might be supposed that feeding it as the amine might be more effective. There is evidence that nicotinamide in the diet can be incorporated unaltered into NAD, but there is also evidence that a proportion of nicotinamide is deaminated in the gut and absorbed as nicotinate. The conversion of nicotinic acid into the amide in the chicken seems to be very efficient and feeding comparisons suggest that any differences in biological activity between the two compounds are quite small. Requirements are therefore discussed on the assumption that the two forms are equally active (their molecular weights are nearly identical).

Growth and freedom from leg lesions have generally been used as criteria for assessing nicotinic acid requirements in young birds; it has sometimes been observed that the requirement is higher to meet the latter criterion. The previous ARC (1974) requirement was 28 mg/kg for young chickens, but more recent information suggests that the current value for modern broilers is much higher. The studies have involved supplementation of maize-soyabean diets containing 21 mg total nicotinic acid/kg (probably about 14 mg available nicotinic acid/kg). In one brief report Ruiz and Harms (1986b) claimed the minimum supplement needed was 33 mg/kg (for a diet containing 2.9 g tryptophan/kg), suggesting a requirement of 47 mg/kg. In a more detailed account of a series of experiments, Waldroup *et al.* (1985) showed consistent responses to supplementation above the 33 mg/kg level. Tryptophan levels in these diets were not reported, but have been calculated to be about 2.3 g/kg for starter diets. This value is somewhat lower than the value reported by Ruiz and Harms (1986b) and the difference, if real, may account for the apparent difference in nicotinic acid requirement. For diets not containing tryptophan in excess of the requirement, it is suggested that an appropriate value for the nicotinic acid requirement of starting broilers is 70 mg/kg.

Quantitative data on turkey responses to nicotinic acid are also few, and complicated by tryptophan interactions. Slinger *et al.* (1953) reported best performance with a supplement of 70 mg/kg to a practical diet, but data of Scott (1953) suggested a supplemental level of 45–50 mg/kg was adequate for a diet containing 27 mg/kg. Zaviezo and McGinnis (1980) reported the requirement to be 40 mg/kg feed with 2.6 g tryptophan/kg and 35 mg/kg with 3.2 g tryptophan/kg. Ruiz and Harms (1988b) have recently reported that 44 mg/kg feed (about 40 mg available/kg feed?) was needed for optimum growth and minimal incidence of leg disorders, in a diet containing a generous amount of tryptophan (3.5 g/kg). With diets just meeting the tryptophan requirement, however, Fisher (1974) reported data suggesting that 60 mg available nicotinic acid/kg was needed for optimum growth and that a slightly higher amount (70 mg/kg) was required for minimum incidence of leg disorders. This value is taken as the preferred requirement for diets not containing excessive tryptophan levels.

The requirement of laying chickens for nicotinic acid seems to be comparatively low. Adams and Carrick (1967) indicated that 6 mg/kg was adequate for egg production but that 10 mg gave better hatchability. Corresponding values from Ringrose *et al.* (1965) are 8 and 10 mg/kg and these are taken as the preferred estimates of requirement for laying and breeding hens respectively. A recent report by Ouart, Harms and Wilson (1987) has not shown any benefit to supplementation of practical maize-soya or wheat-soya layers' diets. Though dementia is a characteristic of nicotinic acid deficiency in humans, nicotinic acid within the normal nutritional ranges does not seem to influence nervousness in battery hens. For breeding turkeys, the only report (Harms *et al.*, 1988) has indicated that a practical maize–soya diet contained sufficient nicotinic acid (23 mg/kg total; 15 mg/kg available?) for maximum egg production, fertility and hatchability. However, about 30 mg available nicotinic acid/kg was needed to maximize egg and body weights. This latter value is taken as the requirement.

PANTOTHENIC ACID

Biochemical functions

Pantothenic acid consists of pantoic acid (dihydroxydimethylbutyric acid) joined to β-alanine by an amide linkage. Only the *d*-form is biologically active. The alcohol derivative, pantothenol, is fully active but other structural changes, such as the addition of another methyl group, give rise either to a loss of vitamin potency or antagonistic activity. Free pantothenic acid occurs only rarely in biological systems; it is most usually present as a component of coenzyme A or acyl carrier proteins. In its isolated form it is a very hygroscopic, unstable oil. The sodium and calcium salts are much more stable. The latter is the most common form used as a feed additive and is stable to air and light if protected from humidity but is slightly sensitive to heat. Pantothenol is sometimes used in liquid preparations.

The metabolic role of pantothenic acid is as a constituent of enzymes and coenzymes. In combination with a nucleotide, it forms part of coenzyme A which plays a central role in many areas of metabolism. As acetyl coenzyme A, it is an essential intermediate in the energy-yielding oxidation of fat, carbohydrate and amino acids. Thus, in the β-oxidation of fatty acids, long chain fatty acyl coenzyme A is successively degraded into molecules of acetyl coenzyme A, with liberation of energy as ATP. Acetyl coenzyme A can also be produced by degradation of carbohydrates and amino acids and itself participates in the generation of energy in the citric acid cycle as a result of its part in the synthesis of citrate by condensation with oxalacetate. Acetyl coenzyme A is also involved in other synthetic pathways. It takes part in acetylation reactions, such as the synthesis of acetyl choline and acetylhexosamines and in the synthesis of cholesterol and steroid hormones. Pantothenic acid also has two roles in lipogenesis: it forms part of acyl carrier protein, a coenzyme for fatty acid synthetase and, as acetyl coenzyme A, is converted to malonyl coenzyme A (by a biotin-dependent mechanism) in the first step of fatty acid synthesis.

Deficiency

In pantothenic acid deficiency, the utilization of energy is impaired and nutrient retention is depressed. Birds seem to be able to obtain almost the full metabolizable energy value from diets, but a greater heat increment and a depressed deposition of carcass energy are indicative of lower productive energy. Nervous disorders occur, probably related to inhibition of acetylcholine synthesis. The endocrine system is also affected: synthesis of steroid hormones is decreased and the activity of the adrenal gland is lowered. These last changes may be responsible for the relationships observed between the vitamin and stress and fertility. Disease resistance is lowered in deficient birds.

   Histological abnormalities include myelin degeneration of nerves, necrosis of lymphocytes, fatty degeneration of the liver and changes in Lieberkuhn's glands in the duodenum. Clinical signs of a deficiency resemble those of biotin in several ways. Growth and feed conversion are depressed and crusty lesions develop at the corners of the beak and around the eyelids. Skin on the feet is affected with hyper- and parakeratosis and acanthosis, and peels off, allowing cracks to develop. Feathering is retarded and colouring is lost. Nervous problems result in locomotor disorders and ataxia. In young turkeys, deformation of the hock joint and slippage of the gastrocnemius tendon have been associated with a lack of pantothenic acid. In adult birds egg laying and, to a greater extent, hatchability are diminished. With a mild deficiency embryonic mortality occurs mainly during the last day or two of incubation but with a more severe deficiency mortality during the first week increases. Embryos show malformations of the head, back and feathers together with fatty liver, hydrocephalus and oedema of the skin.

Requirements

Relationships have been established between pantothenic acid and other nutritional and environmental factors. These include folic acid, biotin (fatty acid synthesis), copper (inhibition of coenzyme A function) and stress but quantitative effects on pantothenic acid requirements have not been established. However, the relationship between pantothenic acid and cobalamin has been investigated in more detail and it has been found that pantothenic acid requirements are about 50–100% higher in hens and growing chickens fed insufficient cobalamin.

   Requirements of growing birds have been assessed by several criteria. Roth-Maier and Kirchgessner (1973) did not observe a growth response in broilers to supplementation of a diet containing 7.8 mg/kg. However, in a subsequent experiment (Roth-Maier and Kirchgessner, 1976) the responses obtained suggested that the maximum amount fed (8 mg/kg) may not have been sufficient. Beagle and

Begin (1976) measured energy utilization in two experiments and found that productive energy derived by broilers was not maximal with a diet containing 10 mg/kg. Dietary levels of 12.5 and 17 mg/kg in the different experiments gave very similar improvements in productive energy relative to the diet with 10 mg/kg. The broiler requirement has been set on the basis of these observations at 12 mg/kg for a diet containing 13 MJ metabolizable energy/kg. Requirements of growing turkeys are about the same. Slinger and Pepper (1954) found that the growth of poults fed a diet calculated to contain 11.5 mg/kg responded to a supplement but Jeroch, Prinz and Hennig (1978) have concluded from a large series of trials that practical diets calculated to contain 13 mg/kg of natural vitamin did not require supplementation, using feed intake, growth and mortality as criteria.

There is little information on the requirements of adult birds of modern strains. Beer, Scott and Nesheim (1963) reported that 1.9 mg/kg was adequate for egg production of White Leghorns laying at a rate of 65%. However, evidence that this level may no longer be adequate, especially for heavier types of bird, has come from the findings of Leeson, Reinhart and Summers (1979) that egg production of Rhode Island Red hens was depressed with a diet calculated to contain about 6 mg/kg of natural vitamin. In view of doubts over calculated dietary levels (discussed later) it is not possible to set with certainty a requirement value for laying hens. However, a value of 7 mg/kg may be appropriate for highly productive heavy birds. In breeding hens Balloun and Phillips (1957) estimated a requirement of 7 mg/kg for hatchability. However, there was evidence that higher levels improved chick viability and growth potential, hence a requirement of 10 mg/kg is considered more satisfactory. The requirement of the breeding turkey is set at 16 mg/kg on the basis of data of Kratzer *et al.* (1955).

PYRIDOXINE

Biochemical function

Pyridoxine or vitamin B_6 is a hydroxymethyl derivative of pyridine. It occurs naturally in plants but in animals it is converted into pyridoxal and pyridoxine. These compounds are then phosphorylated and incorporated into enzyme systems as pyridoxal phosphate. Synthetic pyridoxine is used as its hydrochloride. The lower stabilities of pyridoxal and pyridoxamine mean they are less active than pyridoxine when added to diets.

Pyridoxal phosphate is a cofactor for a number of enzymes, almost all of which catalyse reactions involving amino acids. These reactions involve transamination, decarboxylation or cleavage and include the removal of water or hydrogen sulphide. Thus pyridoxine is important for the metabolism of sulphur-containing amino acids and it is also involved in the synthesis of non-essential amino acids as well as the conversion of tryptophan to nicotinic acid (inhibition of this pathway leads to increased excretion of xanthurenic acid).

Deficiency

Amino acid metabolism is widely impaired in pyridoxine deficiency. Utilization of amino acids such as methionine is impaired, as are the interconversions needed for the efficient incorporation of amino acids into proteins. Both intestinal absorption of amino acids and nitrogen retention are also depressed. There is inadequate synthesis of some proteins and this is reflected in blood where there are changes in the concentrations of individual proteins. A decrease in haemoglobin levels results in microcytic anaemia; decreases in globulins can lower antibody levels and increase the sensitivity of animals to infections. Changes in lipid metabolism, such as depressed fat deposition and elevated plasma cholesterol may be secondary effects.

Clinical signs of severe deficiency in chicks include poor growth and feather development, ataxia and death. Nervous symptoms include tilting of the head and uncontrolled running movements and convulsions. The nervous disorder is not seen in marginal deficiency but growth depression still occurs, accompanied sometimes by leg bone abnormalities. Distortion of the epiphyseal cartilage causes curvature of the leg bone and tendon slippage. Other abnormalities to have been reported are gizzard erosion, pendulous crop and haemorrhages around the follicles of wing feathers. Similar changes can occur in poults, although anaemia has not been reported.

In adult chickens and turkeys deficiency causes decreases in appetite and body weight, especially fat depots. Egg production and hatchability both decrease to zero in severe deficiency and there is regression of combs and wattles. Dermal lesions have been reported around the beaks and on the legs of turkeys.

Requirement

The requirement for pyridoxine is very dependent on the protein level and amino acid composition of the diet. The requirement is elevated when the diet contains excess amino acids that must be catabolized before being excreted. Enzymes involved in the metabolism of sulphur-containing amino acids are very sensitive to the supply of the vitamin. Thus a marginal deficiency of methionine can be overcome by the supply of extra pyridoxine (Kazemi and Daghir, 1971). However, excess methionine exacerbates a deficiency of pyridoxine because the vitamin is also needed to degrade the surplus amino acid. Similar relationships with other amino acids such as serine have also been reported.

There are several criteria by which to assess the requirement in broilers. However, biochemical criteria such as the excretion of xanthurenic acid after dosing with tryptophan or the measurement of glutamate–oxaloacetate transaminase and glutamate–pyruvate transaminase have, in general, not proved so successful in indicating the optimum requirement of the broiler as they have for the pig. Hence most of the estimates or requirement are based upon production data such as growth rate and the retention or accumulation of protein or fat in the carcass.

There have been many studies on the requirements of broilers using both purified and practical diets, but interpretation of results is complicated by the wide differences in the dietary energy and protein levels used. However, extrapolation of results to diets of conventional broiler starter specification (i.e. 13 MJ ME and 230 g CP/kg) suggests that the pyridoxine requirement is in the region of 3.5–4.0 mg/kg (Kirchgessner and Maier, 1968; Gries and Scott, 1972; Daghir and Shah, 1973). The latter value is taken as the preferred requirement.

There are few recent data on which to base the requirement of growing turkeys. Data of Sullivan, Heil and Armintrout (1967) suggested the requirement was 4.5 mg/kg but more recently Waldroup *et al.* (1976) reported that male turkeys showed a growth response to supplementation of a practical maize–soyabean diet. A value of 6 mg/kg may therefore represent a better estimate of the current growing turkey requirement.

In adult birds the most recent studies suggest that the requirements for egg laying and hatchability in chickens are 2 and 4 mg/kg respectively (Abend and Jeroch, 1973; Weiss and Scott, 1979). There is little specific information on breeding turkey requirements.

BIOTIN

Biochemical function

The structure of biotin is based on an imidazole ring with a valeric acid side chain. There are eight optically active forms of this structure but only d-(+)-biotin is biologically active. d-Biotinol and biocytin (a naturally-occurring form of biotin bound to lysine) can be converted to biotin in animal tissues but other structurally-related compounds are only weakly active or strongly antagonistic.

Biotin acts as a cofactor for several enzymes involved in carboxylation reactions. These include pyruvate carboxylase which catalyses the conversion within the mitochondrion of pyruvate to oxaloacetate, an intermediate in the synthesis of glucose. Oxaloacetate is also required by acetyl coA which condenses with it to form citrate before passing into the cytoplasm to enter the lipogenic pathway. Acetyl coA carboxylase, another biotin dependent enzyme, catalyses the conversion of acetyl coA to malonyl coA, the first step in the synthesis of fatty acids. A third enzyme, propionyl coA carboxylase converts propionyl coA into methylmalonyl coA. Biotin-dependent enzymes are thus directly involved in the important processes of gluconeogenesis and lipogenesis and can have effects on other pathways by their influences on many metabolic intermediates. Biotin is covalently bound to its enzymes and a lack of biotin results in decreases in enzyme specific activities. Impairment of lipogenesis is reflected in changes in tissue fatty acids: the proportions of monoenoic fatty acids, especially palmitoleic, at the expense of saturated fatty acids, mainly stearic. There may also be inhibition of the conversion of linoleic to arachidonic acid. Gluconeogenesis via pyruvate is depressed and indirect effects of deficiency depress other metabolic pathways including protein and purine synthesis and oxidative phosphorylation.

Deficiency

Skin and bone abnormalities are the classic lesions of biotin deficiency. In young birds growth is depressed and lesions develop on the foot pads. Hyperkeratosis and acanthosis lead to the formation of cracks and encrustations which may become infected. Dermal lesions may also form at the angles of the beak and around the eyelids and the skin over the rest of the body becomes rough. The foot pad lesions can be exacerbated by poor litter conditions and may lead to the bird spending a longer time sitting, encouraging the formation of breast blisters. Bone lesions are characterized by chondrodystrophy: the most obvious defects are shortened and twisted metatarsal bones, sometimes accompanied by slippage of the gastrocnemius tendon. 'Parrot beak' can also occur.

Fatty liver and kidney syndrome (FLKS) is a condition related to biotin deficiency that specifically affects young growing chickens, mainly broilers. Deficient chicks experiencing a severe loss of activity of pyruvate carboxylase may have difficulty maintaining blood glucose at normal levels. If these birds suffer even a small stress, interruption of feeding or depletion of limited glycogen reserves may lead to them developing a progressive and usually fatal hypoglycaemia. An accompanying mobilization of adipose tissue lipids, perhaps in a vain attempt to counter the metabolic abnormalities, leads to the infiltration of histologically visible lipid not only in liver and kidney, as implied by the name, but in many other tissues as well.

The manifestation of biotin deficiency, whether as classic lesions of FLKS, is influenced considerably by the effect of other nutrients, mainly protein and fat, on the partitioning of biotin between the different biotin-dependent enzymes. The cytoplasmic enzyme acetyl coA carboxylase seems to have preferential access to biotin because when dietary protein or fat levels are low, a situation that creates a high demand for lipogenesis, the activity of pyruvate carboxylase is depressed to a far greater relative extent than that of acetyl coA carboxylase. Under these conditions FLKS can occur, often without the presence of severe classic lesions. However, with higher dietary levels of protein or fat, acetyl coA carboxylase activity is normally lower and is depressed considerably by biotin deficiency. In contrast the depression in pyruvate carboxylase activity is not so great as to prevent gluconeogenesis completely. Under these conditions birds may develop severe classic lesions without the occurrence of FLKS. Deficiency in adult birds rarely results in clinical lesions. Hens can continue to lay eggs when fed diets of very low biotin content but hatchability can be severely impaired. Deficient embryos show stunting of bone growth characteristic of chondrodystrophy.

Requirements

Biotin requirements of growing birds have been assessed by responses in growth and the activity in blood of pyruvate carboxylase, a specific criterion of biotin status. These results (Whitehead and Bannister, 1980) indicated the broiler requirement was 170 µg available biotin/kg but later calculations (Whitehead, 1986) suggested that the requirements of more modern broilers might have increased to 180 µg/kg. Experience has suggested that where other factors predispose birds to FLKS, higher amounts of biotin may be needed. Estimates in the past of growing turkey requirements have ranged from 200 to 450 g/kg but these may have been complicated by interactions between *Mycoplasma* infection and biotin. Requirements do not seem to have been reassessed properly within the last 20 years. It is suggested that a level of 250 µg/kg may represent the current requirement of mycoplasma-free turkeys.

In adult birds the requirement for biotin for egg laying seems to be quite low in both chickens and turkeys: Whitehead (1980) has reported that supplementation of a wheat-based diet estimated to contain 30 µg available biotin/kg did not give any response in egg number over the whole laying year in caged White Leghorn-type hens. However, requirements for hatchability are higher. Recent studies on feed-restricted broiler breeders (Whitehead *et al.*, 1985) have suggested the requirement for hatchability is 100 µg/kg. This amount has also been found to be adequate for Leghorn and Rhode Island Red hens fed *ad libitum* (Bradley,

Atkinson and Krueger, 1976; Leeson, Reinhart and Summers, 1979). Biotin transfer to the egg is important for chick viability but it has been shown that the concentration of biotin incorporated into egg yolk is lower in younger breeding hens. Chicks from younger hens thus hatch with a lower biotin status which can be reflected in poorer viability and growth potential unless these parents are given extra biotin to increase the yolk content (Whitehead *et al.*, 1985). Turkey breeder requirements have been reported to be about 150 μg/kg (Waibel *et al.*, 1969; Arends *et al.*, 1971) but a recent reassessment (Whitehead, unpublished) suggested this level was slightly inadequate for maximum hatchability. A safer requirement value might be 170 μg/kg. Similarly, Robel (1983) has reported that biotin incorporation in turkey eggs increases with age and also (1987) that injecting turkey eggs with biotin can improve hatchability. The latter observation remains to be confirmed, but there is no evidence that very high dietary biotin levels are especially beneficial for hatchability (Whitehead, unpublished).

COBALAMIN

Biochemical functions

A group of compounds known as cobalamins or vitamin B_{12} gives rise to metabolically active cobamide coenzymes. The cobamide is a complex structure in which a cobalt atom is chelated to a substituted corrin ring. The chelate can also involve hydroxyl or cyano anions. The vitamin is usually isolated in this last form, cyanocobalamin, which is quite stable when dry or in acidic environments but is susceptible to oxidation promoted by heat or light when wet.

Cobamide coenzymes participate in reactions involving hydrogen or methyl group transfer. Hydrogen transfer is achieved by methylmalonyl-coA mutase in converting methylmalonyl-coA to succinyl-coA. Methionine synthetase transfers a methyl group provided by a folate derivative in the conversion of homocysteine to methionine. Cobamide enzymes are also thought to be involved in the synthesis of folate polyglutamates. There are thus close metabolic relationships between cobalamin, folic acid and methionine.

Deficiency

Cobalamin deficiency results in a depression of enzyme activities. The decrease in mutase activity results in increased excretion of methylmalonic acid. Loss of activity of the enzymes involved in folate metabolism can induce a functional folate deficiency. Thus, some of the other manifestations of cobalamin deficiency, such as macrocytic anaemia, resemble those of folate deficiency.

Clinical effects of a deficiency are poor growth in chicks and depressed hatchability in hens. Dead embryos show fatty infiltration and haemorrhaging in many tissues.

Requirements

The requirements of poultry for cobalamin are exceedingly small and still ill-defined. They are affected by interactions with protein, fat, methionine, choline and pantothenic acid. Protein and fat appear to increase the requirement whereas choline and methionine have sparing actions but adequate data are not available to quantify these effects. Studies suggest that dietary requirements for cyanocobalamin are about 12 µg/kg for chicks and 15 µg/kg for laying hens (Tada, 1976a, b; Patel and McGinnis, 1977).

FOLIC ACID

Biochemical functions

The structure of folic acid or pteroyl monoglutamic acid is based on a pterin ring and *p*-aminobenzoic acid and glutamic acid units. Folic acid itself does not have direct biological activity but is reduced within the body firstly to dihydrofolic acid and then to tetrahydrofolic acid which is the major metabolically active folate. Other forms of folate involve different 1-carbon units attached to either the 5- or 10- positions of the vitamin, or bridging these positions. Metabolically active folates often involve monoglutamate forms but folates in storage are generally present as polyglutamates with chains of up to 11 or 12 glutamic acid residues linked by the γ-carboxyl unit (unlike the more usual α-linkage in peptides. Folic acid is thus one of the most complex vitamins in terms of its different structural forms.

Folates act as co-substrates for enzymes involved in 1-carbon metabolism. They can accept or donate the 1-carbon units at the oxidation levels of methanol, formaldehyde and formic acid but not CO_2. The folate is not covalently bound to the enzyme and to continue as a catalyst the original folate species must be regenerated by other enzymatic species. Folate metabolism thus involves a highly dynamic series of interconversions between the various folate forms. Folate-dependent pathways occur in many areas of metabolism. These include amino acid interconversions, e.g. serine–glycine, histidine–glutamic acid and homocysteine–methionine, and DNA and uric acid synthesis.

Deficiency

Folate deficiency suppresses the activity of the folate-dependent pathways. Inhibition of DNA synthesis results in the most obvious abnormality, a macrocytic megaloblastic anaemia. Growth is depressed in young birds, pigment is lost from the feathers of coloured strains and leg bone abnormalities, sometimes resulting in tendon slippage, have been observed. In young turkeys a cervical paralysis has also been reported. In adult birds a deficiency depresses hatchability and ultimately egg production. Embryonic mortality is especially high just at the end of the incubation period and embryos show signs of bone abnormalities, especially bending of the tibiotarsus.

Requirements

Folate requirements are influenced by several other nutrients. High protein levels can increase the folate requirement, presumably to meet the need for increased uric acid synthesis. Thus Creek and Vasaitis (1963) found the requirement of chicks to be 0.4 mg/kg with 18% protein but 1.5 mg/kg with 37%. Levels of individual amino acids, such as methionine, serine and glycine can also influence requirements, though quantitative data are limited. Of the vitamins, the interaction with choline has been best studied: inadequate choline levels can increase folate requirements twofold.

Experimental evidence on the folate requirements of growing birds has many limitations: it comes mainly from studies on old strains of birds fed purified diets and growing relatively slowly. The most recent study is that of Wong, Vohra and Kratzer (1977) who obtained different results depending upon the protein source used. With casein–gelatin the requirement did not appear to exceed 0.27 mg/kg, with isolated soya protein it did not exceed 0.44 mg. With casein supplemented with glycine the authors reported the requirement to be between 0.34 and 0.49 mg/kg, although analysis of the dose-response relationship suggests that a requirement of about 0.7 mg/kg would represent a better interpretation of the data. This value is taken as the preferred estimate of requirement on the basis of the existing information, but might still be an underestimate of modern broiler requirements. In the absence of any new information, the requirement for turkey poults of 1.5 mg/kg established by ARC (1975) is retained.

In adult birds there is no new information to use as a basis for revising earlier estimates of folate requirements, 0.3 and 0.5 mg/kg for laying and breeding chickens respectively (ARC, 1975). Experiments on breeding turkeys have shown the importance of maternal carry-over of folate to the chick and also the effect of dietary choline on folate requirement (Miller and Balloun, 1967). At the recommended choline requirement of 1000 mg/kg, the suggested folic acid requirement for breeding turkeys is 1.2 mg/kg.

CHOLINE

Biochemical functions

Choline, or hydroxyethyltrimethylammonium hydroxide, is needed in comparatively large amounts by birds but nevertheless may be considered as a vitamin since it is an essential dietary constituent in young birds. It is involved in several areas of metabolism. It is a constituent of phospholipids, e.g. phosphatidyl choline (lecithin), and sphingomyelin, and acetylcholine, a neurotransmitter. Choline is also involved in transmethylation since it can be a source of labile methyl groups. One such reaction is the combination of methyl groups with homocysteine to form methionine. Conversely, methyl groups from methionine can combine with ethanolamine to form choline. The participation of folate and cobalamin in an alternative methionine synthesis pathway means that nutritional interrelationships exist between these vitamins and choline and methionine.

Deficiency

In chicks and poults choline deficiency causes poor growth and defects in phospholipid metabolism. The latter results in faulty lipid transport and function because of the role of phospholipid *inter alia* in lipoprotein formation. The most obvious consequence of this is the development of fatty livers. Hock joint deformities, including tendon slippage also occur.

Effects of deficiency are seldom seen in older birds since these develop the ability to synthesize choline in amounts sufficient even to permit the high level of inclusion of choline in eggs. There is evidence that the extent of this synthesis can be influenced by dietary supply; thus a bird reared on a high choline diet may show transient signs of deficiency if it is given a diet low in choline. In laying hens these effects include depressed egg production and fatty livers.

Requirements

Because of the various metabolic interrelationships, requirements for choline are highest when dietary levels of methionine, folate or cobalamin are low. Choline and methionine are, to an extent, interreplaceable. On a molar basis methionine provides one labile methyl group to three from choline. Experimental estimates suggest that 1 g choline can be replaced by approximately 2.3–2.7 g methionine (Quillin *et al.,* 1961; Pesti, Harper and Sunde, 1980). However, the limits within which one can replace the other are ill-defined.

In chickens, dietary requirements are highest in young broilers. The requirement suggested by ARC (1975) was 1300 mg/kg and subsequent studies by Pesti, Harper and Sunde (1980) have not suggested that the requirement of broilers up to 3 weeks of age is higher than this, provided that the dietary content of methionine plus cystine is adequate (about 9.5 g/kg). However, the choline requirement rises to about 2000 mg/kg if dietary methionine plus cysteine is only 7.4 g/kg. The results of Derilo and Balnave (1980) are consistent with this conclusion. When responses to methionine or choline with diets deficient in both these nutrients are compared, it has generally been found that responses to methionine are greater (e.g. Baker *et al.,* 1982). The best nutritional philosophy therefore seems to be to supplement with methionine to meet the full sulphur amino acid requirement, then to supplement with choline to meet the choline requirement.

As the ability of the broiler to synthesize choline increases, so its dietary requirement decreases. Thus Molitoris and Baker (1976) have estimated the requirement for growth during the 7th week to be 360 mg/kg with a diet containing 6.2 g methionine plus cystine/kg. They also found that at this age methionine had a minimal effect on choline requirement.

The most recent estimate of the choline requirement of young turkeys is 1600 mg/kg for growth to 4 weeks (Zaviezo, MacAuliffe and McGinnis, 1977). However, there is no information in turkeys on methionine/choline interactions, nor the extent to which choline requirement decreases with age.

In laying hens the relationship with methionine again dominates the question of choline requirement. It seems that when the sulphur amino acid content of a diet is adequate, the choline requirement for egg production is quite low. Thus March (1981) found that supplementing a diet estimated to contain 280 mg choline/kg did not improve egg production but did decrease liver fat content from 40 to 30% of dry liver weight. The higher liver fat content did not cause pathological effects in this experiment, but where excessive liver fat is a problem dietary choline levels of up to 800–900 mg/kg may be helpful. With diets deficient in sulphur amino acids, production responses to higher choline levels (up to 1500 mg total choline/kg) have been reported (Tsiagbe, Kang and Sunde, 1982). When responses to methionine and choline are compared, it has been found that, whereas responses in egg number are quite similar to supplements of either nutrient, responses in egg weight are

Table 4.2 SUMMARY OF METABOLIC ROLES, EFFECTS OF DEFICIENCY AND INTERACTING FACTORS OF VITAMINS

Vitamin	Metabolic roles	Effects of deficiency		Productive	Interacting factors
		Microscopic	Clinical		
Vitamin A	Visual press, glycoprotein synthesis, spermatogenesis	Degeneration of membranes and kidney, hyperkeratosis	Blindness, white deposits in mouth, oesophagus, bone deformities, ataxia, pallor	Decreased growth egg production, hatchability, fertility	Disease, stress, fat, carotenoids
Vitamin D	Calcium and phosphorus uptake, transport and retention	Decreased bone calcification	Rickets, osteomalacia	Decreased growth, egg production, shell quality, hatchability	Calcium, phosphorus vitamin A, vitamin C, rachitogenic agents
Vitamin E	Membrane antioxidant	Degeneration of subcellular membranes, testes, erythrocyte fragility	Encephalomalacia, exudative diathesis, muscular dystrophy	Mortality, depressed egg production, hatchability, carcass storage quality	Selenium, vitamin A unsaturated fatty acids
Vitamin K	Protein metabolism, blood clotting	Decreased bone mineralization	Decreased blood clotting ability, haemorrhaging, anaemia	Mortality, poor carcass quality	Disease, stress, anticoagulants, growth promoters, vitamin A
Thiamin	Carbohydrate metabolism, nerve action	Lesions in duodenum, pancreas	Weakness, nervousness, spasms, head retraction, inability to stand, polyneuritis	Decreased growth, egg production, hatchability	Drugs, thiaminases, stress

Riboflavin	Oxidation of lipids, carbohydrates, amino acids	Inflammation of mucous membranes, nerve degeneration	Dermatitis, curled toe paralysis, clubbed down	Decreased growth, egg production, hatchability	Minerals?
Nicotinic acid	Hydrogen transfer, energy metabolism		Dermatitis, leg bone abnormalities, inflammation	Depressed growth, egg production, hatchability	Tryptophan
Pantothenic acid	Coenzyme A, acyl carrier protein	Myelin degeneration, lymphocyte necrosis, fatty degeneration of liver, hyper- and parakeratosis	Dermatitis, locomotor disorders, ataxia, leg bone abnormalities	depressed nutrient retention, growth egg production, hatchability, fertility	Disease, stress
Pyridoxine	Amino acid metabolism	Microcytic anaemia	Nervous disorders, leg bone abnormalities, gizzard erosion, dermal lesions, ataxia	Depressed growth, body fat deposition, egg production, hatchability	Dietary protein, sulphur amino acids
Biotin	Gluconeogenesis, lipogenesis	Hyperkeratosis, fatty infiltration of tissues	Dermatitis, leg bone abnormalities, fatty liver and kidney syndrome	Depressed growth, hatchability, mortality	Dietary fat, protein
Cobalamin	Hydrogen and methyl group transfer	Macrocytic anaemia	Fatty infiltration	Depressed growth, hatchability	Methionine, choline, folic acid, pantothenic acid
Folic acid	1-carbon metabolism	Macrocytic anaemia, hepatocyte abnormalities	Leg bone abnormalities	Depressed growth, egg production, hatchability, loss of feather colour	Methionine, cobalamin, choline
Choline	Phospholipids, methyl group metabolism, acetyl choline		Fatty liver, leg bone abnormalities	Depressed growth, egg production	Methionine, folic acid, cobalamin

greater for methionine (Keshavarz and Austic, 1985). As in the chick, it is therefore productively more effective to formulate a layer's diet to contain adequate amounts of sulphur amino acids. Under these conditions choline requirements are low, probably not more than 300 mg/kg.

There are few adequate data on which to base a requirement for breeding birds. Nesheim *et al.* (1971) found no benefit in egg production, fertility or hatchability from supplementing a purified diet containing 250 mg choline/kg, though weight gain of the hens was improved. Numbers of eggs involved were not large, but measurements of the choline contents of the eggs showed that supplementation of the diet resulted in only a marginal increase. There is thus little reason to suspect that low dietary choline contents might impair hatchability by depressing the amount of choline in the egg. The requirements of breeding birds are therefore not thought to differ from those of laying hens. In breeding turkeys Balloun and Miller (1964) found that 1000 mg/kg was adequate for egg production and hatchability and growth and viability of poults. Subsequently, Ferguson *et al.* (1975) reported that supplementing a practical diet containing 1230 mg/kg did not improve egg production, fertility or hatchability. The requirement value of 1000 mg/kg thus seems satisfactory.

Choline chloride (70% choline) is the supplemental form of choline usually added to diets. It is deliquescent and though usually absorbed on a carrier such as silica it can depress the stability of other vitamins if feeds or premixes are stored under adverse conditions of temperature or humidity. Betaine is a less aggressive compound and can replace choline in the metabolic pathway involving the conversion of homocysteine to methionine (choline itself is first converted to betaine before participating in this reaction). Betaine can therefore be used to meet part of the choline requirement where a choline supplement is necessary. Quantitative information on extent of replacement is limited, but it would seem to be safe to provide at least 10% of the total choline requirement as betaine. Tables 4.2, 4.3 and 4.4 summarize the effects of deficiency and the vitamin requirements in chickens and turkeys.

Table 4.3   VITAMIN REQUIREMENTS OF CHICKENS (PER kg DIET)

	Broiler starter	Broiler finisher[a]	Chick starter	Chick grower[a]	Laying hen	Breeding hen
Vitamin A (IU)	1400	1400	1300	1300	3600	3600
Vitamin D_3 (ICU)	400	400	200	200	400	400
Vitamin E (mg α-tocopherol)	10	10	10	5	10	10
Vitamin K (mg MSB)	0.4	0.4	0.4	0.4	0.7	0.7
Thiamin (mg)	1.5	1.5	1.2	0.5	0.7	0.7
Riboflavin (mg)	5	4	3	2	2.5	4
Nicotinic acid (mg)	70	45	30	20	8	10
Pantothenic acid (mg)	12	10	10	6	7	10
Pyridoxine (mg)	4	3	3	2	2	4
Biotin (μg)	180	120	120	30	30	100
Cyanocobalamin (μg)	12	10	12	5	12	15
Folic acid (mg)	0.7	0.5	0.5	0.3	0.3	0.5
Choline (mg)	1300	1000	1000	300	300	300

[a] Estimated values

Table 4.4 VITAMIN REQUIREMENTS OF TURKEYS (PER kg DIET)

	0–8 weeks	*9–16 weeks*[a]	*Beyond 16 weeks*[a]	*Breeding hen*
Vitamin A (IU)	2000	2000	2000	3500
Vitamin D$_3$ (ICU)	1200	900	900	2000
Vitamin E (mg α-tocopherol)	10	10	10	10[a]
Vitamin K (mg MSB)	1	1	1	1
Thiamin (mg)	2	2	2	2[a]
Riboflavin (mg)	4	3	3	4[a]
Nicotinic acid (mg)	70	50	40	30
Pantothenic acid (mg)	12	10	9	16
Pyridoxine (mg)	6	4	3	4[a]
Biotin (μg)	250	150	100	170
Cyanocobalamin (μg)	15	5	5	15[a]
Folic acid (mg)	1.5	0.8	0.7	1.2
Choline (mg)	1600	1300	1000	1000

[a] Estimated values

Vitamin contents and bioavailabilities in feed ingredients

The natural vitamin content of feed ingredients can make an important contribution towards the needs of poultry, particularly for B-vitamins. Ingredient vitamin contents are quoted in many tables but, before these can be taken into account in feed formulation, three important factors need to be known about them: accuracy, variability and bioavailability.

Sophisticated methods of assay, such as immunosorbent assay, have now been developed for some vitamins. However, it is probable that values quoted in most tables have been obtained using more traditional methods such as microbiological assay. Though these methods are accurate for some vitamins, for others (e.g. folic acid) this may not be the case. Tables usually only quote single values for each vitamin, whereas analytical surveys can show quite wide variations between different samples of the same ingredient. Whilst some of this variation will be attributable to analytical error, it is undoubtedly true that individual vitamin contents of ingredients do vary. In setting diet standards, therefore, some account must be taken of the likelihood and extent to which individual batches of a feedstuff will have a vitamin content lower than the table average value.

For some vitamins analytically-determined amounts are not fully available to poultry. Good information on bioavailability is available for biotin but for other vitamins where bioavailability is also thought to be limited, information is still scant. Biological assays with birds are ultimately the best measures of vitamin contents. However, these are time-consuming and expensive and do not form the basis for most of the information available on ingredient contents.

VITAMIN A

Retinol can be measured in mixed feeds by chemical methods. Measurement of β-carotene and other carotenoids is more relevant for feed ingredients, though amounts of these are low in relation to the usual retinol supplements.

VITAMIN E

Measurements of vitamin E content of feedstuffs are only relevant if they are based on chromatographic methods that quantify α-tocopherol.

THIAMIN

Chemical and microbiological assay methods can give comparable results. Thiamin in feedstuffs is thought to be fully available but the presence of antagonists and antithiamin compounds in feeds may mean that feed analyses do not give a true indication of the thiamin status of feeds.

RIBOFLAVIN

Information on the riboflavin contents of feedstuffs can be reliably gained by microbiological or chemical assays. The vitamin is thought to be fully available.

NICOTINIC ACID

This vitamin can be measured microbiologically. It is present as its amide in feedstuffs of animal origin and is thought to be fully available in these. However, in cereals a large proportion of nicotinic acid is bound with cellulose and other carbohydrate or proteinaceous macromolecules and is unavailable. Bioavailabilities of 30–35% have been quoted in chickens and rats for wheat, sorghum and maize (Manoukas, Ringrose and Teerie, 1968; Carter and Carpenter, 1980). In contrast, nicotinic acid in soyabean meal is thought to be fully available, but there is no information for other oilseed meals. Measurements of bioavailability are complicated by the ability of chicks to synthesize the vitamin from tryptophan. The accuracy of the limited data on bioavailability must therefore be regarded as uncertain.

PANTOTHENIC ACID

Methods for the most efficient release of this vitamin from feedstuffs seem to vary from one ingredient to another and this may explain the rather wide variation in values reported for this vitamin, whether measured microbiologically or by immunoassay. Chick bioassays have been used to give information on the available pantothenic acid contents of feedstuffs but the results have sometimes been inconsistent with microbiologically-determined values (Latymer and Coates, 1982). In view of the uncertainties over bioavailability and reaction conditions for the release of pantothenic acid in assays, the nutritional relevance of values quoted in tables may be doubtful.

PYRIDOXINE

Reproducible results for feedstuffs can be obtained by microbiological assay or HPLC quantification of the three active forms. Bioavailability is assumed to be high.

BIOTIN

Tabular values are based upon microbiological assays which have been shown to give accurate and reproducible results. Bioavailability is a major problem with this vitamin, but a consistent picture for individual feedstuffs has emerged from chick bioassays based on responses of growth (Frigg, 1976; 1984) and blood pyruvate carboxylase (Whitehead, Armstrong and Waddington, 1982). Bioavailability in wheat is particularly low (5%) and is also low in barley (15%), oats (30%) and sorghum (25%). However biotin is fully available in maize. In the protein sources, it seems to be fully available in animal proteins and soyabean meal. Availability is depressed in other oilseed meals, but since these contain very high levels of the vitamin, they still remain good sources of available biotin.

FOLIC ACID

There is considerable uncertainty over the nutritional values of feedstuffs in relation to this vitamin. One reason is the wide range of folate derivates found in feedstuffs. Separation and identification of individual folates, both mono- and polyglutamates, can be achieved with HPLC but the methodology is not yet adapted to feedstuff analysis. Another chemical method involves isotope dilution using a binding protein.

Traditionally folate analyses have been carried out microbiologically. Micro-organisms such as *Lactobacillus casei* respond to folate monoglutamates but not to polyglutamates. Thus a microbiological assay of a feedstuff buffer extract gives a measure of 'free folate', whereas an assay after enzyme treatment of the extract to degrade polyglutamates to monoglutamates measures 'total folate'. However, recent doubt has been cast on the validity of the usual analytical procedures by the finding that the response of *L. casei* to different monoglutamate forms is not uniform and that the pH of the reaction medium can influence results.

Bioavailability of folates is uncertain. Measurements of 'free folate' are likely to underestimate the nutritional value of feedstuffs since poultry have digestive enzymes capable of degrading polyglutamates (folic acid can only be absorbed as the monoglutamate). The efficiency of this degradation is unknown, but it is probable that 'total folate' measurements will overestimate nutritional values. Animal bioassays have been described but have not been particularly sensitive. Tabular values may thus not give reliable indications of the nutritional folate content of feedstuffs.

COBALAMIN

Microbiological and isotope dilution methods have been used to measure this vitamin but there are large variations in the values reported in feedstuffs. It is uncertain whether this represents analytical error or natural variation. Chick bioassays are complicated by the ability of the hen to pass on reserves to chicks.

CHOLINE

Choline has been measured by several chemical and microbiological methods. It is generally assumed to be largely available. Results from chick bioassays are not

Table 4.5 ASSESSMENT OF THE CURRENT
STATE OF KNOWLEDGE ON AVAILABLE
VITAMIN CONTENTS OF FEEDSTUFFS

Good	Moderate	Poor
Vitamin E	Thiamin	Nicotinic acid
Riboflavin	Pyridoxine	Pantothenic acid
Biotin	Choline	Folic acid
		Cobalamin

thought to be very meaningful because of the interaction of methionine and choline in choline-limiting diets.

The quality of the current information available on feed ingredient vitamin contents is summarized in Table 4.5. A matrix of raw material vitamin contents is presented in Table 4.6.

Natural vitamin contents of mixed feeds

The calculated natural available vitamin contents of practical and hypothetical formulations that might be used in different parts of the world for different classes of poultry are given in Tables 4.7 to 4.10. The formulations list the major vitamin-containing ingredients; they are not intended to be isoenergetic or isonitrogenous and supplements of fat, amino acids, minerals and other additives have been omitted. The calculated vitamin contents represent probable safe minimum levels rather than average contents, but must be interpreted in the light of uncertainties described earlier.

Vitamin contents of broiler prestarter and starter diets are shown in Table 4.7 and can be compared with requirements listed in Table 4.3. There are several noteworthy features. The contents of several vitamins stay remarkably constant over the range of formulations, especially vitamin E, thiamin and riboflavin. However, whereas thiamin is always above the requirement value, riboflavin is consistently below. Nicotinic acid is also generally consistently low, and well below the requirement, though it is higher in those diets that contain sunflower seed meal. This ingredient has a high content of nicotinic acid, although bioavailability has not been established. Pyridoxine level generally meets the requirement except for diets based mainly on wheat. Biotin contents are highly variable, but always below the requirement. The levels are particularly low when diets contain high proportions of wheat or animal proteins. Cassava is also a poor source of biotin and most other vitamins. Cobalamin, of course, is totally lacking in all-vegetable diets, but does not meet the requirement in any of the other diets. Folic acid values should be treated with caution. Choline contents vary about the requirement: they meet the requirement in diets that contain high proportions of wheat or protein provided by low-protein oilseed meals (e.g. sunflower, rape). It is perhaps interesting to observe that practical formulations based on a wide range of ingredients can be little different in vitamin contents from diets of simpler composition.

Broiler grower and finisher diets (Table 4.8) contain generally lower levels of most vitamins. Biotin and pyridoxine remain as two of the most variable vitamins.

Table 4.6 VITAMIN CONTENTS OF FEED INGREDIENTS (per kg)[a]

	Vitamin E (mg)	Thiamin (mg)	Riboflavin (mg)	Nicotinic acid (mg) A	B	Pantothenic acid (mg)	Pyridoxine (mg)	Biotin (µg) A	B	Folic acid (mg)	Cobalamin (µg)	Choline (mg)
Barley	6	4	1	42	14	5	2	130	20	0.2	0	900
Maize	7	4	1	15	5	4	6	65	65	0.2	0	435
Maize gluten 20		1.9	2	15	5	13	8	160	160	0.5	0	250
Oats	7	6	1	12	4	10	1	220	65	0.2	0	800
Sorghum	5	4	1	40	13	8	3	250	65	0.2	0	600
Wheat	7	4	1	36	12	8	0.8	80	4	0.2	0	900
Wheatfeed	35	15	2	90	30	13	10	300	15	0.8	0	950
Cassava	0	1	0.5	10		15	1	50	1		0	450
Peas	0	1.8	0.8	17		7	1.5	140	NA	0.3	0	650
Cottonseed meal	15	6	4	35		9	5	230	NA	0.8	0	2600
Rapeseed meal	15	2	3	160		9	6	900	500	2	0	6500
Soyabean meal	2	6	2	18	18	12	6	235	235	2	0	2500
Sunflower seed meal	8	2	3	176		30	10	1000	280	1.5	0	2600
Fish meal	4	2	5	33	33	7	3	100	100	0.2	88	3500
Herring meal	5	0.4	5	105	105	15	3	130	130	0.2	88	3500
Meat and bone meal	1	0.5	4	44	44	3	2	75	75	1	40	1600
Lucerne	100	3.4	13	20		25	6	540	400	4.2	0	250

[a] Values represent conservative estimates for feed formulation purposes (i.e. are biased towards the lower end of reported ranges). For nicotinic acid and biotin, the A values are total contents and the B values are estimated bioavailable contents. For other vitamins bioavailability is assumed to be high

Table 4.7 NATURAL AVAILABLE VITAMIN CONTENTS OF BROILER PRESTARTER AND STARTER DIETS

	Wheat-based		*Maize-based*					*Mixed*		
Main ingredients (g/kg)										
Maize	500		500	490	460	600	420	300	480	100
Wheat		600					120	70	150	100
Cassava								100		250
Rapeseed meal					50					
Soyabean meal	400	150	400	300	300	150	260	220	230	300
Sunflower seed meal				50	50			40		
Peas						50	50	50		50
Meat and bone meal	50	50		60	50		60	40		50
Fish meal	100	100	60			100	20		40	50
Vitamins (mg/kg)										
Vitamin A (IU)	0	0	1250	1220	1150	1500	1050	750	1200	250
Vitamin E	4	5	4	4	5	5	4	3	5	2
Thiamin	4	3	4	4	4	3.2	3.6	3	3.7	3
Riboflavin	1.5	1.7	1.5	1.5	1.5	1.7	1.6	1.5	1.4	1.3
Nicotinic acid	13	14	10	18	26	11	12	17	9	12
Pantothenic acid	9	7	7	7	7	5	6	7	6	11
Pyridoxine	3	2	5.5	5.4	5.5	4.5	4.5	4	4	3
Biotin (µg)	96	52	130	122	139	90	103	90	90	93
Cobalamin (µg)	0	11	0	5	4	11	4	5	3	5.7
Folic acid	0.9	0.4	0.9	0.5	0.6	0.4	0.7	0.7	0.6	0.7
Choline	1450	1350	1210	1300	1580	1070	1139	1115	1060	1280

Table 4.8 NATURAL AVAILABLE VITAMIN CONTENTS OF BROILER GROWER AND FINISHER DIETS

	Wheat-based		Maize-based			Mixed	
Main ingredients (g/kg)							
Maize			600	640	360	550	550
Wheat	650	690				150	180
Cassava					100		
Peas					150		
Soyabean meal	250	100	300	150	120	180	240
Sunflower seed meal					30		
Rapeseed meal					50		
Meat and bone meal		50		50	60		
Fish meal		60		60	30	40	
Vitamins (mg/kg)							
Vitamin A (IU)	0	0	1500	1600	875	1375	1450
Vitamin E	4	4	5	5	4	5	5
Thiamin	3.7	3	4	3.4	2.6	3.7	4
Riboflavin	1.5	1.5	1.4	1.5	1.3	1.4	1.3
Nicotinic acid	12	14	8	11	19	9	9
Pantothenic acid	8.6	7.6	6	6	7	6	7
Pyridoxine	2	1.5	5.6	5.5	4	4.5	5
Biotin (µg)	60	36	110	88	113	82	95
Cobalamin (µg)	0	8	0	8	5	3	0
Folic acid	0.6	0.3	0.7	0.4	0.5	0.5	0.6
Choline	1210	1160	1010	950	1210	970	1000

The composition of some layers and breeder diets are given in Table 4.9. Features of interest are again the universally low riboflavin levels and the low pyridoxine contents of wheat-based diets. The use of lucerne as a source of yolk pigment has the added advantage of providing considerable amounts of vitamins A and E. When compared to requirements (Table 4.3), the dietary levels show that wheat-based diets can be severely deficient in riboflavin and marginal or deficient in pantothenic acid, pyridoxine, cobalamin and folic acid. The same is broadly true of maized-based diets, with the exception that pyridoxine does not seem to be limiting. For chicken and turkey breeders, the diets are generally deficient in all B-vitamins except thiamin and choline.

When turkey starter diets (Table 4.10) are compared with requirements (Table 4.4), it can be seen that all the formulations presented are deficient in all vitamins save thiamin (adequate) and choline (marginal).

Vitamin supplementation of practical diets

The requirements discussed in the earlier section represented the minimum amounts of each vitamin, as a proportion of the diet, that were judged to be needed to meet the most desirable characteristic of performance or metabolism. The allowance, on the other hand, is the actual amount of the vitamin it is thought necessary to provide in the diet. This is influenced by several factors, such as (a) the degree of uncertainty over the requirement, (b) interactions involving stress, disease, environment and other nutrients, including vitamins, and (c) the advisability of providing a margin of insurance.

IMMUNITY, DISEASE AND STRESS

It is a general characteristic that a deficiency of most vitamins, and also of other nutrients, lowers the disease resistance of birds. This is presumably because of specific metabolic impediments suppressing the synthesis of immunoglobulins and other factors involved in the immune system. This problem is normally solved by providing enough of the vitamin to meet the normal requirement. The point of interest to nutritionists is the extent to which it might be possible to further enhance the immune potential of the bird by providing vitamins in amounts higher than the normal requirement. So far this ability has only been demonstrated convincingly in the cases of vitamins A and E. However, the levels found to give some protection against *E. coli* infection have been so high, 60 000 IU vitamin A/kg or 300 mg vitamin E/kg, as to be well beyond the normal commercial ranges (though 100 mg vitamin E/kg have been reported to give some protection against coccidiosis). Effects of high levels of both vitamins show no signs of synergism (indeed, there is antagonism).

The question of whether to provide additional vitamins to enhance immunity will depend upon the cost of the vitamins, the levels and types of intercurrent diseases encountered, and the types of drugs used in the feed. For instance, the use of antibacterial growth promoters might give adequate, or better, protection against bacterial infections. However, birds fed an 'additive free' diet might benefit from vitamin-induced protection. The choice of vitamin might also depend upon the type of disease. For instance parasitic infections increase the need for vitamin A because

Table 4.9 NATURAL AVAILABLE VITAMIN CONTENTS OF LAYER AND BREEDER DIETS

	Wheat-based		Maize-based			Mixed		
Main ingredients (g/kg)								
Maize			680	600	735	390	570	250
Maize gluten 20								250
Wheat	720	705		200		200	100	
Cassava								200
Peas				100		40		
Soyabean meal	180	50	220		80	60	130	200
Sunflower seed meal		50				100		
Meat and bone meal		70			60	50	30	
Fish meal		25			25	20	20	
Lucerne						40	50	
Vitamins (mg/kg)								
Vitamin A (IU)	0	0	1700	1500	1800	8100	10400	700
Vitamin E	5	5	5	5	5	9	10	4
Thiamin	3.7	3.2	3.8	3.8	3.2	3	3.5	3
Riboflavin	1.2	1.1	1.2	1.3	1.3	1.9	1.9	1.4
Nicotinic acid	12	21	7.5	24	8	26	9	10
Pantothenic acid	8	8	5.6	8	4	8.7	6	10
Pyridoxine	1.7	1.5	5.6	5.8	4.9	4.4	4.8	5
Biotin (µg)	45	33	96	114	75	95	92	103
Cobalamin (µg)	0	9	0	0	9	4	3	0
Folic acid	0.4	0.4	0.6	0.6	0.3	0.5	0.6	0.6
Choline	1100	955	845	1020	700	980	830	761

Table 4.10 NATURAL AVAILABLE VITAMIN CONTENTS OF TURKEY PRESTARTER AND STARTER DIETS

	Wheat-based		Maize-based			Mixed		
Main ingredients (g/kg)								
Maize	420		430	420	380	150	160	120
Wheat		440					240	360
Cassava						150	40	50
Peas						100		
Soyabean meal	400	260	350	280	350	280	420	320
Sunflower seed meal					50	50		
Meat and bone meal	50	100		100		50	80	70
Fish meal	30	100	120	100	120	120	20	10
Vitamins (mg/kg)								
Vitamin A (IU)	0	0	1075	1050	950	375	400	300
Vitamin E	4	4	4	4	5	2.5	3.6	4
Thiamin	3.7	3.2	4	3.6	4	3	3.8	3.5
Riboflavin	1.9	1.8	2	1.4	1.8	1.8	2	1.7
Nicotinic acid	16	18	12	15	21	24	16	15
Pantothenic acid	9.5	8	7	6	8	9	8.6	8
Pyridoxine	3	2.6	5	4	5	3.8	4	3.3
Biotin (µg)	104	80	123	105	108	122	125	98
Cobalamin (µg)	5	12	10	12	10	12	5	3
Folic acid	0.9	0.7	0.8	0.7	0.8	0.7	1	0.8
Choline	1560	1550	1490	1210	1590	1530	1560	1355

of its role in epithelial tissues, though the need for this extra protection in the case of coccidiosis might be constrained by the regular use of specific anticoccidials. Where only a general immune enhancement is desired, the choice of vitamin might be influenced by secondary factors, e.g. high levels of vitamin E enhance stability and keeping quality of poultry meat.

Vitamin needs are also influenced by stress, whether induced by disease or environment. The particular vitamins involved can depend upon the type of disease or stress: vitamin A in the case of coccidial infection, as indicated earlier; some B-vitamins, particularly biotin, for mycoplasma infection in turkeys; pantothenic acid for adrenal function; vitamin C to counter metabolic changes, including immunosuppression, in heat stress. Vitamin E and thiamin have also been implicated in stress. Additional vitamin E has also been reported to be beneficial in malabsorption syndrome, but it is probable that the response was to the addition of one of the more limiting nutrients, rather than being an indication of a specific interaction of the syndrome with this vitamin.

Unfortunately there are few detailed dose-response data to indicate the quantitative effect of stress on vitamin needs. Where the stress causes a decrease in feed intake, the allowance for all vitamins should be increased at least in proportion to the depression in consumption. This is because vitamins are needed for the continuation of metabolic processes even in the absence of feeding.

NUTRITIONAL INTERACTIONS

Specific nutritional interactions involving individual vitamins have been discussed already. Among the fat-soluble vitamins, high dietary levels of vitamin A have been shown to depress the absorption or retention of the others. It is important therefore to ensure an appropriate balance between these vitamins. Results of Frigg and Broz (1984) suggest that when dietary vitamin A is increased five-fold (from 10 000 IU/kg to 50 000 IU/kg), dietary vitamin E needs to be increased three-fold to maintain the same plasma vitamin E level. This quantitative relationship is broadly consistent with a more elaborate equation recently proposed by Weiser *et al.* (1988) to relate dietary levels of the two vitamins.

Interactions also occur amongst the B-vitamins. Specific interactions have already been described but there are others for which there are no biochemical explanations. For instance, biotin status can be depressed by high dietary levels of choline or combinations of other B-vitamins. Quantitative data are not available for these interrelationships. Nevertheless, the importance of maintaining an appropriate metabolic balance between vitamins has been established. The practical consequence of this is that a diet should not be generously fortified with some vitamins whilst others are left at only marginal levels.

REPRODUCTION

An important consideration in the nutrition of breeding hens is the need to provide an adequate carry over a vitamin into the egg and hatching chick. The benefits of this in terms of the health and viability of young chicks have been demonstrated for many vitamins. These responses may be obtained with dietary levels above those needed to maximize hatchability. Since this criterion has been used most frequently to establish the requirement, it is advisable to provide an extra allowance in breeder diets to enhance carry-over to the chick. The amount of a vitamin incorporated into the egg depends upon the dietary level, but there is a limit. This arises because many vitamins are transported into the yolk attached to specific binding proteins. Incorporation of the vitamins in the egg can therefore be limited by the amount of binding protein produced by the bird, rather than the amount of vitamin available from the diet. For instance, saturation of plasma riboflavin-binding protein occurs with a dietary level of about 10 mg riboflavin/kg, giving a yolk concentration of about 4.5 μg/g. This yolk concentration is increased very little by even substantially higher dietary levels. It is thus likely that genetic factors affecting the production of vitamin-binding proteins can influence vitamin carry-over in hens.

Age effects can also influence vitamin carry-over. In both chickens and turkeys it has been shown that the concentration of biotin incorporated into the yolk at a given dietary level increases with age. In consequence, chicks from young breeder flocks can hatch with relatively poorer biotin status. To ensure adequate viability in these chicks it is therefore necessary to provide a larger biotin allowance for the parents when younger. In contrast, turkey egg concentrations of pyridoxine, folic acid and cobalamin have been found to decrease with maternal age. There may therefore be a case for providing higher allowances of these vitamins for 'old' hens.

INDIVIDUAL VITAMINS

Vitamin A

The practical allowance for this vitamin is usually considerably higher than the minimum requirement. This is partially in recognition of the protective effect of the vitamin against disease and stress. However, the supplemental levels fed in the UK are often very high (about 13 000 IU/kg for a broiler starter diet) compared with the minimum requirement (1300 IU/kg) or with levels used in other parts of the world, e.g. 9000–10 000 IU/kg (including contribution from β-carotene in maize-based diets) in North America. The reason for this disparity in commercial practice is not obvious. A large body of commercial data is needed before a more objective assessment of vitamin A allowances can be made on the basis of a cost-benefit analysis.

Vitamin D

As discussed earlier, vitamin D_3 requirements are increased if dietary calcium and phosphorus levels are not optimum. Since levels of these other nutrients may vary in feeds, it is sensible to provide a generous allowance for vitamin D_3. However, in spite of very high dietary levels of D_3 (4000–5000 ICU/kg) and satisfactory levels of calcium and phosphorus, rickets is still seen as a field problem in young broilers and turkeys. Recent studies suggest this problem may be related to impaired vitamin D metabolism in the bird rather than to inadequate nutrition (Bar *et al.*, 1987). The factor responsible for this problem has not yet been identified and there is no information on whether it can be overcome by feeding a vitamin D_3 metabolite.

Until more information becomes available on this problem it is probably justifiable to continue feeding high levels of the vitamin. This procedure has the added benefit of helping to counter the possible effects of some mycotoxins in the feed.

Vitamin E

There are several factors to be considered when setting practical allowances for this vitamin. Polyunsaturated fat content of the diet is one of these. Whilst vitamin E is a natural constituent of fats and oils, the amounts can diminish with processing and storage. In practice it is prudent to assume little contribution from supplemental fat and to provide 10 mg vitamin E/10 g supplemental PUFA as an addition above the requirement. Thus for a diet containing 20 g added PUFA/kg, 30 mg vitamin E should be adequate for normal health and growth.

Somewhat higher levels are needed to maximize the stability of broiler meat after processing. Marusich *et al.* (1975) suggested that the addition of 40 mg vitamin E/kg of diet of relatively low PUFA content was needed for optimum stability. Turkeys are less efficient at incorporating vitamin E into tissues and the corresponding dietary vitamin E value was 200 mg/kg for the last 4 weeks prior to slaughter. If diets contained added PUFA, higher levels of vitamin E would be needed to maintain equivalent meat stability (up to 100 mg vitamin E/kg with 50 g PUFA/kg in broilers).

There has been some indication from field observations that higher vitamin E levels for breeder hens can bring about an improvement in immune status that is passed on to progeny and reflected in lower chick mortality.

Riboflavin

There have been sporadic reports over the last few years in the UK of apparent riboflavin deficiency in broiler breeder flocks. Hatchability has been depressed and clubbed down, curled toes and nerve lesions have been seen in embryos and hatching chicks. However, chemical analyses of diets, eggs and hatching chick plasma have not indicated a lack of riboflavin. If a functional riboflavin deficiency is nevertheless involved, it is possible that it is caused by some factor, a disease or other agent, blocking a specific area of riboflavin metabolism. Until more information is acquired on this problem, it might be prudent to feed sufficient riboflavin to ensure maximum incorporation of the vitamin in the egg.

Vitamin C

Ascorbic acid (AA) is needed for normal metabolic function in birds but it can be synthesized by them. There is no compelling evidence that this synthesis is inadequate to meet the needs of the bird under normal conditions and AA is thus not routinely added to poultry diets. Nevertheless, research is finding new areas of metabolism that involve AA, such as vitamin D_3 hydroxylation, and a nutritional benefit may yet be shown.

In contrast, the nutritional role of AA under conditions of stress is well established. The relatively high concentrations of AA in the adrenal are rapidly depleted during stress and adrenocortical insufficiency leads to many of the metabolic problems associated with heat and other stresses, such as corticosteroid hormone imbalance, immunosuppression and mortality. Inclusion of AA in the diet of heat stressed birds has been shown to limit these abnormalities (Pardue, Thaxton and Brake, 1985). There are various reports that adding AA to diets a rate of 200, 400 mg/kg or higher has decreased mortality and improved growth, egg production and eggshell quality in heat stressed poultry.

SUPPLEMENTS FOR PRACTICAL DIETS

The amount of a vitamin added to a diet as a supplement is dependent upon several factors. Firstly, there is the allowance it is thought desirable for the bird to receive. From this may be subtracted the amount of available vitamin thought to be naturally present in the feed. However, a proportion may be added to take into account any losses that may occur during feed processing and storage. This is a complex matter because vitamin stability depends upon the degree of protection provided by the individual vitamin product or premix and the nature and severity of the feed mixing and storage conditions. It is therefore not possible to give individual values for stabilities of vitamins. Suffice it to say that manufacturers of vitamins and vitamin premixes continue to bring out new products for which superior stabilities are claimed. Finally, a vitamin supplement may contain a margin for insurance, to guard against error in any of the earlier judgements.

The natural vitamin contents of practical diets can vary quite considerably depending on the individual feed ingredients and their levels, as has been shown in Tables 4.7 to 4.10. It is therefore not possible to give a supplement formulation for each type of diet that is optimal under all conditions. Ideally, a vitamin supplement should perhaps be formulated for the specific diet to which it is to be added, but this has the practical disadvantage that supplement formulations would have to change with changes in ingredient contents in least cost formulations. A more realistic approach is to consider the range of ingredient levels regularly used for a particular type of diet and to use the minimum natural level of each vitamin contained over this range as the basis for calculating the appropriate supplemental level.

Broiler and broiler breeder diets

Supplements for broiler diets are given in Table 4.11. These have been divided into two broad categories, namely supplements for diets based mainly on wheat or maize since these represent the two most common types on a worldwide basis. However, there may be many local variations, in the contents of cereals and other ingredients, so these supplements should be taken as guidelines only. For starter/grower diets (0–4 weeks) based on wheat, a supplemental level for vitamin A of 10 000 IU/kg is suggested as a minimum, though higher levels are often provided in the belief that they give extra insurance against stress and disease. Vitamin E at 30 mg/kg should be adequate for most diets, but higher levels might be appropriate if diets contain very high levels of PUFA or if keeping quality of meat is a particular criterion. The high level of vitamin K in relation to the requirement is mainly to guard against processing and storage losses. On the basis of calculated feed contents it would not seem necessary to add thiamin as a supplement, though

Table 4.11 SUPPLEMENTAL VITAMIN LEVELS FOR BROILER AND BROILER BREEDER DIETS

Bird type	Broiler				Broiler breeder							
Diet type	Starter/grower (0–28 days)		Finisher (28 days–end)		Chick (0–6 weeks)		Grower (6–17 weeks)		Prebreeder 17–20 weeks		Breeding hen (20–60 weeks)	
Cereal base	Wheat	Maize	Wheat	Maize	Wheat	Maize	Wheat	Maize	Wheat	Maize	Wheat	Maize
Vitamin (mg/kg)												
Vitamin A (IU)	10000	9000	10000	9000	8000	7000	8000	7000	8000	7000	12000	11000
Vitamin D$_3$ (ICU)	5000	5000	5000	5000	3000	3000	3000	3000	3000	3000	3000	3000
Vitamin E	30	30	30	30	10	10	10	10	15	15	30	30
Vitamin K[a]	3	3	3	3	2	2	2	2	2	2	3	3
Thiamin	–	–	–	–	–	–	–	–	–	–	–	–
Riboflavin	6	6	5	5	3	3	3	3	8	8	12	12
Nicotinic acid	60	65	30	35	20	25	10	15	10	15	12	17
Pantothenic acid	12	14	10	12	5	7	3	5	5	10	12	15
Pyridoxine	3	1	2	1	3	1	2	–	4	1	6	3
Biotin (µg)	180	60[b]	100	20[b]	80	30	–	–	100	70	200[c]	150[c]
Cyanocobalamin (µg)	12	12	10	10	10	10	–	–	10	10	15	15
Folic acid	1.5	1.5	1	1	0.5	0.5	–	–	1	1	2	2
Choline	250	450	–	100	–	200	–	–	–	–	–	200

[a] As MSB
[b] For maize/soyabean diet should be higher if diet contains animal proteins
[c] Can be decreased by 50 µg/kg after 34 weeks

small amounts (1–2 mg/kg) are often added for insurance. A biotin supplement of 180 µg/kg should in general be adequate but larger amounts might be needed if the diet contains a high proportion of animal protein (especially meat and bone meal) or if FLKS occurs. A generous folic acid supplement seems prudent in view of uncertainties over requirements and feed contents.

Some modifications are apparent in supplements for maize-based diets. Vitamin A can be lowered to take account of the carotenoid content of maize. However, nicotinic acid and choline should be increased above the levels added to wheat-based diets, although this would not be necessary if the diet contained high proportions of oilseed meals especially rich in these vitamins (e.g. sunflower, rape). The biotin supplement of 60 µg/kg is appropriate for maize-based diets where the protein is provided mainly by oilseed meals; where protein is provided mainly by animal proteins, 100–120 µg/kg might be a safer supplemental level.

The vitamin specifications for broiler finisher diets are usually lower than for starter diets, though experimental evidence on the quantitative basis for this are limited. Suggested values are set out in Table 4.11. Diets based on maize and oilseed meals probably need little or no biotin supplement during this period; a supplement of 50 µg/kg is recommended for diets containing animal protein sources.

Although broiler breeder replacement stock are fed *ad libitum* for the first few weeks, they are generally given diets of relatively low ME content to avoid very fast growth. Under these circumstances, a chick starter rather than a broiler starter vitamin supplement is adequate. However when feed restriction starts they can be given a grower supplement of lower specification (Table 4.11). Towards the end of the growing period, it may be helpful to give the hens a pre-breeder supplement of enhanced specification to ensure that they enter the breeding stage in good metabolic condition. The contents of the breeding supplements (Table 4.11) are relatively high to allow an adequate carry-over of each vitamin to the chick.

Growing and laying chickens

Supplemental vitamin levels for the diets of laying hens and their replacement chicks are given in Table 4.12. Supplements of biotin and choline are not needed for wheat-based layers diets. A choline supplement has been included for maize-based diets but this may not be necessary if the diet contains adequate methionine and fatty liver is not a problem. The natural vitamin A content of a layer's diet is likely to be quite high if dried grass or lucerne are among the ingredients. However, the amounts are likely to be highly variable depending upon the processing conditions, length of time of storage, etc. hence supplemental vitamin A should always be added.

Turkeys

Supplements for growing and breeding turkeys are given in Table 4.13. As with the supplements for other types of birds, the levels of nicotinic acid, biotin and choline in these supplements can vary depending upon the proportions of some of the oilseed meals rich in these vitamins. In the breeder diet it might be helpful to provide an extra 50 µg biotin/kg during the first month of egg-laying and to increase pyridoxine and folic acid towards the end of the breeding period.

Table 4.12 SUPPLEMENTAL VITAMIN LEVELS FOR LAYING HENS AND REPLACEMENT CHICKS

Bird type	Chicks				Laying hen	
Diet type	Starter (0–6 weeks)		Grower (6–18 weeks)			
Cereal base	*Wheat*	*Maize*	*Wheat*	*Maize*	*Wheat*	*Maize*
Vitamin (mg/kg)						
Vitamin A (IU)	8000	7000	6000	5000	7000	6000
Vitamin D$_3$ (ICU)	3000	3000	1500	1500	3000	3000
Vitamin E	10	10	5	5	5	5
Vitamin K[a]	2	2	2	2	2	2
Thiamin	–	–	1	1	–	–
Riboflavin	3	3	1	1	3	3
Nicotinic acid	20	25	5	10	5	10
Pantothenic acid	5	7	3	5	5	7
Pyridoxine	3	1	1	–	1	–
Biotin (µg)	80	30	–	–	–	–
Cyanocobalamin (µg)	10	10	5	5	10	10
Folic acid	0.5	0.5	–	–	–	–
Choline	–	200	–	–	–	–[b]

[a] As MSB
[b] Choline supplement not needed for egg production but 100–200 mg might be helpful in minimizing liver fat content

Table 4.13 SUPPLEMENTAL VITAMIN LEVELS FOR TURKEY DIETS

Vitamin (mg/kg)	Prestarter/starter (0–8 weeks)		Rearer/grower (8–16 weeks)		Finisher (16+ weeks)		Breeder	
	Wheat	*Maize*	*Wheat*	*Maize*	*Wheat*	*Maize*	*Wheat*	*Maize*
Vitamin A (IU)	12000	11000	10000	9000	10000	9000	12000	11000
Vitamin D$_3$ (ICU)	5000	5000	5000	5000	3000	3000	3000	3000
Vitamin E	25	25	25	25	20	20	30	30
Vitamin K[a]	4	4	3	3	3	3	3	3
Thiamin	2	2	–	–	–	–	2	2
Riboflavin	6	6	4	4	3	3	12	12
Nicotinic acid	70	75	60	65	50	55	50	55
Pantothenic acid	15	18	8	10	6	8	18	20
Pyridoxine	8	6	3	2	3	2	5	3
Biotin (µg)	200	150	150	100	100	50	250	200
Cyanocobalamin (µg)	15	15	10	10	10	10	20	20
Folic acid	2	2	1	1	1	1	2	2
Choline	250	450	200	400	–	200	200	400

[a] As MSB

Table 4.14 ADDITIONAL VITAMIN SUPPLEMENT FOR
STRESSED BIRDS[a]

	Meat birds[b]	Layers	Breeders
Vitamin A (IU/kg)	5000	3000	3000
Vitamin E (mg/kg)	30	30	30
Vitamin C (mg/kg)	500	500	1000
Thiamin (mg/kg)	3	3	3
Pantothenic acid (mg/kg)	10	10	10

[a] These amounts are in addition to the amounts provided by the
normal supplements. If the food intake of the birds is depressed, the
amounts of vitamins provided by the normal supplement should be
increased in proportion to maintain the daily intake of each vitamin
[b] If broilers are affected by FLKS, an additional biotin supplement of
50–150 µg is needed

Stress

The vitamin supplements presented in Tables 4.11 to 4.13 are for high performance
birds housed under temperate conditions with satisfactory feed processing and
storage conditions. If processing or storage conditions are known to be especially
detrimental to vitamins, higher levels should be included in the supplements. If
birds are kept under conditions that are routinely or occasionally stressful
(temperature, etc.) an additional supplemental 'cocktail' such as the one indicated
in Table 4.14 might be provided, although the exact choice of vitamins and their
levels might depend upon the nature and severity of the stress.

FREE RANGE PRODUCTION

The amounts of vitamins provided in diet supplements can be decreased if birds
have access to open ranges. There are two reasons for this. Firstly, the birds will
have access to natural vitamins present on the range and, secondly, requirements
for vitamins may be less because of the lower performance of birds maintained on
ranges (e.g. 'label rouge' meat chickens are grown more slowly than intensively-
housed broilers).

Open ranges can theoretically provide birds with all their vitamin needs. For
example, vitamins A and E can be provided by grass, the B vitamins and vitamin K
from plants, insects, worms and bacterial matter and vitamin D from sunlight.
However, the extent to which the birds' needs are met in practice will obviously
depend upon the quality of the range and the season. A small range largely devoid
of plant or insect life will provide little in the way of vitamins and the vitamin
content of a range will be much lower in winter when plants die back and insects
disappear. Because of these factors, it is not possible to recommend specific
supplemental vitamin levels for diets given to free range birds. An effective
practical procedure may be to vary the inclusion rate in the diet of a vitamin premix
in relation to the perceived quality of the range and to the season.

Future considerations

Though much is known about vitamin nutrition, a lot of information on which requirements are based is out of date. As new information becomes available, the process of updating vitamin feeding practices will continue. Apart from a general revision of vitamin requirements, under both temperate and tropical conditions, there are several specific areas where advances would be welcome.

The feeding of vitamin D_3 metabolites is of growing interest in relation to improving specific areas of poultry performance and research on the different natural and synthetic metabolites could result in important practical benefits to shell and bone quality.

Also in relation to bones, the problem of leg weakness is of ever increasing importance, especially in the fastest growing meat-type birds. The causes seem varied and mysterious but good vitamin nutrition is vital for healthy bone formation. It is important that breeders and nutritionists work closely on this problem. It is also probable that, as in the past, adventitious discoveries involving vitamins will be made in relation to regular or sporadic syndromes. Though vitamins constitute only a relatively small proportion of the cost of a mixed feed, their ubiquity and complexity mean that high levels of scientific and practical expertise will continue to be important for the effective nutrition and production of poultry.

References

ABDULRAHIM, S. M., PATEL, M. B. and McGINNIS, J. (1979). *Poultry Science*, **58**, 858–863

ABEND, R. and JEROCH, H. (1977). *Jahrbuch für Tierernährung und Futterung*, **8**, 361–364

ADAMS, R.L. and CARRICK, C.W. (1967). *Poultry Science*, **46**, 712–718

ARC (1975). *The Nutrient Requirements of Farm Livestock No. 1: Poultry.* Agricultural Research Council, London

ARENDS, L. G., KIENHOLZ, E. W., SHUTZE, J. M. and TAYLOR, D. D. (1971). *Poultry Science*, **50**, 208–214

BAKER, D.H., ALLEN, N.K. and KLEISS, A.J. (1973). *Journal of Animal Science*, **36**, 299–302

BAKER, D.H., HALPIN, K.M., CZARNECKI, G.L. and PARSONS, C.M. (1982). *Poultry Science*, **62**, 133–137

BALLOUN, S.L. and MILLER, D.L. (1964). *Poultry Science*, **43**, 64–67

BALLOUN, S.L. and PHILLIPS, R.E. (1957). *Poultry Science*, **36**, 929–934

BAR, A., ROSENBERG, J., PERLMAN, R. and HURWITZ, S. (1987). *Poultry Science*, **66**, 68–72

BAUERNFIEND, J.C. and DE RITTER, E. (1959). *Proceedings of Poultry Science Convention, Sydney*, p. 310

BEAGLE, W.S. and BEGIN, J.J. (1976). *Poultry Science*, **55**, 950–957

BEER, A.E., SCOTT, M.L. and NESHEIM, M.C. (1963). *British Poultry Science*, **4**, 243–253

BRADLEY, J.W., ATKINSON, R.L. and KRUEGER, W.F. (1976). *Poultry Science*, **55**, 2490–2492

CARTER, E.G.A. and CARPENTER, K.J. (1980). *Federation Proceedings*, **39**, 557

CANTOR, A.H., MUSSER, M.A. and BACON, W.L. (1980). *Poultry Science*, **59**, 563–568

CHARLES, O.W. (1977). *Poultry Science,* **56,** 1701

COUCH, J.R., CREGER, C.R. and CAVEZ, R. (1971). *British Poultry Science,* **12,** 367–371

CREEK, R.D. and VASAITIS, V. (1963). *Poultry Science,* **42,** 1136

DAGHIR, N.J. and SHAH, M.A. (1973). *Poultry Science,* **52,** 1247–1252

DERILO, Y.L. and BALNAVE, D. (1980). *British Poultry Science,* **21,** 479–487

DONOVAN, G.A. (1965). *Poultry Science,* **44,** 1292–1298

EDENS, F.W., VAN KREY, H.P., KELLY, M. and SIEGEL, P.B. (1970). *Poultry Science,* **49,** 295–297

FERGUSON, T.M., ATKINSON, R.L., BRADLEY, J.W. and MILLER, D.H. (1975). *Poultry Science,* **54,** 1679–1684

FISHER, C. (1974). *Proceedings of Roche Symposium,* pp. 11–29. London

FRIGG, M. (1976). *Poultry Science,* **55,** 2310–2318

FRIGG, M. (1984). *Poultry Science,* **63,** 750–753

FRIGG, M. and BROZ, J. (1984). *International Journal of Vitamin and Nutrition Research,* **54,** 125–134

FROST, D.V., PERDUE, H.S. and SPRUTH, H.C. (1956). *Journal of Nutrition,* **59,** 181–196

FRY, J.L., WALDROUP, P.W., DAMRON, B.L., HARMS, R.H. and WILSON, H.R. (1968). *Poultry Science,* **47,** 630–634

GRIES, C.L. and SCOTT, M.L. (1972). *Journal of Nutrition,* **102,** 1259–1268

GRIMMINGER, P.W. (1957). *Poultry Science,* **36,** 1227–1234

GRIMMINGER, P.W. (1964). *Poultry Science,* **43,** 1289–1290

GRIMMINGER, P.W. (1965). *Poultry Science,* **44,** 210–213

GRIMMINGER, P.W. and DONIS, O. (1960). *Journal of Nutrition,* **70,** 361–368

HAMILTON, R.M.G. (1980). *Poultry Science,* **59,** 598–604

HARMS, R.H., RUIZ, N., BURESH, R.E. and WILSON, H.R. (1988). *Poultry Science,* **67,** 336–338

HARMS, R.H., WALDROUP, P.W. and COX, D.D. (1962). *Poultry Science,* **41,** 1836–1839

HENRY, H.L. and NORMAN, A.W. (1978). *Science,* **201,** 835–837

JENSEN, L.S. (1965). *Poultry Science,* **44,** 1609–1610

JENSEN, L.S., FLETCHER, D.L., LILBURN, M.S. and AKIBA, Y. (1981). *Poultry Science,* **60,** 1603

JENSEN, L.S., FLETCHER, D.L., LILBURN, M.S. and AKIBA, Y. (1983). *Nutrition Reports International,* **28,** 171–179

JEROCH, H. (1970). *Archiv für Tierernährung,* **20,** 545–551

JEROCH, H. (1971). *Archiv für Tierernährung,* **21,** 249–256

JEROCH, H., PRINZ, M. and HENNIG, A. (1978). *Archiv für Tierernährung,* **28,** 53–65

KAZEMI, R. and DAGHIR, N.J. (1971). *Poultry Science,* **50,** 1296–1302

KESHAVARZ, K. and AUSTIC, R.E. (1985). *Poultry Science,* **64,** 114–118

KIRCHGESSNER, M. and MAIER, D.A. (1968). *Archiv für Tierernährung,* **18,** 309–315

KRATZER, F.H., DAVIS, P.N., MARSHALL, B.J. and WILLIAMS, D.E. (1955). *Poultry Science,* **34,** 68–72

LATYMER, E.A. and COATES, M.E. (1982). *British Journal of Nutrition,* **47,** 131–137

LEE, D.J.W. (1982). *British Poultry Science,* **23,** 263–272

LEESON, S., REINHART, B.S. and SUMMERS, J.D. (1979). *Canadian Journal of Animal Science,* **59,** 561–567; 569–575

LETH, T. and SØNDERGAARD, H. (1977). *Journal of Nutrition,* **107,** 2236–2243

McNAUGHTON, J.L., DAY, E.J. and DILWORTH, B.C. (1977). *Poultry Science,* **56,** 511–516

MANOUKAS, A.G., RINGROSE, R.C. and TEERIE, A.E. (1968). *Poultry Science,* **47,** 1836–1842

MARCH, B.E. (1981). *Poultry Science,* **60,** 818–823

MARUSICH, W.L., DE RITTER, E., OGRINZ, E.F., KEATING, J., MITROVIC, M. and BUNNELL, R.H. (1975). *Poultry Science,* **54,** 831–844

MARUSICH, W.L., HANSON, L.J., OGRINZ, E.F., EISENBEIS, H. and CAMERLENGO, M. (1983). *Poultry Science,* **62,** 1355

MILLER, D.L. and BALLOUN, S.L. (1967). *Poultry Science,* **46,** 1502

MOLITORIS, B.A. and BAKER, D.H. (1976). *Poultry Science,* **55,** 220–224

NEISHEIM, M.C., NORVELL, M.J., CEBALLOS, E. and LEACH, R.M. (1971). *Poultry Science,* **50,** 820–831

NELSON, T.S. and NORRIS, L.C. (1960). *Journal of Nutrition,* **72,** 137–144

NELSON, T.S. and NORRIS, L.C. (1961a). *Poultry Science,* **40,** 392–395

NELSON, T.S. and NORRIS, L.C. (1961b). *Journal of Nutrition,* **73,** 135–142

NORMAN, A.W., LEATHERS, V. and BISHOP, J.E. (1983). *Journal of Nutrition,* **113,** 2505–2515

OUART, M.D., HARMS, R.H. and WILSON, H.R. (1987). *Zootechnica International,* August, 44–45

PANDA, B., COMBS, G.F. and DE VOLT, H.M. (1964). *Poultry Science,* **43,** 154–168

PARDUE, S.L., THAXTON, J.P. and BRAKE, J. (1985). *Journal of Applied Physiology,* **58,** 1511–1516

PARSONS, A.H. and COMBS, G.F. (1979). *Proceedings of the Cornell Nutrition Conference,* 86–92

PATEL, M.B. and McGINNIS, J. (1977). *Poultry Science,* **56,** 45–53

PEACOCK, R.G. (1970). *Dissertation Abstracts International,* **31,** 983–984

PESTI, G.M., HARPER, A.E. and SUNDE, M.L. (1980). *Poultry Science,* **59,** 1073–1081

POLIN, D., WYNOSKY, E.R. and PORTER, C.C. (1962a). *Journal of Nutrition,* **76,** 59–68

POLIN, D., WYNOSKY, E.R. and PORTER, C.C. (1962b). *Proceedings of the Society for Experimental Biology and Medicine,* **110,** 844–846

POLIN, D., WYNOSKY, E.R. and PORTER, C.C. (1963). *Proceedings of the Society for Experimental Biology and Medicine,* **114,** 273–277

PRINZ, M., JEROCH, H., MOCKEL, P. and HENNIG, A. (1979). *Tierernahrung und Futterung,* **11,** 269–278

QUILLIN, E.C., COMBS, G.F., CREEK, R.D. and ROSOMER, G.L. (1961). *Poultry Science,* **40,** 639–645

RAMBECK, W.A. (1988). *Proceedings of the 6th European Symposium on Poultry Nutrition,* D3–11

REDDY, V.R., PANDA, B. and RAO, P.V. (1977). *Indian Journal of Animal Science,* **47,** 148–152

REID, B.L., HEYWANG, B.W., KURNICK, A.A., VAVICH, M.G. and HULAN, B.J. (1965). *Poultry Science,* **44,** 446–452

RINGROSE, R.C., MANOUKAS, A.G., KING KSON, R. and TERRI, A.E. (1965). *Poultry Science,* **44,** 1053–1065

ROBEL, E.J. (1983). *Poultry Science,* **62,** 1751–1756

ROBEL, E.J. (1983). *Comparative Biochemistry and Physiology,* **86B,** 265–267

ROBENALT, R.C. (1960). *Poultry Science,* **39,** 354–360

ROTH-MAIER, D. and KIRCHGESSNER, M. (1973). *Archiv für Geflügelkunde,* **37,** 205–207

ROTH-MAIER, D. and KIRCHGESSNER, M. (1976). *Archiv für Geflügelkunde*, **40**, 120–123

RUIZ, N. and HARMS, R.H. (1986a). *Poultry Science*, **54** (Suppl. 1), 191

RUIZ, N. and HARMS, R.H. (1986b). *Poultry Science*, **65** (Suppl. 1), 190–191

RUIZ, N. and HARMS, R.H. (1988a). *Poultry Science*, **67**, 794–799

RUIZ, N. and HARMS, R.H. (1988b). *Poultry Science*, **67**, 760–765

SCOTT, M.L. (1953). *Poultry Science*, **32**, 670–677

SHEN, H., SUMMERS, J.D. and LEESON, S. (1981). *Poultry Science*, **60**, 1485–1490

SINGH, S.P. and DONOVAN, G.A. (1973). *Poultry Science*, **52**, 1295–1301

SLINGER, S.P. and PEPPER, W. F. (1954). *Poultry Science*, **33**, 633–637

SLINGER, S.P., PEPPER, W.F., MORPHET, A.M. and EVANS, E.V. (1953). *Poultry Science*, **32**, 754–762

SOARES, J.H., SWERDEL, M.R. and OTTINGER, M.A. (1979). *Poultry Science*, **58**, 1004–1006

STEVENS, V.I., BLAIR, R. and RIDDELL, C. (1983). *Poultry Science*, **62**, 2073–2082

STEVENS, V.I., BLAIR, R., SALMON, R.E. and STEVENS, J.P. (1984a). *Poultry Science*, **63**, 760–764

STEVENS, V.I., BLAIR, R. and SALMON, R.E. (1984b). *Poultry Science*, **63**, 765–774

STOEWSAND, G.S. and SCOTT, M.L. (1961). *Poultry Science*, **40**, 1255–1262

SULLIVAN, T.W., HEIL, H.M. and ARMINTROUT, M.E. (1967). *Poultry Science*, **46**, 1560–1564

TADA, M. (1976a). *Japanese Poultry Science*, **13**, 1–7

TADA, M. (1976b). *Japanese Poultry Science*, **13**, 8–13

TENGERDY, R.P. and BROWN, J.C. (1977). *Poultry Science*, **56**, 957–963

TENGERDY, R.P. and NOCKELS, C.F. (1975). *Poultry Science*, **54**, 1292–1296

THORNTON, P.A. and SHUTZE, J.V. (1960). *Poultry Science*, **39**, 192–199

TINTE, P.J. and AUSTIC, R.E. (1974). *Poultry Science*, **53**, 2125–2136

TSIAGBE, V.K., KANG, C.W. and SUNDE, M. L. (1982). *Poultry Science*, **61**, 2060–2064

VALINIETSE, M.Y. and BAUMAN, V.K. (1982). *Nutrition Abstracts and Reviews*, **52B**, Abstract 5236

VELTMAN, J.R., JENSEN, L.S. and ROWLAND, G.N. (1983). *Poultry Science*, **62**, 1518

WAIBEL, P.E., KRISTA, L.M., ARNOLD, R.L., BLAYLOCK, L.G. and NEAGLE, L.M. (1969). *Poultry Science*, **48**, 1979–1985

WALDROUP, P.W., AMMERMAN, C.B. and HARMS, R.H. (1963). *Poultry Science*, **42**, 982–989

WALDROUP, P.W., HELLWIG, H.M., SPENCER, G.K., SMITH, N.K., FANCHER, B.I., JACKSON, M.E., JOHNSON, Z.B. and GOODWIN, T.L. (1985). *Poultry Science*, **64**, 1777–1784

WALDROUP, P.W., MAXEY, J.F., LATTER, L.W., JONES, B.D. and MESHEW, M.L. (1976). Bulletin, University of Arkansas Experimental Station, No. 805

WALDROUP, P.W., STEARNS, J.E., AMMERMAN, C.B. and HARMS, R.H. (1965). *Poultry Science*, **44**, 543–548

WEISER, H. and SALKELD, R.M. (1977). *Acta Vitamina Enzymologica (Milano)*, **31**, 143–155

WEISER, H. and TAGWERKER, F.J. (1981). *Papers Dedicated to Professor Johannes Moustgaard*, The Royal Danish Agricultural Society, Copenhagen

WEISER, H., BACHMANN, H. and BIEBER-WLASCHNY, M. (1988). *Proceedings of XVIII World's Poultry Congress, Nagoya*, pp. 834–836

WEISS, F.G. and SCOTT, M.'L. (1979). *Journal of Nutrition*, **109**, 1010–1017

WHITEHEAD, C.C. (1980). *British Journal of Nutrition*, **44**, 151–159

WHITEHEAD, C.C. (1986). In *Nutrient Requirements of Poultry and Nutritional Research*, pp. 173–190. Ed. Fisher, C. and Boorman, K.N. Butterworths, London

WHITEHEAD, C.C. and BANNISTER, D.W. (1980). *British Journal of Nutrition,* **43**, 541–549

WHITEHEAD, C.C., ARMSTRONG, J. and WADDINGTON, D. (1982). *British Journal of Nutrition,* **48**, 81–88

WHITEHEAD, C.C., DEWAR, W.A. and DOWNIE, J.N. (1972). *British Poultry Sciencee,* **13**, 197–200

WHITEHEAD, C.C., PEARSON, R.A. and HERRON, K.M. (1985). *British Poultry Science,* **26**, 73–82

WONG, P.C., VOHRA, P. and KRATZER, F.H. (1977). *Poultry Science,* **56**, 1852–1860

YEN, J.T., JENSEN, A.H. and BAKER, D.H. (1977). *Journal of Animal Science,* **45**, 269–278

YOSHIDA, M., HOSHII, H. and MORIMOTO, H. (1966). *Poultry Science,* **45**, 736–744

ZAVIEZO, D. and McGINNIS, J. (1980). *Poultry Science,* **59**, 1675

ZAVIEZO, D., MacAULIFFE, T. and McGINNIS, J. (1977). *Poultry Science,* **56**, 1772

5

EFFECT OF PELLET QUALITY ON THE PERFORMANCE OF MEAT BIRDS

E. T. MORAN, Jr
Poultry Science Department and Alabama Agricultural Experiment Station, Auburn University, AL 36849-5416, USA

Introduction

Pelleting is a processing procedure that is employed by the feed manufacturing industry to improve farm animal performance. Poultry generally responds to the greatest extent, and pelleted feed is in widespread commercial use in the poultry industry.

Changes occur to feed ingredients when a mixture of them is subjected to severe physical compression. Preconditioning with steam generally increases the extent of these changes. Disruption of cell walls that encapsulate nutrients is advantageous when this would not have been accomplished by the digestive system itself. Such an advantage can be expected to be more extensive with the young than the old as dietary use of feedstuffs having encapsulated nutrients increases. The forage meals and grain milling by-products are the primary feedstuff sources that consistently have improved nutrient availability, regardless of bird age (Bayley, Summers and Slinger, 1968; Olsen and Slinger, 1968; Saunders, Walker and Kohler, 1969; Summers, Bentley and Slinger, 1968). Protein denaturation and starch gelatinization also occur, but benefits from these changes largely relate to young fowl with immature gastrointestinal systems (Moran, 1985).

The pellet itself has benefits beyond any associated improvement in nutrient digestibility. Realization is wholly dependent upon maintaining pellet integrity from the point of manufacture to consumption. The following is a short description of the pelleting process, a rationalization of the changes which occur that influence binding, and an examination of the significance of pellet quality to meat bird performance.

Pelleting conditions

The feed mixture may be either conditioned with steam prior to compaction or untreated. Steam treatment predominates commercially, and recognition of the fundamentals of conditioning, extrusion, drying and cooling is paramount to any rationalization of binding and pellet stability. More thorough descriptions may be obtained by referring to Behnke and Fahrenholz (1986), Robinson (1976), and Wellin (1976).

CONDITIONING

Conditioning involves the direct injection of steam into feed that is being rapidly stirred. Moisture and heat are continuously transferred to the ingredients over an average of 10–30 seconds as feed enters the conditioning chamber until ingredients leave for extrusion. Steam pressure determines the extent of moisture increase from condensation and the associated temperature change. Generally, moisture level in grain-based feeds increases about 5%, with the final temperature at about 70–90°C.

DIE EXTRUSION

During pelleting, feed is physically forced through holes in a metal die by a series of rollers. Hole diameter varies with the desired pellet size, while its length in the die is usually 7- to 10-fold that of the diameter. Associated friction increases feed temperature. Decreasing hole diameter increases friction heat as does extending the hole length. Similarly, holes having alteration in relief from entry to exit also influence the degree of friction.

Moisture level during die extrusion can be a critical element to the extent of binding by acting as medium for chemical changes as well as a lubricant and moderator of friction. Skoch *et al.* (1981) reported that pellet temperature when leaving the die increased approximately 5°C when feed was steam conditioned, but was in excess of 25°C with dry pelleting.

COOLING AND DRYING

Immediately following extrusion, pellets are subjected to counter or cross-directional air flow. As a result, pellet temperature is generally reduced to 6–8°C above that of the air. Moisture in the pellets is progressively decreased to a marginally higher level than in the original mash.

Feed components and pellet stability

GRAIN

Grain usually comprises the greatest part of poultry feed and has the most impact on pellet stability. About 80% of the kernel is endosperm, and the associated starch, protein and fibre are responsible for binding capacity. Each component appears to act independently in this respect with their respective contributions varying substantially between the grains.

Starch

Amylose and amylopectin are organized in granular form such that birefringence occurs with the passage of plane polarized light. Amylose is thought to exist as a helix, while amylopectin represents a collection of these helices that are connected to one another. Most helices are placed against each other to form a crystalline network (Moran, 1982a). Orientation of these helices is perpendicular to the granule surface, thereby permitting the entry of water.

Heating starch in the presence of moisture greatly accentuates the rate at which water enters the granule. Accumulation of water causes a progressive swelling and loss of crystallinity (Figure 5.1). By definition, gelatinization occurs when the granule has swollen to its maximum, yet remains intact, and birefringence nearly disappears (Goering, Fritts and Allen, 1974).

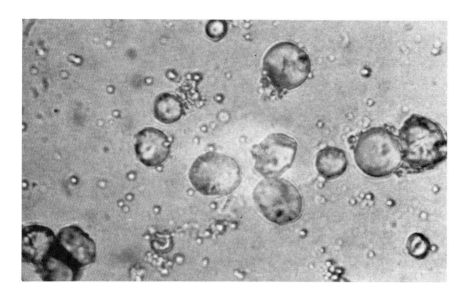

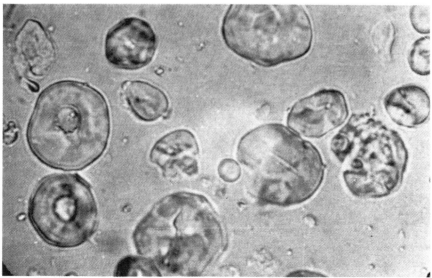

Figure 5.1 Microscopic views of maize starch. (Top) Granules from maize ground to pass through a 2-mm screen. (Bottom) Illustration of the extensive swelling of granules which occurs after the ground maize is exposed to steam at atmospheric pressure for 10 min. (From Mercier and Guilbot, 1974, courtesy of the Institut de Naturel Recherche Agronomique)

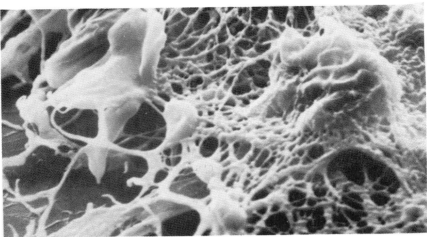

Figure 5.2 Scanning electron micrograph of dried wheat starch granules after moist heating for 45 min at 95°C. The exudate is presumed to represent the release of free amylose and amylopectin molecules as the granule 'melts'. (From Miller, Derby and Trimbo, 1973, courtesy of the American Association of Cereal Chemists)

Additional heat beyond that necessary to cause gelatinization is required for the release of polymers and 'melting' of the granule (Figure 5.2). Starches can differ in their proportions of each polymer, and as amylopectin increases so also does the extent that swelling and heat are needed for disintegration (Hill and Dronzek, 1973).

Free amylopectin is probably responsible for most of the pellet binding attributable to starch. Schwartz and Zelinskie (1978) examined the effectiveness of starches in stabilizing pills where the pharmaceutical preparation was based on minerals. Binding was better using cooked rather than raw starch and when the cooked starch had an increasing proportion of amylopectin.

Steam pelleting substantially gelatinizes grain starch, but, the extent of granule melting and free polymer release appears to be marginal. Mercier and Guilbot (1974) examined the susceptibility of corn starch to alpha-amylase digestion after each stage of the pelleting process. Although conditioning alone increased susceptibility to amylase, digestibility was far greater after die extrusion. Minimizing moisture at this time to that which can be absorbed by the granule allows for the generation of additional heat. The additional heat and physical forces have been shown to play a major role in starch solubilization (Bhattacharya and Hanna, 1987; Gomez and Aguilera, 1983). Pelleting conditions in practice, do not favour granule solubilization. Moisture exceeding the capacity for granule uptake acts as a lubricant and is commercially advocated to reduce the consumption of electric power (Wellin, 1976).

Protein

By definition, the grain endosperm protein that is soluble in alcohol is termed a prolamine. Prolamines are microscopically distinctive because they appear as small

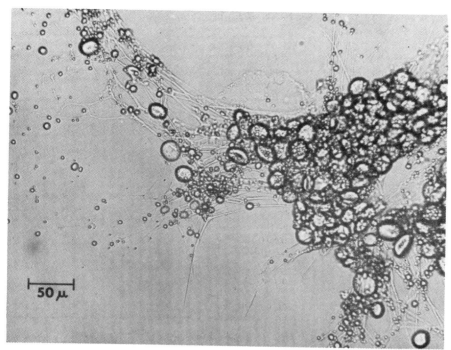

Figure 5.3 Microscopic view of wheat after wetting of its flour. Moisture is rapidly absorbed by the matrix protein with a near immediate increase of volume about 20-fold. Resultant fibrils are translucent, long and stick to the array of suspended particulates. (From Bernardin and Kasarda, 1973, courtesy of the American Association of Cereal Chemists)

spheres associated with the starch granules. In some grains, especially wheat, they lose their distinctiveness and form part of the cell matrix during the final stage of development (Adams, Novellie and Liebenberg, 1976; Bechtel, Gaines and Pomeranz, 1982; Pernollet and Camilleri, 1983). Moistening of wheat flour has been shown to transform rapidly this matrix protein into long 'sticky' fibrils that interconnect starch granules with other particulates (Figure 5.3; Bernardin and Kasarda, 1973; Seckinger and Wolf, 1970). Triticale, rye and barley also produce doughs having viscoelastic properties, but maize, sorghum, millet and oats do not.

Fibril formation is believed to occur during steam conditioning, with binding being stabilized upon heat denaturation as the pellet is extruded. Dahle (1971) reported that wheat dough loses its viscoelastic and sticky properties at 90–100°C.

Fibre

Solubilized fibre contributes to the viscoelastic properties of those grains which can form doughs. This fibre almost exclusively comes from the weak endosperm cell walls, and represents a composite of glucan and pentosan hemicelluloses. Substantial differences exist between the grains in the amount of endosperm fibre that is soluble, the contributions of glucans and pentosans, and molecular sizes

Table 5.1   AMOUNT OF FIBRE SOLUBILIZED IN WATER FROM VARIOUS GRAINS, THE MOLECULAR NATURE OF EACH CONTRIBUTION AND THEIR VISCOSITIES

Grain	Soluble fibre		Glucans[b]		Pentosan[b] molecular weight
	(% kernel)	*Relative viscosity*[a]	*Soluble fibre (%)*	*Molecular weight*	
Maize	0.32	123	12	NA	NA
Wheat	0.64	296	10	33 700	38 000
Rye	1.05	304	17	64 200	173 000
Barley	1.34	247	59	55 800	58 800
Oats	1.65	115	21	26 800	34 200

[a] Preece and MacKenzie (1952); viscosities are given for 0.5% solutions where water = 100
[b] D'Appolonia (1973) citing Podrazky (1954)

representing each type polymer (Table 5.1; D'Appolonia, 1973; Henry, 1985; Preece and Mackenzie, 1952).

Solubilized grain fibre creates viscous solutions, with the extent of viscosity being related to the nature of component sugars, their connection with one another and total size of the molecule. Wheat, rye and barley create fibre solutions which are particularly viscous. Such conditions are thought to participate in dough viscoelasticity and enhance the baking performance of bread (D'Appolonia and Gilles, 1971; Jelaca and Hlynka, 1971; Patil, Tsen and Lineback, 1975). These same polysaccharides are also likely to contribute to pellet binding.

High quality grain

The pellet binding ability of any grain that is of good quality can be attributed to the starch, protein and fibre. While starch is similar among the grains, the abilities to form protein fibrils and viscous fibre solutions differ markedly. In this respect, maize and wheat have substantial differences for both characteristics.

Wheat is known to enhance pellet stability better than maize. An example of this difference is given using turkey feed (Table 5.2). Having feeds where wheat was progressively substituted for maize and steam condition temperature was held constant led to a reduction in the percentage of fines. Increasing conditioning temperature with feeds having maize or wheat reduced the percentage of fines further but maize was more responsive to temperature than wheat. In summary, protein fibrils and soluble fibre are of primary advantage to binding, while starch is secondary.

Poor quality grain

Animal feeds do not always contain reasonable quality grain, and many alterations associated with down-grading may affect pellet binding. Interrupting grain development before the seed matures is known to cause kernel shrivelling and

Table 5.2 INFLUENCE OF MAIZE VERSUS WHEAT
AND CONDITIONING TEMPERATURE ON FINES
CONTENT OF STEAM PELLETED TURKEY FEED

Maize/wheat (% ratio diet)[a]	*Conditioning temperature* (°C)	*Fines (%)*	
		In feed	*Reduction*
60:0	80	67.8	Control
	88	50.8	25.0
45:15	80	55.1	18.7
30:30	80	46.0	32.1
0:60	80	26.4	61.1
	88	23.5	65.4

[a] Total grain was 60%; other major ingredients were soyabean
meal, 19.4% and added fat, 4.0%
(From W. F. Krueger, Cuddy Foods Ltd, Marshville, North
Carolina, personal communication)

diversion from human use to animal feedstuff. In this situation, the matrix proteins
do not readily respond to moisture (Crozet and Guilbot, 1979; Pernollet and
Camilleri, 1983) and the starch granules have increased resistance to gelatinization
(Baruch *et al.*, 1979; Karlsson, Olered and Eliasson, 1983), while the amount and
viscosity of soluble fibre increases (Gohl, Nilsson and Thomke, 1978). Such
changes have been shown to affect adversely raising and baking performance of
wheat doughs (Dexter *et al.*, 1985; Tipples, 1980), and binding characteristics
during pelleting are expected to respond similarly.

Germination of mature kernels results in a progressive degradation of
endosperm walls, proteins and starch (Gram, 1982; Mikola and Mikola, 1980;
Palmer, 1987; Selvig, Aarnes and Lie, 1986). These changes also adversely affect
wheat doughs and bread formation (Ibrahim and D'Appolonia, 1979; Kulp,
Roewe-Smith and Lorenz, 1983, Lorenz *et al.*, 1983; Lorenz and Kulp, 1981;
Lorenz and Valvano, 1981; Lukow and Bushuk, 1984; Morad and Rubenthaler,
1983). Again, decreased pellet binding can be expected.

Heating of grain in storage because of excess moisture or artificial drying prior to
storage also creates alterations that decrease baking performance of wheat
(Kastankova, 1983; Kirleis *et al.*, 1982; Slusanchi *et al.*, 1973; Westermarch-
Rosendahl, 1978; Westermarch-Rosendahl and Ylimaki, 1978). Decreased pellet
binding with wheat might be expected because of protein denaturation and loss of a
dominant contributor. However, a certain amount of starch gelatinization could
prove beneficial when this factor dominates binding, for example with maize.

HIGH SUGAR FEEDSTUFFS

Feedstuffs containing free sugars may improve the extent of binding by virtue of
chemical changes that happen during pelleting. While pellet stability is desirable,
the chemical changes may adversely affect nutritional value.

Caramelization

Behnke and Fahrenholz (1986) noted that the inclusion of sugar in feed increased the power requirement for pelleting when the conditioning from steam was as low as 65°C. Caramelization is the loss of water from the sugar molecule itself, and temperature to initiate the process is about 120°C. Presumably, this temperature is attained at the pellet-die interface during extrusion, and heat generated from resistance to flow, in turn, increases the extent of starch granule dissolution.

Caramelization is independent of moisture conditions, and this change may not be the primary factor predisposing to improved binding. Aumaitre *et al.* (1978) examined the effect of dry pelleting a feed high in sucrose and when part of the sucrose was substituted with a glucose-fructose mixture (Table 5.3). The high sucrose feed could not be pelleted rapidly and led to a poor pellet although energy consumption increased. Addition of a small amount of water to the high sucrose feed or its partial substitution with glucose-fructose improved all measurements, but not markedly.

Table 5.3 EFFECT OF DRY PELLETING A FEED HIGH IN VARIOUS SUGARS ON MANUFACTURING YIELD AND PELLET CHARACTERISTICS

Item	*14% Sucrose*		*Glucose/fructose*	
	No added water	*+2% added water*	*5%*	*10%*
Feed analyses (%)[a]				
Dry matter	88.6	—	89.7	90.0
Total reducing sugar	2.4	—	5.6	7.8
True glucose	1.0	—	2.2	2.9
Fructose	3.1	—	3.9	3.5
Sucrose	14.0	—	10.3	6.7
Pelleter operation				
Energy (Wh/kg)	29.7[x]	27.9[y]	27.2[y]	21.8[z]
Yield (kg/h)	—[b]	66[x]	77[y]	102[z]
Pellet characteristics				
Temperature (°C)	58[x]	60[x]	54[y]	52[y]
Moisture after pellet (%)	7.1	8.4	8.4	10.4
after cooling (%)	6.5	7.5	8.7	9.1
Hardness (bar/cm)[c]	3.6[x]	3.4[x]	3.2[y]	1.9[y]
Durability (%)	4.8[x]	0.9[y]	0.7[y]	0.9[y]

[a] Feed contained about 32% wheat, 21% soyabean meal, 4% dried skim milk, 26% refatted skim milk and 3% vitamin-mineral premix
[b] Feed difficult to pellet as a dry meal
[c] High values related to increased hardness
[x,y,z] Means of any one row without a common letter are statistically different (P<.01)
(From Aumaitre *et al.*, 1978)

Browning reaction

Moisture provided by steam conditioning permits the browning reaction to occur in feed while the normal 70–90°C temperature can greatly accentuate the rate of reaction. Free aldehyde and amino groups initially combine, then the resulting Schiff's base undergoes oxidative changes to form complex melanoides that darken and become viscous as they polymerize (Motai, 1976). Amino groups from supplementary free amino acids are particularly reactive, while the free amino groups in protein, such as the epsilon amino group of lysine, are secondary by comparison (Wolf *et al.*, 1981).

The browning reaction does not occur to any extent when most feeds are pelleted because free aldehyde groups are usually present in negligible amounts. Several feedstuffs are known to be high in this respect, but they are not ordinarily incorporated into poultry feed. Milk by-products having lactose, molasses high in monosaccharides, and unstabilized fat that has undergone oxidative rancidity are all capable of supplying the necessary second reactant. Dried bakery by-product has a variable composition (Waldroup, Whelchel and Johnson, 1982), and the inclusion of confectioner's sugar represents an unknown threat.

The enhancement of binding attributable to high sugar feeds is thought largely to occur because of the viscous melanoides. Melanoid formation is thought to start during conditioning and increases at the die-feed interface, in turn, promoting resistance to extrusion. While improved pellet stability is a desirable result, possible losses of supplemental amino acids are not.

COMMERCIAL BINDERS

Feed additives are commercially available solely for the purpose of enhancing pellet quality. Each may be considered as acting in parallel with starch, protein and fibre as previously described for grain. The following is a short description of these binders and their possible modes of action.

Lignosulphonates

The paper industry incorporates either ammonium or calcium bisulphite with wood pulp, then moisture and heat disintegrates the lignin leaving free cellulose. Dehydrating the resulting liquor yields a powder composed of ammonium or calcium lignosulphonates and an array of complex sugars (Table 5.4). Incorporation of these products amounts to about 1–2% of the feed, and binding is attributed to both the lignosulphonate and carbohydrate fractions. The lignosulphonate binders are independent of conditioning temperature once sufficient steam condenses to effect their solubilization (Winowski, 1985).

Although the carbohydrate fraction largely comprises B-linked sugars which are not expected to be digested as such, they can readily enter the caeca for microbial fermentation. Proudfoot and DeWitt (1977) noted that the caeca of broilers given calcium lignosulphonate progressively enlarge as the inclusion level increases while the contents darken and thicken. The metabolizable energy (ME) value is probably the result of retrieved volatile fatty acids. Morrison *et al.* (1968) reported an ME value of 13.4 MJ (3220 kcal)/kg for a calcium lignosulphonate having 44% total soluble carbohydrate while Kivimae (1978) suggested 6.0 MJ (1460 kcal)/kg for another source having 22% carbohydrate.

Hemicellulose extract

The pressed wood industry treats wood at high steam pressure but without chemicals that would affect the lignin. A portion of the hemicelluloses and sugars is solubilized from the pulp, and this carbohydrate complex when dried is used as a binder. Recommended use is similar to the lignosulphonates but the higher proportion of soluble carbohydrates would suggest an advantage in ME content (Table 5.4).

Fibre is an organization of cellulose fibrils held together by a hemicellulose-pectic acid complex and stabilized by lignin (Albersheim, 1978). The hemicelluloses are hydrogen bonded to the fibril surface and pectic acids connect hemicelluloses between fibrils. heat treatment is believed to detach the hemicelluloses from the fibrils permitting some solubilization of the hemicellulose-pectic acid complex. Dissolution of these carbohydrates occurs during steam conditioning of feed, and binding between particulates is probably established during die compaction upon resumption of H-bonding of compatible surfaces, namely cellulose and starch.

Hemicellulose extract has a particularly dark colour, suggesting that the wood treatment formed browning reaction melanoides which may also participate in the binding. The high proportion of reducing sugars remaining indicates that a certain amount of browning reaction may also occur during pelleting as well. Although the lignosulphonates also have reducing sugars present, browning may not be of any significance because sulphites are inhibitors ofthe reaction.

Table 5.4 APPROXIMATE COMPOSITION OF COMMERCIALLY AVAILABLE PELLET BINDERS DERVIED FROM THE WOOD AND PAPER INDUSTRY (%)

Analyses	Lignosulphonates		Hemicellulose extract[c]
	Ammonium[a]	Calcium[b]	
Approximate			
Moisture	5.0	8.0	3.0
Protein	29.4	0.9	<0.1
Fat	0.1	0.1	<0.1
Fibre	0.1	<0.1	1.5
Ash	<0.1	20.4	8.0
Organic			
Lignin	65.0	39.1	<0.1
Total sugar	17.1	39.6	84.0
Reducing sugar	17.0	<3.0	15.0
Inorganic			
Sulphur	5.5	5.2 Nil	
Calcium	<0.1	8.7	0.5

[a] Tembind A 002, Temfibre Inc., Temiscaming, Quebec, Canada
[b] Peltex 1050, Reed Lignin Inc., Rothschild, Wisconsin, USA
[c] Masonex, Masonite Corp., Laurel, Mississippi, USA

Carboxymethylcellulose (CMC)

Treating cellulose fibrils with alkali disrupts the H-bonding holding the individual polymers together to cause a swelling. Reacting the exposed hydroxy groups with sodium monochloroacetate results in the addition of carboxymethyl groups, and the molecule becomes a weak acid (Ganz, 1973). This circumstance leads to an attraction for water, which aggregates in layers around the molecule and creates viscous solutions when dissolved.

The level of CMC advocated for pellet binding is 0.1%. Presumably, hydration occurs during steam conditioning with associated heat permitting the water layers to be more mobile than usual. Subsequent compaction would bring particulates close together and allow the CMC to develop ionic attractions with particulates upon displacement of water.

Bentonite

Trilayered aluminium silicate having sodium or calcium as its exchangeable cation is the only effective form of bentonite for binding purposes. Feed inclusion is about 1–2%, and this mineral must be hydrated to be functional.

The sodium form is best, and hydration of the mineral results in a five-fold increase in weight. During this change, the aluminium silicate layers become separated, and water is attracted to their ionic surfaces creating a 12- to 15-fold increase in volume.

Pellet binding with bentonite probably resembles CMC. Moisture is necessary for binder hydration; heat permits mobility of bound water; and attractions with particulates are formed during extrusion. Bentonite effectiveness has been reported to be dependent upon conditioning temperature (Table 5.5); however, this dependence may better relate to need for moisture from steam condensate than temperature. The increased consumption of power to process feed indicates that resistance to die extrusion is occurring, and binding can be further facilitated by additional starch dissolution.

Table 5.5 EFFECT OF SODIUM BENTONITE AND CONDITIONING TEMPERATURE ON PELLET STABILITY OF TURKEY FEED AND ENERGY REQUIREMENT FOR PROCESSING[a]

Temperature increase (°C)	Bentonite (%)	Fines (%)	Electricity (kWh/tonne)
17	0	29.2	17.7
	2.5	23.4	17.9
34	0	19.3	11.4
	2.5	13.0	11.2
51	0	13.1	8.8
	2.5	7.9	10.2

[a] Feed contained maize, sorghum, soyabean meal, and added fat as the primary ingredients
(From Pfost and Young, 1973; Pfost, 1976)

Starch

Grain that has been heat processed such that the constituent starch granules would have initiated dissolution has also been marketed as a binder. Levels employed may vary from 0.5 to 5.0%, with the advantage that binder may be substituted for the grain portion of the feed without nutritional alteration. Given that amylopectin is the most active participant in binding, then use of the waxy type grains where this polymer represents 99% of the starch would offer the greatest advantage.

Binding by amylopectin can be rationalized in terms of its structure and ability to H-bond. As mentioned previously, the molecule is believed to occur as a series of amylose helices connected in a progressive manner to resemble a tree or bush. In the granule, these helices H-bond to each other and create crystalline regions just as the cellulose polymers aggregate to form fibrils. Pellet binding by amylopectin probably occurs when the free helices at either end of this very large molecule H-bond to compatible starch or fibre surfaces on different particulates.

Gelatin

Skin and bone have a high proportion of collagen, and rendering these by-products of the meat industry yields gelatins. Gelatin failing to meet standards for human use has been marketed as a pellet binder. Recommended usage varies between 0.1 and 0.25% of the feed.

Pellet binding with gelatin may be different from the adherence occurring with those grains forming protein fibrils. Fibril sticky characteristics only exist in the native state, with adherences established at this time persisting after denaturation. Such contacts and stability have been partially attributed to disulphide cross-linking involving cystine (Dreese, Faubion and Hoseney, 1988). Collagen fibrils in the native state are essentially triple helices of protein polymers held together by a combination of covalent and H-bonding. Denaturation destroys all organization such that the resulting gelatin is readily soluble in water and forms gels. Accordingly, ionic attractions seem to be the likely basis of pellet binding.

Inadequate removal of moisture during pellet drying and rewetting after manufacture may represent a threat to gelatin binding. Salt linkages are viewed as suffering from moisture, while heat would be necessary to disrupt H-bonds proportionate to their extent of involvement. Covalent bonds would not be affected.

FAT

All of the previously mentioned factors associated with pellet binding depend upon compatibility with water to be effective. Protein, fibre and starch when involved had to be solubilized before they could act. Similarly, attachment sites on particulates also presented a hydrophilic surface. Fat is viewed as creating pellet stability problems by reducing the formation of active binder and interfering with subsequent attachments.

Table 5.6   EFFECT OF ADDING FAT TO A CHICKEN
BROILER FEED BEFORE AND AFTER STEAM PELLETING ON
THE PROPORTION OF FINES AND USAGE OF ELECTRICITY

Added fat (%)		Fines[a] (%)	Production (tonnes/h)	Power (kWh/tonne)
Mixer	*Post-cooling*			
1.00	4.66	18.0	11.55	10.97
2.00	3.66	22.0	12.10	9.72
3.00	2.66	29.2	13.20	8.78
4.00	1.66	31.6	13.20	7.89
5.33	0.33	50.8	—	—

[a] Percentage passing through a 2360 mm screen (From Richardson and
Day, 1976)

Starch

Added fat has been shown to coat starch granules and reduce their uptake of water
such that additional heat is required for gelatinization (Eliasson *et al.*, 1981;
Eliasson, Larsson and Miezis, 1981). This coating of starch as well as other
particulates also lubricates die extrusion to further reduce the likelihood of granule
dissolution. Richardson and Day (1976) examined the significance of fat added to
broiler feed before and after pelleting in terms of resulting fines and power usage.
Progressively diverting the amount of fat added into the mixer, its addition after
steam pellet cooling, improved stability but increased energy consumption for
processing (Table 5.6).

Commercial binders

One of the reasons commercial binders are effective relates to their ready solubility
compared with the extensive treatment required by grain to reach the same stage.
Thus, the advantage of using a commercial binder would best occur when
conditions do not permit the grains to realize their binding potential. This
advantage is particularly evident when fat is mixed in the feed prior to pelleting.

Adding 5.25% fat to a chicken broiler feed consisting primarily of maize and
soyabean meal reduced the proportion of intact pellets by over 50% (Table 5.7).
Use of 1.25% calcium lignosulphonate provided little additional benefit when the
fat had been omitted, but the proportion of intact pellets improved over 40% when
fat was present.

The extent of improvement from any of the commercial binders may be expected
to vary with their level of inclusion, amount of fat and grain source. Salmon (1985)
added fat to a turkey feed based on wheat in 3% increments from 0 to 9% (Table
5.8). The proportion of intact pellets decreased with fat inclusion. As expected, the
improvement from added binder with wheat was far less than previously mentioned
for maize. Supplementing the feed with 2.5% sodium bentonite was of no benefit
when fat was not added, but stability improved with the 3 and 6% levels. A further
increase to 9% fat counteracted bentonite's binding advantage.

Table 5.7 EFFECT OF ADDED FAT ON YIELD OF INTACT
PELLETS AND PARTICULATE SIZE DISTRIBUTION OF A
MAIZE-SOYABEAN MEAL CHICKEN BROILER FEED AND
THE ADVANTAGE OF SUPPLEMENTAL CALCIUM
LIGNOSULPHONATE

Screen (μm)	No added fat		5.25% added fat[a]	
	Control	+CaLS[b]	Control	+CaLS[b]
4750	90.8	92.8	43.9	62.9
3360	6.6	5.2	16.4	13.3
2360	1.2	0.9	11.2	7.2
1700	0.5	0.4	9.4	12.3
1180	0.2	0.2	7.2	4.8
850	0.1	0.1	3.9	1.8
600	0.1	0.1	2.7	1.4
425	0.1	0.1	1.8	1.1
Residue	0.3	0.3	4.0	2.5

[a] Added fat displaced an equal proportion of maize
[b] Calcium lignosulphonate was added to the complete feed (12.5 kg to
1000 kg feed)
(From Moran, unpublished data)

Table 5.8 EFFECT OF LEVEL OF ADDED FAT
ON PELLET QUALITY OF TURKEY FEED
BASED ON WHEAT AND REALIZATION OF
IMPROVEMENT FROM SUPPLEMENTAL
SODIUM BENTONITE

Added fat (%)[a]	% Retention (2 mm screen)	
	Control	2.5% sodium bentonite
0	97.3	96.2
3	93.0	94.9
6	89.7	93.0
9	84.4	84.2

[a] Wheat and soyabean meal were the primary
ingredients with both being altered to maintain a
constant energy-protein ratio as the level of added fat
increased
(From Salmon, 1985)

High fat feedstuffs

Meat meal, fish meal and full-fat soyabean meal are examples of feedstuffs that in
themselves are high in fat. All high fat containing feedstuffs have generally been
heat processed, and the fat which was localized to specific tissues or areas of seeds
is now uniformly redistributed throughout the product, particularly on particulate
surfaces. In turn, pellet cohesion can be expected to suffer as their combined
dietary level of inclusion increases.

Pellet stability and bird performance

Fowl have been shown to prefer feed in particulate form rather than mash (Schiffman, 1969). This preference is not for a fixed size, but changes to conform with dimension of the oral cavity. Such change in preference decreases the work of prehension, time spent standing at the feeder, and competition for food (Reddy *et al.*, 1962; Jensen *et al.*, 1962; Savory, 1974). Pelleted feed would also appear to reduce the extent of motility and other work associated with the gastrointestinal tract. Choi *et al.* (1986) observed that various parts of the gastrointestinal system in broiler chickens given pelleted feed were reduced in weight relative to the body. These alterations were greater with older than younger birds (Table 5.9).

The mouth of the fowl is particularly well suited to benefit from pellets. Immobility of the beak creates problems in prehending finely divided feeds, while the absence of teeth further prevents the reduction of particulates larger than the oral cavity. Should beak trimming be imposed, then the advantage of pellets can be changed to a disadvantage depending on how the trimming was accomplished (Lonsdale, Vondell and Ringrose, 1957; Vondell and Ringrose, 1957; Deaton *et al.*, 1987).

Saliva produced by mammals represents a mixture of mucous and serous secretions which can be altered in their proportions to accommodate feed characteristics and optimize mastication. Fowl only produce a mucous saliva which is used to coat feed as it is rotated in the oral cavity and lubricates the bolus (Moran, 1982b). Feed as a mash requires more mucus to enable it to be swallowed than if pelleted.

The mucus may react with some of the same components in grain that enhance pellet quality, in turn, creating problems in beak movement. Wheat is an extreme example. Sticky fibrils can form with mucus and accrue at the corners of the mouth where large salivary ducts are located. High dietary levels of wheat, particularly when finely ground, are known to cause beak malformations and extensive feed wastage (Poley, 1938). Moran (1987) suggested that these birds would consume

Table 5.9 CONTRIBUTION OF VARIOUS PARTS OF THE GASTROINTESTINAL TRACT WITH BROILER CHICKENS GIVEN MASH AND PELLETED FEEDS

| Feed form | Gain (g) | Feed/gain | % of body weight | | | Small intestine length (m/kg body weight)[b] |
			Liver	GI tract[a]	Gizzard	
0–4 weeks of age						
Mash	696[x]	1.67	2.26	7.54	1.73[x]	1.42
Crumbles	729[y]	1.66	2.36	6.74	1.52[y]	1.35
4–8 weeks of age						
Mash	983[x]	2.69	2.36	4.97[x]	1.62[x]	0.70
Pellets	1101[y]	2.67	2.47	4.35[y]	1.24[y]	0.68

[a] Proventiculus through cloaca including the caeca
[b] Duodenum to ileum
[xy] Different letters between values in a column, respective of age, corresponds to a statistical difference ($P<0.01$)
(From Choi *et al.*, 1986)

additional water for relief which may cause secondary stresses. Feed-contaminated water elicits bacterial growth, and an excess intake of water leads to wet litter. Differences in water consumption between birds given mash and pellets were not observed by Eley and Hoffman (1949) nor Savory (1974); however, the feed circumstances in these studies would not be expected to create oral problems.

Early performance

The benefits occurring with pelleting that can be attributed to particulate integrity, denaturations and physical disruptions are not expected to be of constant proportions. Feedstuffs incorporated in any feed represent an infinite array in itself. Furthermore, the use of dry pelleting as opposed to steam pelleting superimposes another variable. Summers, Bentley and Slinger (1968) reported a particularly diverse response in chick growth, feed conversion and ME when maize and wheat bran represented one-half of a complete feed and the feeds were either left untreated or pelleted using dry and steam procedures (Table 5.10).

The extent of performance response to pelleting can be expected to decrease as the energy level of the feed increases. McNaughton and Reece (1984) reported that feed-grain ratios of broiler chicks given crumbled feeds having 12.9 (3100), 13.1 (3150) and 13.3 (3200) MJ (kcal) ME/kg decreased by 0.088, 0.078 and 0.042 compared with the respective mash. Similarly, Auckland and Fulton (1972) observed that broiler chicks fed crumbles grew 9 and 6% faster than those fed mash when given low and high energy feeds, respectively. Reductions in feed intake and the associated work of prehension are seen as being the primary factors in these observations.

Late performance

Once the gastrointestinal system matures, mechanisms are in place to cope with anti-digestive enzyme factors and to improve availability of nutrients encapsulated by fibrous cells (Moran, 1982b). Sibbald (1977) measured the metabolizable energy of 10 different practical poultry diets using adult roosters, and steam pelleting failed to improve any of them.

An increased beak size now permits consumption of pellets and the overall efficiency of eating improves. Proudfoot and Sefton (1978) using broiler chickens and Proudfoot and Hulan (1982) with small-type turkeys observed that final body weight and feed conversion suffered as the proportion of fines in pelleted feed increased (Table 5.11). The productive advantage from intact pellets was measured by Wenk and Van Es (1979) in broiler chickens as an increased retention of energy, protein and fat which they attributed to reduced flock activity.

One means of capitalizing on the reduced maintenance expenditure occurring with stable pellets is to decrease dietary ME accordingly. Shen, Summers and Leeson (1985) substituted barley for maize in broiler feeds containing the same amount of balanced protein and obtained equivalent growth after steam pelleting. Failure to consider the productive energy increase from pellets can create adverse circumstances should energy already be high relative to protein. Jensen *et al.* (1965) examined the effect of pelleting on large turkeys from 8 weeks of age to marketing. When feeds having protein and lysine submarginal to the requirement and recommended energy levels were pelleted, the deficiency was accentuated.

Table 5.10 INFLUENCE OF FEEDSTUFF AND PELLETING PROCEDURE ON CHICK RESPONSE AND METABOLIZABLE ENERGY CONTENT

Feedstuff		Chick response[a]		ME content of food	
Ingredient	Treatment[b]	Weight (g)	FCR[c]	(ME/kg)	(MJ ME/kg)
Wheat bran	Mash	246	2.82	1.46	6.11
	Dry crumbled	309	2.33	1.48	6.19
	Steam crumbled	335	2.09	2.05	8.58
Maize	Mash	276	2.24	3.45	14.43
	Dry crumbled	294	1.85	3.58	14.98
	Steam crumbled	309	1.82	3.61	15.10

[a] Response by chicks to a complete feed where 50% was either wheat bran or maize and the remainder was of constant proportions of maize, soyabean meal, minerals, and vitamins
[b] Only the wheat bran or maize was treated. The remaining portion of the complete feed was as a mash
[c] Food conversion ratio (kg feed/kg liveweight gain)
(From Summers *et al.*, 1968)

Table 5.11 EFFECT OF PROPORTION OF FINES IN FEED ON THE LIVE PERFORMANCE OF TURKEY BROILERS GROWN TO 14 WEEKS OF AGE

Diet texture[a]		Body weight (g)		Combined FCR[b]
		Male	Female	
Mash		6872	4947	2.50
Fines,	0%	6951	5219	2.31
	7.5	6942	5173	2.35
	15	7000	5281	2.37
	30	6966	5156	2.37
	60	6895	5134	2.40

[a] All textures had feeds of common composition, and were based on wheat, maize, soyabean meal, fish meal, and added fat. Feed 1–3 weeks was crumbled for all treatments involving fines and as a mash when not pelleted. Texture treatments of pelleted feed were imposed on the grower (3–10 weeks of age) and finisher (10–14 weeks)
[b] Feed conversion ratio (kg feed/kg liveweight gain) for mixed sexes
(From Proudfoot and Hulan, 1982)

Feed particulates should be sized relative to the oral cavity as the birds progress from day old through to marketing. Moran and Ferket (1984) provided large turkeys with small crumbles, large crumbles, standard pellets (around 45 mm diameter), and large pellets (around 60 mm diameter) from 0–4, 4–8, 8–12 and 12–20 weeks, respectively. The feeds employed were based on either a maize, soyabean meal, wheat and fat mixture to satisfy National Research Council (1977) requirements or one where barley was substituted for all the fat and progressively greater amounts of the maize and soyabean meal such that crude protein and essential amino acids would be similar but ME fell below recommendations. Each

Table 5.12 EFFECT OF DECREASING DIETARY ME ON THE RESPONSE OF LARGE TURKEYS IN TERMS OF LIVE PERFORMANCE AND CARCASS QUALITY WHEN THE FEEDS ARE PRESENTED AS A MASH *VERSUS* PELLET AND REGROUND PELLET

Energy[a]	Form	16–20 weeks of age			FCR (kg feed/kg liveweight gain 0–2 weeks)	Finish (% grade A)[b]	Meat[c]	Skin[c]
		Weight (g)	(g)	F/G				
NRC	Mash	14 099	3864	3.92	2.99	99.0	3360	778
	Ground pellet	14 175	3803	3.88	3.01	96.6	3358	784
	Whole pellet[d]	14 973	4148	3.30	2.63	99.1	3690	826
Low	Mash	13 153	3402	4.41	3.27	68.9	3188	487
	Ground pellet	13 577	3599	4.43	3.33	83.9	3391	580
	Whole pellet[d]	14 436	3904	3.88	2.92	85.6	3674	610
Statistics								
Energy		✶✶	✶✶✶	✶✶	✶✶		NS	✶✶
Form		✶✶	✶✶	✶✶	✶✶	NS	✶	NS
ExF		NS	✶✶	NS	NS	NS	NS	NS
SEM		110.7	91.0	0.116	0.034	NS	85.5	48.2

[a] NRC relates to the suggested ME values of the National Research Council (1977) and corresponds to 13.3 MJ/kg (3200 kcal/kg) during the 16–20 week period. Low was when barley was incorporated into the same feed (at 15, 20, 30 and 40% during the 0–4, 4–8, 8–12, 12–16 and 16–20 week periods, respectively, at the expense of all added fat and progressive amounts of maize and soyabean meal that would maintain protein and essential amino acids at levels suggested by the National Research Council (1977)
[b] Subjective evaluation of fat along the main feather tracts with the chilled carcass
[c] Total meat and skin removed from the breast of the chilled carcass
[d] Refers to feeds that were in fine crumbles, coarse crumbles, normal pellet (45 mm), and large pellet (60 mm) during the 0–4, 4–8, 8–12, 12–20 week intervals, respectively
(From Moran and Ferket, 1984)

feed was not only compared in particulate form, but as a mash and when the particulates were ground to mash consistency.

Live performance in response to the particulate form of feed and the imposition of barley was greatest during the 4-week period preceding marketing (Table 5.12). During this time the feed necessary to realize gain maximized as did the extent of barley usage. Inclusion of barley led to an absence of fines and increased pellet length. Fines and short pellets were prominent with the feed dominated by maize and having added fat even though 15% wheat was included. Gain from 16 to 20 weeks with birds given the barley based feed as pellets was equivalent to when the maize-soyabean meal feed was in the same form while feed conversions were similar to when the maize-soyabean meal pellets had been ground.

Although final body weight of turkeys that had received the pelleted barley feeds was not as great as from maize-soyabean meal pellets, breast meat content of the carcasses was the same whereas skin and its fat content were reduced.

Conclusion

Pelleting of feed given to meat birds usually improves their performance. Processing first involves a steam treatment to condition the feed, then the hot mass is extruded through a die. The resulting physical changes may improve nutrient availability, but this advantage largely relates to the young. Having feed as a pellet permits birds of all ages to decrease their energy expenditures for food seeking and eating.

Pellet binding mostly involves starch, certain of the proteins and some fibre types that are contributed by the grains. While starch is common to all of the grains, gluten-forming proteins and soluble hemicelluloses of variable viscosity differ among the grains. Wheat readily forms doughs and is a good pellet binder, but maize is the reverse. Binders must first be solubilized to be effective, and dietary fat is believed to interfere in this respect while also coating particulates to destabilize attachments.

Pellet stability assumes progressive importance with beak growth and as the feed necessary to realize weight gain increases. Secondary hazards from water contaminated with feed and wet litter arising from undue water intake may also be minimized. These advantages magnify prior to marketing when the pen environment deteriorates.

Benefits from pellet stability occur because additional productive energy is recovered from existing feed metabolizable energy. Full realization of this improvement depends on the presence of other nutrients to complement its use, particularly balanced protein if meat formation is to be the beneficiary rather than fat.

References

ADAMS, C.A., NOVELLIE, L. and LIEBENBERG, N.W. (1976). *Cereal Chemistry*, **53**, 1–12

ALBERSHEIM, P. (1978). In *Biochemistry of Carbohydrates II*, pp. 217–237. Ed. Manners, D.J., University Park Press, Baltimore, MD

AUCKLAND, J.N. and FULTON, R.B. (1972). *Poultry Science*, **51**, 1968–1975

AUMAITRE, A., MELCION, J.P., VAISSADE, P. and SEVE, B. (1978). *Annales de Zootechnie*, **27**, 409–421

BARUCH, D.W., MEREDITH, P., JENKINS, L.D. and SIMMONS, L.D. (1979). *Cereal Chemistry*, **56**, 554–558

BAYLEY, H.S., SUMMERS, J.D. and SLINGER, S.J. (1968). *Poultry Science*, **47**, 931–939

BECHTEL, D.B., GAINES, R.L. and PROMERANZ, Y. (1982). *Cereal Chemistry*, **59**, 336–343

BEHNKE, K. and FAHRENHOLZ, C. (1986). In *Proceedings of the Maryland Nutrition Conference*, pp. 67–72, University of Maryland, College Park

BERNARDIN, J.E. and KASARDA, D.D. (1973). *Cereal Chemistry*, **50**, 529–536

BHATTACHARYA, M. and HANNA, M.A. (1987). *Journal of Food Science*, **52**, 764–766

CHOI, J.H., SO, B.S., RYU, K.S. and KANG, S.L. (1986). *Poultry Science*, **65**, 594–597

CROZET, N. and GUILBOT, A. (1979). *Annales de Technologie Agriculture*, **28**, 211–222

DAHLE, L.K. (1971). *Cereal Chemistry*, **48**, 706–714

D'APPOLONIA, B.L. (1973). In *Industrial Uses of Cereals*, pp. 138–160. Ed. Pomeranz, Y. American Association Cereal Chemists, Minneapolis, Minnesota

D'APPOLONIA, B.L. and GILLES, K.A. (1971). *Cereal Chemistry*, **48**, 427–436

DEATON, J.W., LOTT, B.D., BRANTON, S.L. and SIMMONS, J.D. (1987). *Poultry Science*, **66**, 1552–1554

DEXTER, J.E., MARTIN, D.G., PRESTON, K.R., TIPPLES, K.H. and MacGREGOR, A.W. (1985). *Cereal Chemistry,* **62**, 75–80

DREESE, P.C., FAUBION, J.M. and HOSENEY, R.C. (1988). *Cereal Foods World,* **33**, 225–228

ELEY, C.P. and HOFFMANN, E. (1949). *Poultry Science,* **28**, 215–222

ELIASSON,A.-C., CARLSON, T.L.-G., LARSSON, K. and MIEZIS, Y. (1981). *Starch,* **33**, 130–134

ELIASSON, A.-C., LARSSON, K. and MIEZIS, Y. (1981). *Starch,* **33**, 231–235

GANZ, A.J. (1973). *Cereal Science Today,* **18**, 398–416

GOERING, K.J., FRITTS, D.H. and ALLEN, K.G.D. (1974). *Cereal Chemistry,* **51**, 764–771

GOHL, B., NILSSON, M. and THOMKE, S. (1978). *Cereal Chemistry,* **55**, 341–347

GOMEZ, M.H. and AGUILERA, J.M. (1982). *Journal of Food Science,* **48**, 378–381

GRAM, N.H. (1982). *Carlsberg Research Communications,* **47**, 173–185

HENRY, R.J. (1985). *Journal of the Science of Food and Agriculture,* **36**, 1243–1253

HILL, R.D. and DRONZEK, B.L. (1973). *Starch,* **25**, 367–372

IBRAHIM, Y. and D'APPOLONIA, B.L. (1979). *Bakers Digest,* **53**, 17–19

JELACA, S.L. and HLYNKA, I. (1971). *Cereal Chemistry,* **48**, 211–222

JENSEN, L.S., MERRILL, L.H., REDDY, C.V. and McGINNIS, J. (1962). *Poultry Science,* **41**, 1414–1419

JENSEN, L.S., RANIT, S.O., WAGSTAFF, R.W. and McGINNIS, J. (1965). *Poultry Science,* **44**, 1435–1441

KAROSSON, R., OLERED, R. and ELIASSON, A.-C. (1983). *Starch,* **35**, 335–340

KASTANKOVA, J. (1983). *Rostlinna Vyroba,* **29**, 771–781

KIRLEIS, A.W., HOUSLEY, T.L., EMAN, A.M., PATTERSON, F.L. and OKOS, M.R. (1982). *Crop Science,* **22**, 871–876

KIVIMAE, A. (1978). *Archiv fur Geflugelkunde,* **42**, 238–245

KULP, K., ROEWE-SMITH, P. and LORENZ, K. (1983). *Cereal Chemistry,* **60**, 355–359

LONSDALE, M.B., VONDELL, R.M. and RINGROSE, R.C. (1957). *Poultry Science,* **36**, 565–571

LORENZ, K. and KULP, K. (1981). *Starch,* **33**, 183–187

LORENZ, K. and VALVANO, R. (1981). *Journal of Food Science,* **46**, 1018–1020

LORENZ, K., ROEWE-SMITH, P., KULP, P. and BATES, L. (1983). *Cereal Chemistry,* **60**, 360–366

LUKOW, O.M. and BUSHUK (1984). *Cereal Chemistry,* **61**, 336–339

McNAUGHTON, J.L. and REECE, F.N. (1984). *Poultry Science,* **63**, 682–685

MERCIER, C. and GUILBOT, A. (1974). *Annales de Zootechnie,* **23**, 241–251

MIKOLA, L. and MIKOLA, J. (1980). *Planta,* **149**, 149–154

MILLER, B.S., DERBY, R.I. and TRIMBO, H.G. (1973). *Cereal Chemistry,* **50**, 271–280

MORAD, M.M. and RUBENTHALER, G.L. (1983). *Cereal Chemistry,* **60**, 413–417

MORAN, E.T. (1982a). *Poultry Science,* **61**, 1257–1267

MORAN, E.T. (1982b). *Comparative Nutrition of Fowl and Swine – The Gastrointestinal System,* 253 pp. University of Guelph, Canada

MORAN, E.T. (1985). *Journal of Nutrition,* **115**, 665–674

MORAN, E.T. (1987). *Poultry-Misset International,* **3**(3), 30–31

MORAN, E.T. and FERKET, P.R. (1984). In *Proceedings of the University of Guelph Nutrition Conference for Feed Manufacturers,* pp. 12–17. University of Guelph, Canada

MORRISON, H.L., WALDROUP, P.W., GREENE, D.E. and STEPHENSON, E.L. (1968). *Poultry Science,* **47**, 592–597

MOTAI, H. (1976). *Agricultural and Biological Chemistry,* **40**, 1–7

NATIONAL RESEARCH COUNCIL (1977). *Nutrient Requirements of Poultry,* 7th edition. National Academy of Science, Washington, DC

OLSEN, E.M. and SLINGER, S.J. (1968). *Canadian Journal of Animal Science,* **48**, 35–39

PALMER, G.H. (1987). *Journal of the Institute of Brewing,* **93**, 105–107

PATIL, S.K., TSEN, C.C. and LINEBACK, D.R. (1975). *Cereal Chemistry,* **52**, 44–56

PERNOLLET, J.-C. and CAMILLERI, C. (1983). *Physiologie Vegetale,* **21**, 1093–1103

PFOST, H.B. (1976). In *Feed Manufacturing Technology,* pp. 332–335. Ed. Pfost, H.G., American Feed Manufactuer's Association, Arlington, VA

PFOST, H.B. and YOUNG, L.R. (1973). *Feedstuffs,* **45**(49), 26

PODRAZKY, V. (1954). *Chemistry and Industry,* **2**, 712

POLEY, W.E. (1938). *Poultry Science,* **12**, 331–335

PREECE, I.A. and MacKENZIE, K.G. (1952). *Journal of the Institute of Brewing,* **58**, 457–464

PROUDFOOT, F.G. and DEWITT, W. F. (1977). *Poultry Science,* **55**, 629–631

PROUDFOOT, F.G. and HULAN, H.W. (1982). *Poultry Science,* **61**, 327–330

PROUDFOOT, F.G. and SEFTON, A.E. (1978). *Poultry Science,* **57**, 408–416

REDDY, C.V., JENSEN, L.S., MERRILL, L.N. and McGINNIS, J. (1962). *Journal of Nutrition,* **77**, 428–433

RICHARDSON, W. and DAY, E.J. (1976). *Feedstuffs,* **48**(23), 24

ROBINSON, R. (1976). In *Feed Manufacturing Technology,* pp. 103–110. Ed. Pfost, H.B., American Feed Manufacturers' Association, Arlington, VA

SALMON, R.E. (1985). *Animal Feed Science and Technology,* **12**, 223–232

SAUNDERS, R.M., WALKER, H.S. and KOHLER, G.O. (1969). *Poultry Science,* **48**, 1497–1503

SAVORY, C.J. (1974). *British Poultry Science,* **15**, 281–286

SCHIFFMAN, H.R. (1969). *Journal of Comparative and Physiological Psychology,* **67**, 462–464

SCHWARTZ, J.B. and ZELINSKIE, J.A. (1978). *Drug Development and Industrial Pharmacy,* **4**, 463–483

SECKINGER, H.L. and WOLF, M.J. (1970). *Cereal Chemistry,* **47**, 236–343

SELVIG, A., AARNES, H. and LIE, S. (1986). *Journal of the Institute of Brewing,* **92**, 185–187

SHEN, H., SUMMERS, J.D. and LEESON, S. (1985). *Animal Feed Science and Technology,* **13**, 57–67

SIBBALD, I.R. (1977). *Poultry Science,* **56**, 1686–1688

SKOCH, E.R., BEHNKE, K.C., DEYOE, C.W. and BINDER, S.F. (1981). *Animal Feed Science and Technology,* **6**, 83–90

SLUSANCHI, H., FILIPESCU, H., OPROIU, E. and CSERESNYES, Z. (1973). *Annales de Technologie Agriculture,* **22**, 417–426

SUMMERS, J.D., BENTLEY, H.U. and SLINGER, S.J. (1968). *Cereal Chemistry,* **45**, 612–615

TIPPLES, K.J. (1980). *Canadian Journal of Plant Science,* **60**, 357–369

VONDELL, R.M. and RINGROSE, R.C. (1957). *Poultry Science,* **36**, 1310–1312

WALDROUP, P.W., WHELCHEL, D.L. and JOHNSON, Z.B. (1982). *Animal Feed Science and Technology,* **7**, 419–421

WELLIN, F. (1976). In *Feed Manufacturing Technology,* pp. 329–335. Ed. Pfost, H.B., American Feed Manufacturers Association, Arlington, VA

WENK, C. and VAN ES, A.J.H. (1979). *Archiv fur Geflugelkunde,* **43**, 210–214

WESTERMARCH-ROSENDAHL, C. (1978). *Acta Agriculture Scandinavica,* **28**, 159–168
WESTERMARCH-ROSENDAHL, C. and YLIMAKI, A. (1978). *Acta Agriculture Scandinavica,* **28**, 151–158
WINOWISKI, T. (1985). *Feed Management,* **36**(7), 28–33
WOLF, J.C., THOMPSON, D.R., WARTHESEN, J.J. and REINECCIUS, G.A. (1981). *Journal of Food Science,* **46**, 1074–1078

6

NUTRITION OF RABBITS

G. SANTOMÁ*, J.C. DE BLAS, R. CARABAÑO and M.J. FRAGA
Departamento de Producción Animal, Universidad Politécnica de Madrid, Spain

Introduction

Rabbits are bred almost all over the world for many different purposes. They are used as laboratory animals in many research institutes and pharmaceutical industries, for example, for the production and development of new vaccines or polyclonal antibodies, for immunological and technological studies, for drug testing, etc. (e.g. in Japan this is practically the only use for this species). In Germany, the UK, the USA and Canada rabbits are mainly bred as pets or for exhibitions, whereas in other countries there is an increasing interest in rabbit wool. The angora rabbit is raised with this aim in China (80% of total Angora rabbit production) and Italy is the main market (50% of a 6000-tonne total market) (Zina, 1988). Furthermore, the Rex breed gives a high quality fur for hats, gloves, coats, etc.

However, the main use of the rabbit as an agricultural species is meat production. As a result of its peculiar digestive physiology, which permits it to use farm and industrial by-products, and its small size, short reproductive cycle and meat quality, the rabbit is very suitable for both small-scale farms and intensive production.

As rabbits were used as laboratory animals for a long time, some basic principles about their nutrition were well known at a very early stage (one of the first publications on the nutrient requirements of rabbits was produced by the NRC in 1966). However research on rabbit nutrition in intensive animal production is recent, and suffers from a number of shortcomings which need further study.

RABBIT MEAT PRODUCTION AND STRUCTURE

There is a lack of official statistical information concerning rabbit meat production in most countries. This is probably due to the fact that rabbit meat represents less than 5% of meat production, even in rabbit producing countries, and because of the production structure, which is still largely based on very small farms that are

*Present address: Cyanamid Ibérica SA, Aptdo 471, Madrid.

Table 6.1 STRUCTURE OF RABBIT MEAT PRODUCTION IN EUROPE

Country	Rabbit meat production (1987) (tonnes/year)	Production trend	Rabbit meat trade		Rabbit meat consumption		Production structure
			Imports	Exports	(kg/head/year)	Trend	
Italy	220 000	↑	25 000	–	3.8	↑	65% production from farms > 50 does
France	150 000	=	13 000	3000	3.1	=	→ familiar farms (3–50 does) ↑ industrial farms
Spain	123 000	=	–	–	3.3	=	→ familiar farms ↑ industrial farms. 55% production from farms > 20 does
West Germany	22 000		4800	–	0.3	←	
Belgium	16 500	↑ ↑	7700	2400	2.48	↑ ↑	40% production from farms < 10 does
United Kingdom	15 000	↑	7400	–	0.40	↑ ↑	↑ farms 50–200 does 35% production from those farms
Netherlands	9500		3000	3000	0.65		
Portugal	6000	=	–	–	0.61	=	
Greece	6000		–	–	0.63		
EEC	570 000	← =	45 500	–	1.8	↑	
Hungary	37 000	=	–	25 000	1.2		Large farms
Czechoslovakia	30 000		–	3000	1.76		Small farms and large for export
Poland	25 000	=	–	10 000	0.42		Small farms (3–50 does)

↑ increasing trend
↓ decreasing trend
= net balance
(From Magdelaine, 1985; Holdas, 1988; Doria, 1988; Maertens and Peeters, 1988; Colin, 1988)

Table 6.2 STRUCTURE OF RABBIT MEAT PRODUCTION IN THE EEC

	Traditional farms	Part-time specialized rabbitries	Intensive farms
No. of does	< 30	30–250	> 250
Feeding method	Farm products and by-products + feed	Feed	Feed
Production system	Extensive	Semi-intensive	Intensive
Labour	Part-time	Part-time	Full-time
Housing quality	Low	New or transformed housing	Good

Table 6.3 DISTRIBUTION OF RABBIT MEAT PRODUCTION (%) ACCORDING TO THE TYPE OF FARM

Type of farm	Italy	France	Spain	Average
Traditional	26	31	44	32
Part-time specialized	46	46	37	44
Intensive	28	23	19	24

(From Colin, 1988)

very difficult to control. Despite this, Ouhayoun (1985) estimated a world rabbit meat production of 1 million tonnes per year. Approximately 80% is produced in Europe (including the USSR), mainly in Italy, France and Spain which represent almost 90% of the EEC rabbit meat production which is estimated to be 550 000 tonnes (Table 6.1).

The main areas of rabbit meat production and consumption are the north of Italy (Venetto region), the west of France and the Mediterranean coast of Spain. In these three countries the average rabbit meat consumption is 3–4 kg/head/year, while in the rest of the EEC it is less than 0.5 kg.

There is a shortage of rabbits in the EEC and despite increased production in recent years, the Community imports around 46 000 tonnes per year. Countries like China and Hungary specialize in exporting rabbits to the EEC while others like Poland and Czechoslovakia consume rabbit but export part of their production to the EEC (Table 6.1).

In Italy, France and Spain rabbit meat production is based on three types of farms (Table 6.2) ranging from traditional to intensive. They have all undergone intensification in recent years, but rabbit meat production from traditional farms still represents one-third of total production (Table 6.3). This means that rabbit meat for home consumption is still an important market (75%) in Spain (Doria, 1988); more than 40% in France (Ouhayaoun, 1985). This is changing very fast and 50% of rabbit meat sold in France is now as portions.

PRODUCTION CHARACTERISTICS OF RABBITS

The cycle of production and performance criteria of semi-intensive and intensive rabbit systems are shown in Figure 6.1 and Table 6.4.

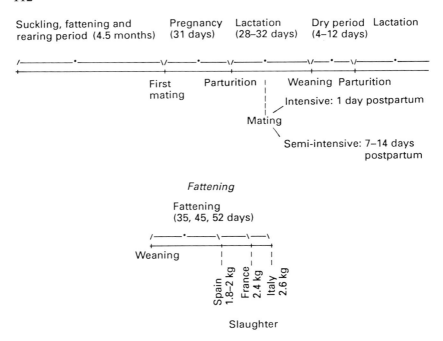

Figure 6.1 Simplified cycle of intensive rabbit meat production (figures are approximate)

Table 6.4 MINIMUM PERFORMANCE FOR OPTIMUM MEAT RABBIT PRODUCTION

Production parameters	
Breeding period	
Weaned rabbits/cage/year[a]	40
Mating acceptability	70%
Birth/matings	60%
Born alive/litter	7.5
Litters/cage/year	6.5
Interval between parturitions	60 days
Mortality birth-weaning (max.)	20%
Weaned rabbits/litter	6
Rabbit weight at weaning (30 days)[a]	500 g
Total feed intake/kg of weaned rabbit[a]	4.3 kg
Replacement rate per year	60–140%
Fattening period (slaughter at 2.5 kg)	
Liveweight gain[a]	33 g/day
Feed conversion ratio (max.)[a] (20% ADF diet)	3.5 g/g
Mortality (max.)[a]	10%

[a] Key parameters
(Adapted from Surdeau and Henaff, 1983)

A 4-kg liveweight doe can produce almost 70 weaned rabbits per year (550 g live weight). Compared with pigs on liveweight basis, a 150 kg liveweight sow would have to produce about 200 weaned piglets each weighing 7 kg. In practice does produce between 25 and 60 weaned rabbits per year, depending on production factors, which is an extraordinary reproductive yield. At 30 days weaning young rabbits reach a weight which is 13 times their birth weight. This reflects a high milking capacity of the doe (4–6 kg in 30 days of lactation). During the fattening period growth is also very high with rabbits increasing weaning weight 3.6 times in 30 days.

The most important difference among countries in relation to rabbit meat production is the final slaughter weight. It varies from 1.8–2 kg in Spain up to 2.6–2.8 kg in Italy. This situation makes trade difficult among the three most important countries in rabbit production.

Digestive physiology

Some herbivorous animals have developed several types of digestive reservoirs in order to increase the efficiency of utilization of their highly fibrous diets. Small herbivorous species like rabbits have followed a different route from ruminants because their high nutritive requirements per unit of body weight can only be satisfied through a high flow of nutrients from the digestive tract.

Rabbits have achieved this through a high voluntary feed intake (approximately four times higher than a 250 kg steer, and twice as much as a 40 kg growing pig on liveweight basis), associated with a low retention time of the digesta in the gut (Table 6.5). However, the characteristics of rabbit digestion permit a selective retention of both easily digestible feed particles and microorganisms together with the re-utilization of part of their gut contents by the reingestion of caecal material.

Table 6.5 MEAN RETENTION TIMES OF DIGESTA IN THE GUT OF DIFFERENT ANIMAL SPECIES

Species	*Mean retention time* (h)
Cattle	68.8
Sheep	47.4
Pigs	43.3
Horse	37.9
Rabbit	17.1

(adapted from Warner, 1981)

MAIN CHARACTERISTICS OF RABBIT DIGESTION

The digestive enzyme system of the rabbit is similar to other monogastric species, so that it can be assumed that the digestibility of the non cell-wall constituents at the small intestine is comparable. However, on reaching the ileocaecal-colonic junction the digesta is subject to a specific separation mechanism (Bjrnhag, 1987) which permits the rabbit to retain the small partices and soluble substances in the caecum, whereas the large ones progress towards the anus and are excreted as hard faeces (Figure 6.2).

Rabbits produce another type of faeces, the so-called soft faeces, which are excreted and systematically reingested from the anus, mainly in the early morning. They come from the caecum to which their chemical composition is similar (Figure 6.3) but different from the hard faeces (Table 6.6).

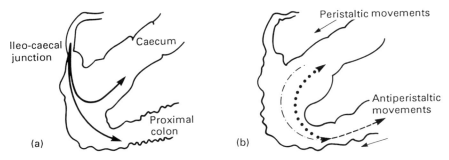

Figure 6.2 Mechanism of selective separation of ileal digesta in the hindgut. (a) Ileal digesta enter the caecum and proximal colon at the ileo-caecal junction. (b) Caecum contractions push its contents to the proximal colon. Antiperistaltic contractions move small particles and fluids backwards into the caecum, but permit large particles to progress towards the anus. —, ileal digesta; – · – ·, caecal content; · · · ·, small particles; – – –, large particles

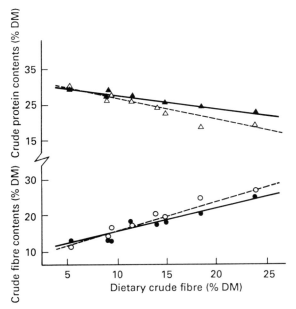

Figure 6.3 Effect of crude fibre (CF) level of the diet on CF levels of caecum contents (○) and soft faeces (●) and on crude protein levels of caecum contents (△) and soft faeces (▲) (Carabaño *et al.*, 1988)

Table 6.6 CHEMICAL COMPOSITION OF HARD AND SOFT FAECES (MEAN VALUES AND RANGE OF VARIATION)

	Hard faeces	Soft faeces	References
Dry matter (g/kg)	603 (464–671)	349 (276–427)	4, 6, 7, 8, 9, 10
Crude protein (g/kg DM)	126 (54–189)	289 (218–427)	2, 4, 5, 6, 7, 8, 9, 10
Crude fibre (g/kg DM)	322 (194–428)	184 (131–273)	2, 4, 5, 6, 7, 9, 10
Ash (g/kg DM)	90 (77–167)	125 (95–168)	2, 4, 5, 7, 8, 9, 10
Na^+ (mmol/kg DM)	40	120	3
K^+ (mmol/kg DM)	95	280	3
PO^{2-}_4 (mmol/kg DM)	10	110	3
Nicotinic acid (µg/g)	39.7	139.1	1
Riboflavin (µg/g)	9.4	30.2	1
Pantothenic acid (µg/g)	8.4	51.6	1
Cyanocobalamine (µg/g)	0.9	2.9	1

1. Kulwich *et al.* (1953), 2. Huang *et al.* (1954), 3. Bonnafous (1973), 4. Proto (1976). 5. Portsmouth (1977), 6. Hörnike and Bjrnhag (1980), 7. Fekete and Bokori (1985). 8. Fraga *et al.* (1984), 9. Carabaño *et al.* (1988), 10. Lorente *et al.* (1988)

ROLE OF FIBRE IN RABBIT DIGESTION

Fibre is the main dietary component responsible for proper digestion in rabbits, but this differs from ruminants because its role is related to both physical and chemical properties. Thus rabbit diets which are not satisfactory in terms of fibre favour digestive disorders.

Fibre level

The determination of the optimum dietary fibre level has been one of the main aims of the research on rabbit nutrition. Rabbits fed low fibre diets show a high incidence of digestive disorders usually manifest as diarrhoea and associated with high mortality. This effect can be explained by low fibre levels promoting a higher retention time of the digesta in the gut (Hoover and Heitmann, 1972; Lebas and Laplace, 1977; Fraga *et al.*, 1984; Gidenne, 1987). Furthermore, an increase in the caecal contents is observed with diets lower than 12% crude fibre (Figure 6.4), which can be related to a lower renewal of the caecal contents (Carabaño *et al.*, 1988). These situations could favour both, an undesirable fermentation pattern in the caecum, and a proliferation of pathogenic microorganisms. In conclusion, fibre acts as ballast in rabbit diets by maintaining an adequate transit time of the digesta in the gut.

Fibre type

Many different chemical compounds are included as fibre. Depending on the relative amount of each of them and their relationships in a complete diet, different ballast behaviour of the fibre can be expected. Gidenne, Poncet and Gómez (1986) showed that substituting lucerne hay with beet pulp in isofibrous diets led to a

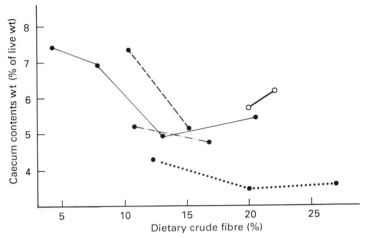

Figure 6.4 Effect of dietary crude fibre level on caecum contents weight. ●—●, De Blas *et al.*, 1986; ●–·–·●, Dehalle, 1981; ●···●, Lebas *et al.*, 1982; ○—○, Gidenne and Lebas, 1984; ●––––●, Blas, 1986

higher retention time of the digesta, and Bouyssou, Candau and Ruckebush (1988) reported a decreased intestinal motility with such a change. Pérez de Ayala and Fraga (unpublished) have also found higher retention times and higher caecal contents in isofibrous diets on substituting lucerne hay with both beet pulp and citrus pulp, whereas when grape cake was used the opposite effect was found (Table 6.7).

Table 6.7   EFFECT OF SOURCE OF DIETARY FIBRE ON RETENTION TIME AND WEIGHT OF CAECUM CONTENTS

	Diet			
	51% Citrus pulp	*29% Beet pulp*	*27% Lucerne hay*	*25% Grape-cake*
Dietary crude fibre (%)	8.3	9.4	8.2	8.5
Dietary ADF (%)	15.4	11.9	11.1	17.2
Lignin (%)	0.85	1.95	1.91	7.74
Retention time (h)	17.9	17.2	16.5	9.3
Weight of caecum contents (% LW)	9.1	7.4	5.6	4.4

(Pérez de Ayala, P. and Fraga, J. J., unpublished data)

Fibre particle size

The degree of grinding of fibre affects the intestinal motility (Pairet *et al.*, 1986; Bouyssou, Candau and Ruckebush, 1988). Finely grinding causes a higher retention time of the diet in the gut (Lebas and Laplace, 1977) and an increase of the caecal contents (Candau *et al.*, 1986). There is not complete agreement among authors about the minimum screen diameter to avoid these adverse conditions, since it depends on the fibre source and the factors which influence the particle size distribution. However a practical minimum of 2 mm seems reasonable (Lebas and Frank, 1986).

Fibre and caecal metabolism. Fibre as an energy source

The digestive function of fibre, means that a low digestibility of cell-wall constituents in rabbits can be expected compared with other animals (Table 6.8). As a result energy supply from crude fibre is poor in conventional diets (less than 5% of total dietary DE) where an average crude fibre digestibility of around 17% can be found (De Blas *et al.*, 1986; Maertens, Moermans and De Groote, 1987).

However, fine fibre particles and soluble material enter the caecum and are fermented by its microbial population mainly to volatile fatty acids (VFA). According to Carabaño *et al.* (1988), energy is the limiting factor for the optimal growth of caecal microorganisms. Both total concentration and relative proportions of VFA in the caecum are different from the rumen. In this way propionic acid represents a very low percentage (8% of total) with acetic acid being dominant (73%) and also with a high level of butyric acid (17%). Reports on the total VFA concentration in the caecum have been very variable, ranging from 34.5 μmol/g DM (Morisse, Boilletot and Maurice, 1985) to 351 μmol/g DM (Candau *et al.*, 1986). However it can be deduced that the above mentioned factors which favour an increase in retention time of the feed in the gut, also favour an increase of the VFA concentration in the caecum, especially of acetic acid with highly digestible fibre sources (beet pulp, citrus pulp) and butyric acid with low fibre diets (<14% CF on DM basis); consequently the caecal pH decreases.

Different authors (Prohászka, 1980; Borriello and Carman, 1983; Rolfe, 1984; Toofanian and Hammen, 1986) have shown a relationship between these metabolic changes in the caecum and the growth of different pathogenic microorganisms responsible for digestive disorders. However, more systematic studies are necessary to understand the influence of the diet on energy metabolism in the caecum. Chemical compounds in the digesta other than crude fibre can play an important role in VFA production since a first approach shows some discrepancies between the energy supply from crude fibre (around 5% of dietary digestible energy) and the energy supply from the VFA which is between 12 and 40% of energy requirements for maintenance in adult rabbits (Hoover and Heitmann, 1972 and Marty and Vernay, 1984 respectively).

Table 6.8   AVERAGE DIGESTIBILITIES (%) OF CELL-WALL CONSTITUENTS IN DIFFERENT ANIMAL SPECIES[a]

	Cellulose	*Hemicellulose*	*Lignin*
Sheep	30.0	40.3	5.1
Pigs	30.4	46.4	2.0
Chicken	9.6	4.2	–5.6
Rabbits	16.1	24.7	–7.4
Sheep[b]	28.7–55		
Rabbits[b]	25.0–55		
Pigs[b]	1.0–34.3		
Chicken[b]	1.0–41.9		

[a] Cell-wall contents of diets (compounded feed) as NDF (% DM)
[b] (Fonnesbeck *et al.*, 1974)

Fibre and soft faeces

A portion of the caecal contents is excreted daily as soft faeces. Although there are some differences between authors, which probably reflect the different methods used to determine the soft faeces production, it can be deduced that for conventional diets the production of soft faeces is not related to dietary fibre level (Dehalle, 1981; Carabaño *et al.*, 1988). Nevertheless Carabaño *et al.* (1988) and Gidenne (1987) obtained lower and higher values with dietary fibre levels below 10% CF and above 16% CF respectively. This suggests that only a relatively small proportion of the caecal material is removed each day with low fibre diets, and therefore the risk of digestive disorders is increased.

The contribution of soft faeces to total dry matter intake with practical diets is fairly constant (around 14%, ranging from 9 to 15%, Proto, Gargano and Gianani, 1968; Dehalle, 1981; Carabaño *et al.*, 1988; Lorente *et al.*, 1988). Only when high levels of forages or indigestible by-products are included in the diet, can soft faeces reach 20–23% of total dry matter consumption (Gidenne, 1987; Falcao e Cunha and Lebas, 1986).

An increase in dietary fibre level promotes an increase of fibre level of soft faeces (Figure 6.3) but not to the same extent, which demonstrates the efficiency of the separation mechanism by avoiding the entry of large amounts of fibre into the caecum.

PROTEIN DIGESTION

According to Lebas, Corring and Courtot (1971) and Henshell (1973) the proteolytic enzyme capacity of rabbits is practically complete at 4 weeks of age and its evolution depends mainly on the development of the endocrine system and much less by diet (Corring, Lebas and Courtot, 1972).

The protein digestibility (CPD) of adult rabbits is related to protein source (De Blas *et al.*, 1984b; Maertens and De Groote, 1984). In this way, protein from protein concentrates and cereals is well digested (higher than 70%) whereas protein more or less linked to fibre shows lower values (55–70%) but higher than in other monogastric species (e.g. CPD of lucerne hay and grass meal is between 30 and 50% in pigs and poultry; Just, Jorgensen and Fernández, 1985; Green, 1987).

There is a lack of data showing the degree of protein digestion before the caecum. Recently Gidenne (1988) has developed an ileal cannula for rabbits and his first results show that only 35% of total protein digestion of lucerne hay takes place in the small intestine, although higher values can be expected for protein concentrates. Therefore, caecum metabolism seems to be important for protein utilization in rabbits, at least for protein coming from forages.

Nitrogen metabolism in the caecum

Ammonia is the main end product of the nitrogen catabolism in the caecum as well as the main nitrogenous source for microbial protein synthesis. As in ruminants an important source of the caecal ammonia comes from the catabolism of the blood urea (approximately 25% of the total caecal ammonia, Forsythe and Parker, 1985) and the rest comes from the degradation of both dietary and endogenous nitrogenous compounds by the caecal microorganisms which show a high proteolytic activity (Makkar and Singh, 1987).

Ammonia concentration in the caecum is between 6 and 8.5 mg NH_3-N/100 ml caecum contents in practical diets (Carabaño *et al.,* 1988) which seems adequate for a proper protein microbial synthesis when compared with ruminants (Satter and Slyter, 1974) and supports the view that energy is more limiting for optimum microbial growth in the caecum as it is in pigs (Just, 1983). In situations where ammonia concentration in the caecum could be the limiting factor for caecal microbial growth (e.g. very low protein diets) urea supply has not proved to be satisfactory (Olceste and Pearson, 1948; King, 1971, and later confirmed by other authors) because urea is hydrolysed and absorbed as ammonia before reaching the caecum resulting in an increase of urinary nitrogen. Furthermore an increase in caecal ammonia favours a higher pH than optimum and therefore can promote digestive disorders (Lebas, 1984).

Soft faeces and protein digestibility

The main contribution of soft faeces consumption (caecotrophy), as far as nutrient supply is concerned is protein. Chemical composition of soft faeces is similar to that of caecal contents, although soft faeces has a slightly higher protein content and lower fibre (Huang, Ulrich and McCay, 1954; Carabaño *et al.,* 1988) (Figure 6.3). These differences are related to the high nitrogen content of the mucosal envelope which covers the caecal contents in the last sections of the large intestine to produce the final soft faeces, before excretion. Soft faeces also include less moisture than the caecum contents because of the water absorption at the final sections of the large intestine although absorption is lower than with hard faeces (Table 6.6).

Once they are consumed by the rabbit, soft faeces remain in the stomach for 6–8 h, due to the mucosal cover which protects them from the digestive attack. Meanwhile microorganisms continue their fermentative processes producing appreciable amounts of lactic acid because the phosphate buffer contained in the soft faeces permits a suitable pH. Eventually the mucosal cover is destroyed and soft faeces are subject to the normal digestive process (Griffiths and Davies, 1963).

Protein supply from soft faeces varies from 10% (Spreadbury, 1978) to 55% (Falcao e Cunha and Lebas, 1986) of total protein intake depending on the raw materials used. Low digestibility diets with a high percentage of nitrogen coming from forages or from low digestibility by-products increased that percentage. In practical diets the protein supply from soft faeces is around 18% of total protein intake.

One of the main advantages of caecotrophy is its positive effect on protein digestibility of the diet. According to Fraga and De Blas (1977) and Stephens (1977) CP digestibility increases between 5 and 20% as a result of caecotrophy, rising to the highest values when protein coming from protein concentrates decreases. This observation explains why rabbits have a better utilization of protein from forages than other non ruminant species.

Although there are few data available on amino acid composition and contribution of soft faeces, they appear to be a good source of lysine and methionine (Proto, 1976) which are generally the most limiting amino acids in rabbit diets. That is probably why Kennedy and Hershberger (1974) found that caecotrophy can overcome poor protein quality diets, though in conventional ones the amino acid supply from soft faeces is not enough to dispense with amino acid supply from the diet (see later).

Apart from protein (see Table 6.6) soft faeces represent an important source of B vitamins, vitamin K and a reutilization of some minerals like iron (Salze, 1983). Although the B vitamin supply can be enough for traditional rabbit production (Harris, Cheeke and Patton, 1983) it is necessary to supply extra B vitamins and minerals to rabbits intensively raised.

STARCH DIGESTION

As a result of the intensive production, rabbit feeds have a high nutrient density and therefore generally include higher level of cereals and starch sources than traditional feeds. This change induced Cheeke and Patton (1980) to propose the hypothesis that high starch levels in rabbit diets associated with a fast transit time of the digesta of these diets, could provide an important supply of starch to the caecal microorganisms, which on fermentation could promote digestive disorders.

Even with high inclusion levels of cereals (>60% of barley), starch is well digested by rabbits (>95%, Wolter, Nouwakpo and Durix, 1980; Eggum *et al.,* 1982; Blas, 1986) but unfortunately there are few available data showing starch digestibility at each part of the gut especially the small intestine and caecum.

Wolter, Nouwakpo and Durix (1980) found that about 70% of dietary starch reaches the small intestine without any degradation, despite the high activity of the salivary amylase in rabbits reported by Blas (1986). It seems that low stomach pH makes this enzyme unstable. At first starch digestion capacity in the small intestine is high because pH in this section is optimum and the caecal starch concentration reported by Fraga *et al.* (1984) is low (between 1 and 1.9% DM) even with high dietary starch contents (30%). However, caecal microorganisms have an important amylase activity (Blas, 1986) which is even higher than ruminants (Makkar and Singh, 1987) and makes the caecal starch concentration irrelevant as an indicator of starch flow to the caecum.

Some results support the Cheeke and Patton (1980) hypothesis. Wolter, Nouwakpo and Durix (1980) determined an 85% starch digestibility before the caecum with diets including 35% cereals but this result needs confirmation because of the procedure used. A 15% available starch diet in the caecum could be significant in developing undesirable fermentation. Weanling rabbits seem to be more sensitive to a starch overload in the hind-gut because their pancreatic enzyme system is still immature and shows a rapid development from 3–4 weeks of age (Corring, Lebas and Courtot, 1972). In this way Blas (1986) found that in 28-day-old rabbits a starch content in the terminal ileum of around 4% with diets including 30% starch, while in adult rabbits this value was lower than 0.5%. This observation is of practical importance since the highest incidence of digestive disorders appears during the first week post-weaning (about 28 and 40 days of age) and suggests that it is necessary either to put some constraints in feed formulation (minimum lactose and fibre, maximum starch) or to follow some specific feeding practices (feed restriction, straw supplementation).

Some authors (Morisse, 1982; Lee *et al.,* 1985) have suggested that starch digestibility depends on its source as well as on treatment. However Santomá *et al.*

(1985) did not find any differences in terms of mortality, growth, feed conversion rate, and dry matter, organic matter and protein digestibilities when using high levels (>33%) of different cereals (wheat, maize, oats, barley) and Seroux (1986) found no differences using barley, maize and wheat either ground or as flakes.

In conclusion further research is necessary to confirm the possible role of starch in promoting digestive disorders in rabbits and it appears that starch ileal digestibility will be a crucial determination.

FAT DIGESTION

As rabbit diets normally have a fibrous nature, fats show potential for increasing energy concentration. There is a lack of information on fat digestion in the rabbit but digestibility results indicate that it is probably similar to other monogastric species. In this way Maertens, Huyghebaert and De Groote (1986) and Santomá *et al.* (1987) found a positive relationship between degree of unsaturation of fats and their digestibility similar to pigs and poultry. These authors also found a negative relationship between dietary fat level and fat digestibility for saturated fats.

In addition Santomá *et al.* (1987) and Fraga *et al.* (1988) detected an extracaloric effect of fats similar to that in poultry (Mateos and Sell, 1981) and this has been explained by a 5.8 percentage unit increase in digestibility of the non-fat components of the diet when fat is added at levels above 3%. From comparisons with poultry this effect would be expected at lower fat levels but has not been studied in rabbits.

Parigi-Bini and Chiericato (1974) found that fat digestibility decreased about 6% when fat content of faeces was analysed with an initial acid hydrolysis and Maertens, Huyghebaert and De Groote (1986) found that this difference increased according to the saturation of fats and their inclusion level in the diet. These results suggest that soap formation in the gut is a factor to be considered when studying fat digestibility in rabbits as in other species.

Energy utilization by rabbits

As in all animals energy is the main nutritional component which determines rabbit performance. Thus, important factors which affect its utilization are discussed.

DIGESTIBLE ENERGY

Energy losses in faeces generally range from 25% to 45% of dietary gross energy. Because of the digestive physiology of the rabbit, cell-wall constituents are the main chemical components involved in the digestion of feed energy. Different authors have shown a negative relationship between energy digestibility (ED) and fibre level in the diet, with slightly better correlations using acid detergent fibre (ADF) than crude fibre (CF) (Table 6.9). The best estimation of feed digestible energy has been obtained through the more digestible nutrients (crude protein, ether extract and nitrogen-free extract) (Maertens, Moermans and De Groote, 1988).

Table 6.9 PREDICTION OF ENERGY VALUE IN COMPOUND FEEDS

---

(1) Digestibility of energy
$ED(\%) = 84.77 - 1.16\,ADF\,(\%\,DM)$ $R^2 = 0.82; P<0.001; n = 73$
   (De Blas *et al.*, 1984a)
$ED(\%) = 87.34 - 1.28\,ADF\,(\%\,DM)$ $R^2 = 0.66; P<0.001; n = 29$
   (Battaglini and Grandi, 1984)
$ED(\%) = 84.93 - 1.13\,ADF\,(\%\,DM)$ $R^2 = 0.80; P<0.001; n = 175$
   (Ortiz, 1986)
$DE(MJ/kg) = 13.5 - 18.12\,ADF(g/g)$ $R^2 = 0.76; RSD = 4.72$
   (Maertens *et al.*, 1988)
$DE(MJ/kg) = -7.54 + 29.71\,CP\,(g/g) + 50.25\,EE\,(g/g) + 23.34\,NFE\,(g/g)$ $R^2 = 0.90; RSD = 3.22$
   (Maertens *et al.*, 1988)

(2) Urine energy losses
$ME/DE = 0.94$
   (Spreadbury and Davidson, 1978; Ortiz, 1986)

(3) Efficiency of ME utilization
$-$Growth: $k_g = 0.958 - 0.0122\,ADF\,(\%\,DM)$ $n = 5$ $R^2 = 0.89$ $P<0.05$
   (Ortiz, 1986)
$-$Lactation: $k_l = 0.76 - 0.83$
   (Partridge *et al.*, 1983, 1986; Lorenie, 1987)

The results obtained with these equations underestimate rabbit feeds which include a high level of digestible fibre (e.g. beet pulp, citrus pulp) (De Blas *et al.*, 1984a; Maertens, Moermans and De Groote, 1988) and diets with added fat. As mentioned previously, added fat has a synergistic effect on diet digestibility because it increases the digestibility of the rest of the diet (Santomá *et al.*, 1987; Fraga *et al.*, 1988).

RELATIONSHIP BETWEEN ME AND DE

According to Spreadbury and Davidson (1978) urinary energy losses vary between 4 and 8% of DE in diets with a high variation in both fibre and protein level (4–33% ADF and 16–28% CP). These energy losses are not related to dietary fibre level in diets balanced for energy/protein (Ortiz, 1986) so that 6% of DE may be assumed as an average urinary energy loss in rabbits.

There is a lack of information on energy losses as methane in this species, but diets which include a large amount of digestible fibre, can have values high enough to be considered in the same way as pigs.

HEAT LOSSES

Growth

Different authors (Parigi-Bini and Rive, 1978; De Blas *et al.*, 1985; Parigi-Bini and Xiccato, 1986) have determined the energy efficiencies for protein and fat synthesis in rabbits, with values ranging from 38 to 45% and from 64 to 70% respectively. De Blas, Fraga and Rodríguez (1985) proposed an average energy efficiency for growth (k_g) of 56% using the comparative slaughter technique. As happens in other species k_g values are higher with calorimetric techniques. In this way Ortiz (1986) estimated through indirect calorimetric techniques that $k_g = 69\%$ with a maintenance heat production (523 MJ/kg 0.75) similar to the one determined by De Blas, Fraga and Rodríguez (1985).

Total heat production increases with dietary fibre at a rate of 0.37 to 1.5% per unit of ADF increase (De Blas, Fraga and Rodriguez, 1985; Ortiz, 1986). This relationship is probably due to an increase of both digestion costs and fermentative activity in the caecum, when using high fibre diets.

Thus k_g can be estimated through dietary fibre level and Ortiz (1986) obtained the following regression equation by using diets between 11 and 27% ADF.

$$k_g = 0.958 - 0.0122 \times \text{ADF} \,(\% \text{ DM}) \qquad (6.1)$$
$$n = 5; \text{R}^2 = 0.89; P<0.05$$

Heat production also depends on the crude protein content of the diet. De Blas, Fraga and Rodriguez (1985) found a positive relationship between these parameters (0.33% increase in heat production/1 unit of crude protein increase) using crude protein levels between 12 and 18%. This effect is probably due to a greater amount of catabolized protein when its level in the diet is increased.

There are no data relating dietary fat level to k_g but a positive relationship between them can be expected on the basis of information in other species.

Milk production

Using calorimetric methods Partridge, Fuller and Pullar (1983) and Partridge, Lobley and Fordyce (1986) determined an efficiency of ME utilization for milk production (k_l) of 76 and 87% with diets having 5% added maize oil and 5% added maize oil plus 2.5% soyabean oil respectively. The same authors estimated an efficiency of milk synthesis from tissue mobilization of 94%. A slightly lower k_l value of 70% has been reported by Lorente (1987).

COMPARISON WITH OTHER SPECIES

Although pigs use higher dietary energy concentrations than rabbits, the energy efficiencies discussed are very similar to the values reported recently by Henry, Vogt and Zoiopoulos (1988) for pigs.

If MAFF (1975) estimations are used for cattle and the above equations for rabbits, digestion of fibrous diets appears better for cattle. However, rabbits compensate their lower feed digestibility through lower heat and methane losses than cattle, so that the NE values for growth and lactation are similar for both species.

ENERGY EVALUATION UNIT

It can be deduced that DE as energy unit will overestimate high fibre diets and will underestimate high added fat diets. However, fat addition is limited by technological factors (rabbit feed must be pelleted) and the error associated with using DE instead of NE in practical diets with extreme fibre contents is low (around 5%), although it can be higher when using highly digestible fibre sources.

At present the state of knowledge DE seems to be a sufficiently precise unit, so that for simplicity it has been accepted world-wide for expressing the energy value of feeds, feedstuffs as well as the energy requirements of rabbits.

DIGESTIBLE ENERGY CONTENT OF FEEDSTUFFS

The information available on DE content of feedstuffs for rabbits is poor. Table 6.10 gives a summary of recent papers which have studied this subject in different ways. Among the different authors agreement is quite remarkable despite it being

Table 6.10 ENERGY, FIBRE AND CRUDE PROTEIN DIGESTIBILITIES OF SOME INDIVIDUAL FEEDSTUFFS

Feedstuffs	Energy digestibility			Fibre digestibility			Crude protein digestibility		
	(1)	(2)[a]	(3)	(1)[b]	(2)[b]	(3)[c]	(1)	(2)	(3)
Barley	–	0.82	0.77	–	0.67	0.54	–	0.75	0.59
Oats	0.65	0.71	0.65	–0.06	0.26	0.08	0.73	0.78	1.00
Sorghum	0.78	0.90	–	–0.03	0.76	–	0.55	0.70	–
Wheat bran	–	0.74	0.59	–	0.18	0.10	–	0.76	0.67
Maize gluten feed	0.68	–	0.65	0.44	–	0.28	0.80	–	0.61
Beet pulp	0.77	0.77	–	0.60	0.70	–	0.45	0.62	–
Beef tallow	0.66	–	0.70	–	–	–	–	–	–
Pork lard	0.79	–	0.75	–	–	–	–	–	–
Soyabean oil	0.93	–	0.90	–	–	–	–	–	–
Soyabean (full-fat)	0.90	–	0.82	0.71	–	0.16	0.88	–	0.85
Soyabean meal 44	0.76	0.81	0.83	–0.02	0.56	0.41	0.79	0.82	0.83
Sunflower meal 28–32	0.52	–	0.53	0.05	–	0.17	0.75	–	0.73
Lucerne meal	0.42	–	–	0.15	0.38	–	0.64	0.55	–
Grape cake	0.15	–	0.30	0.12	–	–	0.45	–	0.15

(1) Maertens and de Groote (1984)
(2) Fekete and Gippert (1986)
(3) Villamide and De Blas, unpublished data
[a] Organic matter digestibility
[b] Crude fibre digestibility
[c] Acid detergent fibre digestibility

difficult to include a high level of some raw materials in the feed without either disturbing digestive physiology (unbalanced ingredients), or affecting feed quality (e.g. fats). This is why it is necessary to use many replicates as well as to use different basal diets according to the raw material under study.

There is also a remarkable similarity between DE values of feedstuffs for rabbits and pigs for almost all ingredients, even the forages.

ENERGY REQUIREMENTS. FACTORIAL METHOD

Energy requirements according to the factorial method are summarized in Table 6.11.

Table 6.11 ENERGY REQUIREMENTS. FACTORIAL METHOD

(1) Maintenance
    growing animals
        for RE = 0: 552 kJ DE/kg$^{0.75}$ day
           (De Blas *et al.*, 1985; Ortiz, 1986; Partridge, 1986)
        for LWG = 0: 485 kJ DE/kg 0.75 day
           (Evans, 1982; De Blas *et al.*, 1985; Parigi-Bini and Ziccato, 1986)
    lactating does
        531 kJ DE/kg$^{0.75}$ day
           (Partridge *et al.*, 1986; Lorente, 1987)
    pregnant does
        356 kJ DE/kg$^{0.75}$ day
           (Partridge, 1986)
        452 kJ DE/kg$^{0.75}$ day
           (Lorente, 1987)
(2) Milk production
    13.4 MJ DE/kg milk
        (Coates *et al.*, 1964; Davies *et al.*, 1964; Partridge *et al.*, 1983–1986; Lorente, 1987)
(3) Pregnancy (last 10 days)
    135 kJ DE/kg$^{0.75}$ day
        (Lebas, 1979)
(4) Growth
    kJ DE/day
        (De Blas *et al.*, 1985)

Market weight (kg)	*Growth rate* (g/day)		
	30	*35*	*40*
2.00	348	421	495
2.25	380	454	528
2.50	414	489	561

Maintenance

As shown in Table 6.11 energy requirements for maintenance in growing rabbits differ by about 14% depending on the determination method. Higher values are obtained when estimating these requirements through retained energy (RE=0) from regression equations relating RE to DE intake, than when they are estimated from rabbits neither losing nor gaining weight.

As far as rabbit does are concerned, the maintenance requirements for pregnancy reported by Partridge, Lobley and Fordyce (1986) are surprisingly low and 21% lower than those obtained by Lorente (1987). In lactating does there is a closer agreement between these authors.

Milk production

Energy content of doe milk varies between 8.36 and 10.25 MJ/kg according to stage of lactation, rabbit breed, milk production, etc. If 9.6 MJ/kg is assumed as an average, with $K_1 = 76\%$, an energy requirement of 13.4 MJ DE/kg of milk is obtained.

Pregnancy

As in other species there is an increase of energy requirements around during the last third of pregnancy. Lebas (1979) estimated a 30% increase over maintenance requirements during the last 10 days of pregnancy.

Growth

Average energy requirements for the fattening period vary between 348 and 561 kJ DE/day (Table 6.11) depending on growth rate (30–40 g/day) and slaughter weight (2–2.5 kg) (De Blas, Fraga and Rodríguez, 1985).

Nutritive allowances for rabbits in practical feed formulation

When recommending nutrient allowances for practical feed formulation consideration should be given to the production system to be used. There is a large variety of production systems for rabbits so that traditional farms generally use just one low density feed for both breeding and growing rabbits, with low incidence of digestive disorders as the main objective. However some intensive farms use up to six different feeds (three for breeding and three for growing animals) in order to minimize the production cost. Therefore the allowances discussed later must be adjusted to the circumstances of each particular case.

ENERGY CONCENTRATION IN THE DIET

Rabbits can achieve high levels of performance on fibrous diets as a result of their peculiar digestive physiology. In this way they reach maximum growth rates with energy concentrations from about 10.5 MJ DE/kg DM (Figure 6.5).

On the other hand they are different from other species in that low dietary fibre levels do not favour better carcass yields. As shown in Figure 6.4 these diets promote an increase of caecum contents when levels are lower than 12% CF, but this effect is compensated by a reduction of other parts of the gut contents (De Blas *et al.*, 1986).

As far as lactating does are concerned, Méndez, de Blas and Fraga (1986), Lebas, Viard and Coudert (1988) and Fraga *et al.* (1988) have reported high performances using diets ranging from 9.5 to 11.4 MJ DE/kg DM without significant differences between productive parameters. However, Fraga *et al.* (1988) found that a 3.5% lard addition to a 18% ADF diet increased feed intake (12%), doe milk production (21%), litter weight at 21 days (18%) as well as a decrease in mortality of suckling rabbits during lactation (12%), especially in those litters with more than nine (20%). This effect has been confirmed by Maertens and De Groote (1988) and also reported in sows (e.g. Coofey, Seerley and Mabry, 1982).

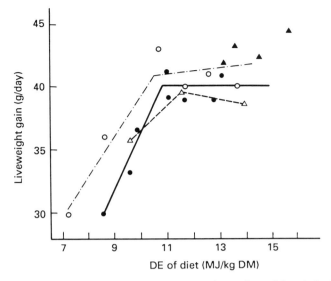

Figure 6.5 Effect of dietary DE concentration on liveweight gain in the fattening period. —, Partridge, 1986; – – –, Lebas *et al.*, 1982; – · – · – ·, De Blas *et al.*, 1986; ▲, Partridge *et al.*, 1986

In conclusion, a minimum of 10.5 MJ DE/kg DM is enough to allow high performances for growth and lactation. Responses to fat addition in lactating does may be of practical interest and need further research.

MINIMUM DIETARY FIBRE CONTENTS

Low dietary fibre levels favour the incidence of digestive disorders in rabbits. Therefore, it is necessary to include a minimum dietary fibre level to prevent these problems. But, as fibre is a predisposing agent more than a producing one, and it is also a chemical compound which can show very different characteristics depending on its source, the minimum requirements determined by different authors and institutions vary between 5 and 14% CF for growing rabbits (Davidson and Spreadbury, 1975; INRA, 1984).

Practical recommendations for growing rabbits vary between 10 and 14% crude fibre, or 14–18% ADF and for lactating does, these requirements are lower and vary between 10–12% crude fibre or 14 to 16% ADF (NRC, 1977; INRA, 1984 and De Blas *et al.*, 1986).

Even the highest minimum dietary fibre recommendations are not safe in practical conditions. Diets that contain this minimum can include high amounts of either digestible fibre or starch which can reach the caecum, particularly in weanling rabbits, and promote digestive disorders.

This situation occurs when using high levels of beet or citrus pulp, or with crude fibre concentrates (sunflower hulls, rice bran, olive pulp, grape-cake).

In order to take into account a part of this variation INRA (1984) proposed utilization of indigestible fibre as unit for expressing the minimum dietary fibre requirements. However, the information available on fibre digestibility of raw materials is very scarce and its determination frequently shows erratic results probably as a result of methodological problems (Table 6.10).

Before finding the best way to prevent digestive problems through dietary means it seems reasonable to establish maximum levels of some digestible nutrients (e.g. starch in weanling rabbits) and a minimum level of fibre. In this way a minimum CF or ADF level in rabbit diets coming from traditional sources (alfalfa hay, straw, wheat bran, sunflower, cereals, etc.) can be recommended. In this context fibre supplied should not be considered as fibre in raw materials rich in digestible fibre (beet and citrus pulps) and constraints should be imposed on less known by-products.

The minimum dietary fibre requirements established indirectly a maximum energy concentration in the diet as it can be inferred from the equations shown in Table 6.9. However this maximum could be increased by the addition of fat. In this way, at fat levels lower than 3%, Santomá *et al.* (1987) estimated approximately a 230 kJ DE/kg DM increase in the diet for each 1% of added fat, when considering the extra supply of fat energy as well as the digestibility increase of the rest of the diet. Thus, taking a minimum of 16% ADF/kg DM and a 2% of fat addition, the maximum dietary energy concentration in rabbits is around 13.0 MJ DE/kg DM.

PROTEIN REQUIREMENTS

Protein requirements are generally expressed as crude protein. However as mentioned earlier and shown in Table 6.10 there are significant differences in protein digestibilities between protein concentrates (approximately 80%), cereals and brans (65–70%) and forages (45–65%). In this way, CP digestibility differs among diets more according to their raw materials than to chemical composition (Fraga *et al.*, 1984; De Blas *et al.*, 1984b). Therefore DCP is more precise than CP for protein evaluation of both rabbit feeds and feedstuffs.

Since energy concentration of rabbit diets may vary between extreme values (10.5–13.0 MJ DE/kg DM) it is advisable to recommend protein requirements as a DE/DCP ratio.

The effect of energy/protein variation on the performance of growing rabbits is shown in Figure 6.6. The best results in terms of growth rate, feed conversion ratio, mortality, protein and fat retention and DE intake were obtained using about 98 kJ DE/g DCP. Thus if energy concentrations vary between 10.5 and 13.0 MJ DE/kg DM, DCP should vary between 10.7 and 13.3% DM. Nevertheless, in countries where rabbits are raised up to 2.5–2.8 kg live weight, and using two or three different growing feeds, a decrease in protein levels should be considered in the later stages.

Protein requirements for lactating does are higher than for growing rabbits. The results of Partridge and Allan (1982), Sánchez *et al.* (1985) and Méndez, de Blas and Fraga (1986) suggest that energy/protein ratios (as kJ DE/g DCP) higher than 80 lead to a lower milk production and growth of suckling rabbits and when this ratio is higher than 90–100 there is a significant decrease of rabbit birth weight, doe weight and doe conception rate. Considering the same practical variation of energy concentration of doe feeds as growing rabbits, the minimum DCP should range between 13.3 to 16.5% DM.

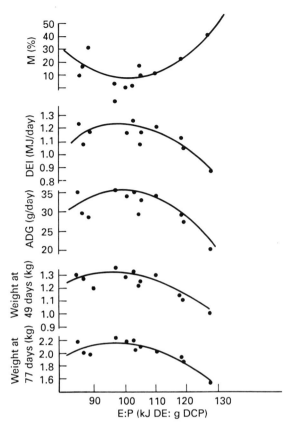

Figure 6.6 Relationship between energy to protein ratio (E:P) and digestible energy intake (DEI), average daily gain (ADG), weight at 49 and 77 days, and mortality (M) (De Blas *et al.*, 1981)

Amino acid requirements

For years it was considered that protein quality was of no concern in rabbit nutrition because of the caecotrophy practice. However, as mentioned earlier soft faeces only represent about 14% of the total dry matter intake and 17–18% of the total protein intake. Consequently despite soft faeces being a good source of the most frequently limiting amino acids (methionine and lysine) (Table 6.12) the quantities are not enough to overlook the amino acid pattern of the diet.

Different authors and institutions have proposed an amino acid patterns for rabbits (Table 6.13). As with other species there are more data for growth than for lactation, as well as considerable variation among the different recommendations. This is due to the different experimental conditions in terms of raw materials used, energy concentration, performance, housing, etc.

Considering the amino acid composition of the most commonly used raw materials in rabbit feeds, the first limiting essential amino acid is generally methionine + cystine followed by lysine. Colin (1978) showed that cystine can meet the total methionine + cystine requirements in a range from 35 to 65%. Some authors have pointed out high requirements for arginine (Table 6.13) but others such as Cheeke and Amberg (1972) have found that rabbits are able to synthesize some arginine, and an antagonism between lysine and arginine has been discarded. Recently Mougham, Schultze and Smith (1988) and Schultze, Smith and Mougham (1988) have started a research programme for the determination of the amino acid requirements of growing rabbits similar to the ideal protein model proposed by Cole (1979) in pigs. These authors recommend the amino acid composition of the whole body of the rabbit (Table 6.12) as a pattern of the balanced protein for this species in the growing period, and suggest that probably the present recommendations overestimate the requirements of several amino acids relative to lysine (methionine + cystine, phenylalanine + tyrosine, threonine, isoleucine, leucine and valine) or that the lysine recommendation is too low. Anyway the utilization of

Table 6.12 AMINO ACID COMPOSITION OF WHOLE BODY AND SOFT FAECES OF THE RABBIT

	Whole body (mg/g N)	Relative to lysine	Soft faeces (mg/g N)	Relative to lysine
Total nitrogen (g/100 g DM)	10.29		4.2	
Lysine	383	100	326	100
Methionine	77.5	20.2	117	36
Cystine	158	41.3	112	34
Arginine	415	108	229	70
Histidine	193	50.4	89	27.3
Threonine	245	64	390	120
Leucine	429	112	416	128
Isoleucine	194	50.7	275	84.4
Valine	239	62.4	363	111
Phenylalanine	249	65	253	77.6
Tyrosine	192	50.1	241	73.9

(Proto, 1976; Mougham *et al.*, 1988)

Table 6.13 AMINO ACID REQUIREMENTS (%) OF RABBITS

	Cheeke (1971) Growth	Adamson and Fisher (1971) Growth	Colin (1975) Growth	Davison and Spreadbury (1975) Growth	NRC (1977) Growth
Methionine + cystine	0.45	0.60	0.6	0.55	0.6
Lysine	0.93	0.70	0.6–0.65	0.9	0.65
Arginine	0.88	1.00	0.8	0.7	0.6
Threonine	–	0.5	0.55	0.6	0.6[a]
Tryptophan	–	0.15	0.15	0.2	0.2[a]
Valine	–	0.7	0.7	0.7	0.7[a]
Leucine	–	0.9	1.05	1.1	1.1[a]
Isoleucine	–	0.7	0.6	0.6	0.6[a]
Histidine	–	0.45	0.35	0.3	0.3[a]
Phenylalanine + tyrosine	–	0.6	1.2	1.1	1.1[a]
Glycine	–	–	–	0.5	–[b]
DE (MJ/kg)	–	–	–	–	10.45

Table 6.13 (continued)

	Spreadbury (1978) Growth	Colin (1988)		INRAd (1984)			AEC (1988)[d]	
		Growth	Lactation	Growth (4–12 weeks)	Lactation	Does and litters fed one diet	Growth (4–11 weeks)	Lactation
Methionine + cystine	0.62	0.62	0.8	0.6	0.6	0.6	0.6	0.65
Lysine	0.94	0.68	0.73	0.65	0.75	0.7	0.7	0.75
Arginine	0.56[c]	0.69	0.88	0.9	0.8	0.9	0.9	0.9
Threonine	–	–	–	0.55	0.7	0.6	0.6	0.65
Tryptophan	–	–	–	0.18	0.22	0.2	0.2	0.22
Valine	–	–	–	0.7	0.85	0.8	0.7	0.85
Leucine	–	–	–	1.05	1.25	1.2	1.1	1.3
Isoleucine	–	–	–	0.6	0.7	0.65	0.6	0.65
Histidine	–	–	–	0.35	0.43	0.4	0.3	0.4
Phenylalanine + tyrosine	–	–	–	1.2	1.4	1.25	1.1	1.3
Glysine	–	–	–	–	–	–	–	–
DE (MJ/kg)	13	10.25	10.67	10.45	10.9	10.45	10.9	10.45

[a] May not be minimum but known to be adequate
[b] Quantitative requirement not determined but dietary need demonstrated
[c] Spreadbury and Davison (1978)
[d] Practical allowances

the amino acid pattern of the rabbit body composition as a model for the amino acid feed composition does not take into account the amino acid requirements for maintenance and the amino acid supply of the caecotrophes, which may change this pattern. This is an attractive and tempting area which needs further research.

In the future it would be desirable, as in pigs, to express amino acid requirements in terms of ileal digestible amino acids relative to energy concentration, but so far no results are available in rabbits.

Summary on nutritive value of feedstuffs for rabbits

Suggested DCP and DE values for the most commonly used raw materials in rabbit feeds are shown in Table 6.14.

Recommendations on maximum inclusion levels of raw materials are also given. These values are only indications since they depend on many factors and have been based on the following criteria:

(1) Negative effect on pellet quality. Pellets cannot be too soft and prone to disintegration (fats, wheat bran, gluten feed) or too hard (molasses) because rabbits decrease feed intake and there is more refused feed.
(2) Variability of chemical composition between batches or suppliers (e.g. distillers dried grains and solubles, alfalfa hay, grape-cake). This criterion depends on the quality control of the feed mill.
(3) Fat oxidation risk (grape cake, oleins).
(4) Prevention of digestive disorders, even when reaching minimum dietary fibre requirements because of both high digestible fibre sources (beet and citrus pulps) and highly indigestible fibre sources (hulls from oats, rice or sunflower).

Table 6.14 NUTRITIVE VALUE OF SOME FEEDSTUFFS FOR RABBITS

Feedstuff	DM (g/kg)	ADF (g/kg)	DCP (g/kg)	DE (MJ/kg)	Maximum level of inclusion (%)
Maize	860	30	68	13.2	20–25
Wheat	860	33	85	12.9	20–25
Barley	890	55	75	12.8	25–30
Oats	900	140	75	11.2	30–35
Cassava	900	[a]	18	12.9	8–10
Citrus pulp	900	[a]	56	11.7	10–15
Beet pulp	900	[a]	50	10.9	15–20
Tallow	995	[a]	–	28.0	4–5[b]
Lard	995	–	–	30.5	4–5[b]
Oleins	995	–	–	28.4	1–2[b]
Molasses (beet)	760	–	54	10.9	3–7[b]
Soyabean (full-fat)	900	89	295	18.4	20–25
Soyabean meal 44	890	102	352	14.1	no limitation
Sunflower meal 38	900	192	304	13.2	no limitation
Sunflower meal 32	900	280	234	9.3	no limitation
Canola meal	880	185	253	11.9	10–12
Maize gluten meal 60	890	69	520	18.0	no limitation
Wheat shorts	870	88	119	10.5	25–30
Wheat bran	880	129	105	10.0	25–30
Maize gluten feed	890	107	140	11.3	15–20
DDGS[c]	890	148	180	13.8	5–10
Lucerne meal 18	900	310	109	7.9	25–30
Grape cake	880	500	18	5.0	10–15
Rice hulls	890	575	–	2.9	2–5
Wheat straw (NaOH treated)	880	415	–	5.4	10–15

[a] Not considered to meet fibre requirements
[b] Using proper technology
[c] Distillers' dried grains and solubles

Future role and importance of rabbit meat production

The main advantages and limitations of rabbit meat production are summarized in Table 6.15. According to Holdas (1988) production trends are on the increase, especially in some developing countries, where the rabbit has been chosen as an alternative for meat production in small-scale family farms. In this context there are some ambitious projects in Africa, Asia and South America.

As far as rabbit production in the EEC is concerned, despite rabbit meat being more expensive than others (pork and poultry) the main problems of its development are probably cultural ones. Even in the consumer countries, rabbit meat is not well known. In a recent public opinion poll undertaken in France, 43% of the people questioned had not tasted rabbit meat, and Doria (1988) points out that only 34% of Spanish families have consumed rabbit meat one or more times. In some circles rabbit meat is rejected for social reasons. However rabbit meat has some physical and chemical characteristics which look very favourable in relation to the present and future consumer demands (Table 6.16).

Table 6.15   FUTURE OF RABBIT PRODUCTION. ADVANTAGES AND LIMITATIONS

Advantages	Limitations

Meat quality:
Present and future consumer demands:
  low fat content
  low in cholesterolaemic effect
  low in sodium
  high in protein
	Economy:
	Developing countries:
	  financial support
	  high temperature
	  infrastructure
	High feed conversion ratios (low concentration diets)

EEC:
Constraints to pig, poultry and milk production
  rabbit is a good alternative
Relatively good research basis
  85% of rabbit research is undertaken in EEC
    but negligible compared to the other species
  Good infrastructure basis
Trend to change traditional meat consumption
  habits
Fish price increase
	Low meat production per labour unit
	Feed characteristics
	Marketing: badly developed marketing channels
	
	Culture, social:
	Not known meat.
	Rejected by some social circles

Developing countries:
Efficient use of by-products
Regular meat supply
Possible part-time activity
	Technical problems:
	Pathology:
	  respiratory problems
	  enteritis, heat susceptibility
	Genetics:
	  low development of pure breeds and hybrids
	Management:
	  low professional standards

Table 6.16   CHEMICAL CHARACTERISTICS AND LIPID COMPOSITION OF THE EDIBLE MEAT FROM DIFFERENT ANIMAL SPECIES

	Energy (MJ/100 g)	Protein (g/100 g)	Lipids (g/100 g)	Phospholipids (%)	% Total fatty acids			
					Unsaturated fatty acids (%)	C 18:2	C 18:3	C 20:4
Lean beef	0.82	20	12	2.17	47.8	2.7	0.3	–
Lean lamb	0.88	18	14.5	2.42	52.94	3.8	–	–
Lean pork	1.09	17	21	1.14	61.5	9	0.5	0.5
Chicken	0.84	19.5	12	1.79	68	19	–	–
Rabbit	0.67	21	8	9.67	63.8	20.5	6	5

(after Ouyahoun, 1985)

Compared with most consumed meats, rabbits have one of the lowest fat contents, with a high degree of unsaturation, as well as high hydrophylic properties. This is due to the high level of phospholipids (9.5% of total lipids versus 1–2% for broiler, pork or beef) which include a high proportion of arachidonic acid and a high level of linolenic acid which mainly comes from the galactolipids of lucerne (Ouhayoun, 1985). Furthermore these characteristics favour a low cholesterolaemic effect (Beynen, 1984). Rabbit meat has a high protein content with a similar amino acid pattern to other meats, and a low sodium level (Ouhayoun, 1985).

From a producer's point of view, rabbits offer an opportunity for development, at a time when difficulties arise because of EEC constraints on the production of milk and cereals and difficulties in some countries (e.g. Netherlands) if starting new poultry or pig farms. Nevertheless some technical problems, particularly pathology, genetics, housing conditions and quality of raw materials should be solved together with an improvement in the quality of labour to increase the profitability of these farms.

Acknowledgements

We are very grateful to Aurora Martín, D. Quintana, J.L. Barderas, P. Pérez de Ayala and Cyanamid Ibérica SA for their help in preparing the manuscript.

References

ADAMSON, I. and FISHER, H. (1971). *Reports International,* **4**, 59

AEC TABLES (1987). *Recommendations for Animal Nutrition*, 5th Edition. *Rabbits*, pp. 49–50. Commentry, France

BATTAGLINI, M. and GRANDI, A. (1984). *Proceedings of the III World Rabbit Science Association Congress*, pp. 252–264, Rome

BEYNEN, A.C. (1984). *Journal of Applied Rabbit Research,* **7**, 133–134

BJÖRNHAG, G. (1987). *Deutsche Tierärztliche Wochenschrift,* **94**, 33–36

BLAS, E. (1986). Doctoral Thesis. Faculty of Veterinary. Zaragoza

BONNAFOUS, R. (1973). Doctoral Thesis. Toulouse. France

BORRIELLO, S.P. and CARMAN, R.J. (1983). *Journal of Clinical Microbiology,* **17**, 414–418

BOUYSSOU, T., CANDAU, M. and RUCKEBUSH, Y. (1988). *Reproduction Nutrition, Developments,* **28**, 181–182

CANDAU, M., AUVERGNE, A., COMES, F. and BOUILLIER-OUDOT, M. (1986). *Annales de Zootechnie,* **35**, 373–386

CARABAÑO, R., FRAGA, M.J., SANTOMÁ, G and DE BLAS, J.C. (1988). *Journal of Animal Science,* **66**, 901–910

CHEEKE, P.R. (1971). *Nutrition Reports International,* **3**, 123

CHEEKE, P.R. and AMBERG, J.W. (1972). *Nutrition Reports International,* **5**, 259–266

CHEEKE, P.R. and PATTON, N.M. (1980). *Journal of Applied Rabbit Research,* **3**, 20–23

COATES, M.E., GREGORY, M.E. and THOMPSON, S.Y. (1964). *British Journal of Nutrition,* **18**, 583–586

COLE, D.J.A. (1979). In *Recent Advances in Animal Nutrition*, pp. 59–72. Ed. Haresign, W. and Lewis, D., Butterworths, London

COLIN, M. (1975). *Information Technique de les Services Vétérinaires,* **47**, 51–54

COLIN, M. (1978). *Annales de Zootechnie,* **24**, 465

COLIN, M. (1988). Personal communication

COOFEY, M.T., SEERLEY, R.W. and MABRY, J.W. (1982). *Journal of Animal Science,* **55**, 1388–1394

CORRING, T., LEBAS, F. and COURTOT, D. (1972). *Annales de Biologie Animale, Biochimie, Biophysique,* **12**(2), 221–231

DAVIDSON, J. and SPREADBURY, D. (1975). *Proceedings of the Nutrition Society*, **34**, 75

DAVIES, J.S., WIDDOWSON, E.M. and McCANCE, R.A. (1964). *British Journal of Nutrition*, **18**, 385–392

DE BLAS, J.C. and VILLAMIDE, M.J. (1988). *La alimentación del conejo*. 2nd edition. Mundiprensa, Madrid

DE BLAS, J.C., PÉREZ, E., FRAGA, M.J., RODRÍGUEZ, J.M. and GÁLVEZ, J.F. (1981). *Journal of Animal Science*, **52**, 1225–1232

DE BLAS, J.C., RODRÍGUEZ, J.M., SANTOMÁ, G. and FRAGA, M.J. (1984a). *Journal of Applied Rabbit Research*, **7**, 72–74

DE BLAS, J.C., FRAGA, M.J., RODRÍGUEZ, J.M. and MÉNDEZ, J. (1984b). *Journal of Applied Rabbit Research*, **7**, 97–100

DE BLAS, J.C., FRAGA, M.J. and RODRÍGUEZ, J.M. (1985). *Journal of Animal Science*, **60**, 1021–1028

DE BLAS, J.C., SANTOMÁ, G., CARABAÑO, R. and FRAGA, M.J. (1986). *Journal of Animal Science*, **63**, 1897–1904

DEHALLE, C. (1981). *Annales de Zootechnie*, **30**, 197–208

DORIA, R. (1988). *Boletín de Cunicultura*, **42**, 13–20

EGGUM, B.O., CHWALIBOG, A., JENSEN, N.E. and BOISEN, S. (1982). *Archiv für Tierernährung*, **32**, 539–549

EVANS, E. (1982). *Journal of Applied Rabbit Research*, **5**, 89–91

FALCAO E CUNHA, L. and LEBAS, F. (1986). *IV Journées de la Recherche Cunicole*. Communication no. 8. Paris

FEKETE, S. and BOKORI, J. (1985). *Journal of Applied Rabbit Research*, **8**, 68–71

FEKETE, S. and GIPPERT, T. (1986). *Journal of Applied Rabbit Research*, **9**, 103–108

FONNESBECK, P.V., HARRIS, L.E. and KEARL, L.C. (1974). *Journal of Animal Science*, **39**, 182

FORSYTHE, S.J. and PARKER, D.S. (1985). *British Journal of Nutrition*, **53**, 183–190

FRAGA, M.J. and DE BLAS, J.C. (1977). *Anales del Instituto Nacional de Investigaciones Agrarias*, **8**(5), 43–47

FRAGA, M.J., BARRENO, C., CARABAÑO, R., MÉNDEZ, J. and DE BLAS, J.C. (1984). *Anales del Instituto Nacional de Investigaciones Agrarias*, **21**, 91–110

FRAGA, M.J., LORENTE, M., CARABAÑO, R. and DE BLAS, J.C. (1989). *Animal Production* (in press)

GIDENNE, T. (1987). *Annales de Zootechnie*, **36**, 85–90

GIDENNE, T. (1988). *Proceedings of the IV World Rabbit Science Association Congress, Budapest*, **2**, 345–352

GIDENNE, T. and LEBAS, F. (1984). *Proceedings of the III World Rabbit Science Association Congress*. Roma. pp. 494–501

GIDENNE, T., PONCET, C. and GÓMEZ, L. (1986). *4emes Journées de la Recherche Cunicole*. Communication no. 4. Paris

GREEN, S. (1987). In *Digestibilities of Amino Acids in Foodstuffs for Poultry and Pigs*. AEC Rhone Poulenc Ntr. Labs, Commentry, France

GRIFFITHS, M. and DAVIES, D. (1963). *Journal of Nutrition*, **80**, 171–180

HARRIS, D.J., CHEEKE, P.R. and PATTON, N.M. (1983). *Journal of Applied Rabbit Research*, **6**, 15–17

HENRY, Y., VOGT, M. and ZOIOPOULOS, P.E. (1988). *Livestock Production Science*, **19**, 299–354

HENSCHELL, M.J. (1973). *British Journal of Nutrition*, **30**, 351–359

HOLDAS, S. (1988). *Proceedings of the IV Congress of the World Rabbit Science Association.* Opening Ceremony, Budapest, pp. 1–17

HOOVER, W.H. and HEITMANN, R.N. (1972). *Journal of Nutrition,* **102**, 375–380

HÖRNICKE, H. and BJÖRNHAG, G. (1980). In *Digestive Physiology and Metabolism in Ruminants,* pp. 707–730. MTP Press Limited, Lancaster

HUANG, T.C., ULRICH, H.E. and McCAY, C.M. (1954). *Journal of Nutrition,* **54**, 621–630

INSTITUT NATIONAL DE LA RECHERCHE AGRONOMIQUE (1984). *L'alimentation des animaux monogastriques: Porc, lapin, volailles,* pp. 77–84. Institut National de la Recherche Agronomique, Versailles

JUST, A. (1983). *IV International Symposium of Protein Metabolism and Nutrition,* pp. 289–309. Clermont-Ferrand

JUST, A., JORGENSEN, H. and FERNÁNDEZ, J.A. (1985). *Livestock Production Science,* **12**, 145–159

KENNEDY, L.G. and HERSHBERGER, T.V. (1974). *Journal of Animal Science,* **39**, 506–511

KING, J.O.L. (1971). *British Veterinary Journal,* **127**, 523–528

KULWICH, R., STRUGLIA, L. and PEARSON, P.B. (1953). *Journal of Nutrition,* **49**, 639–645

LAPLACE, J.P. and LEBAS, F. (1977). *Annales de Zootechnie,* **26**, 413–420

LEBAS, F. (1979). *Cuniculture,* **6**, 67–68

LEBAS, F. (1984). *IX Symposium de Cunicultura.* Opening session, Figueres

LEBAS, F. and FRANCK, T. (1986). *Reproduction, Nutrition, Development,* **26**, 335–336

LEBAS, F. and LAPLACE, J.P. (1977). *Annales de Zootechnie,* **26**, 83–91

LEBAS, F., CORRING, T. and COURTOT, D. (1971). *Annales de Biologie Animale, Biochimie, Biophysique,* **11**, 399–413

LEBAS, F., LAPLACE, J.P. and DROUMENQ, P. (1982). *Annales de Zootechnie,* **31**, 233–256

LEBAS, F., VIARD, F. and COUDERT, P. (1988). *Proceedings of the IV World Rabbit Science Association Congress, Budapest,* **3**, 53–58

LEE, P.C., BROOKS, S.P., KIM, O., HEITLINGER, L.A. and LEBENTHAL, E. (1985). *Journal of Nutrition,* **115**, 93–103

LORENTE, M. (1987). Doctoral Thesis. University of Madrid

LORENTE, M., FRAGA, M.J., CARABAÑO, R. and DE BLAS, J.C. (1988). *Journal of Applied Rabbit Research,* **11**, 11–15

MAERTENS, L. and DE GROOTE, G. (1984). *Proceedings of the III World Rabbit Science Association Congress,* pp. 244–251. Rome

MAERTENS, L. and DE GROOTE, G. (1988). *Proceedings of the IV World Rabbit Science Association Congress, Budapest,* **3**, 42–52

MAERTENS, L. and PEETERS, J.E. (1988). *Proceedings of the IV World Rabbit Science Association Congress, Budapest,* **1**, 5–8

MAERTENS, L., HUYGHEBAERT, G. and DE GROOTE, G. (1986). *Cuni Sciences,* **3**, 7–14

MAERTENS, L., MOERMANS, R. and DE GROOTE, G. (1987). *Revue de l'Agriculture,* **40**(5), 1205–1216

MAERTENS, L., MOERMANS, R. and DE GROOTE, G. (1988). *Journal of Applied Rabbit Research,* **11**, 60–67

MAGDELAINE, P. (1985). *Proceedings de la Association pour la promotion Industrie-Agriculture,* pp. 43–60. Toulouse

MAKKAR, H.P.S. and SINGH, B. (1987). *Journal of Applied Rabbit Research,* **10**, 172–174

MARTY, J. and VERNAY, M. (1984). *British Journal of Nutrition,* **51**, 265–277

MATEOS, G.G. and SELL, J.L. (1981). *Poultry Science,* **60**, 1925–1930

MÉNDEZ, J., DE BLAS, J.C. and FRAGA, M.J. (1986). *Journal of Animal Science,* **62**, 1624–1634

MINISTRY OF AGRICULTURE, FISHERIES AND FOOD (1975). *Energy allowances and feeding systems for ruminants.* Technical Bulletin no. 33. HMSO. Ministry of Agriculture, Fisheries and Food, London

MORISSE, J.P. (1982). *L'eleveur de lapins,* **20**, 43–46

MORISSE, J.P., BOILLETOT, E. and MAURICE, R. (1985). *Recueil de Medicine Vetérinaire,* **161**, 443–449

MOUGHAM, P.J., SCHULTZE, W.H. and SMITH, W.C. (1988). *Animal Production,* **47**, 297–301

NRC (1966). *Nutrient Requirements of Domestic Animals. No. 9. Nutrient Requirements of Rabbits,* 1st edn. National Academy of Science, Washington DC

NRC (1977). *Nutrient Requirements of Domestic Animals. No. 9. Nutrient Requirements of Rabbits,* 2nd edn. National Academy of Science, Washington DC

OLCESTE, O. and PEARSON, P.B. (1948). *Proceedings of the Society for Experimental Biology and Medicine,* **69**, 377–379

ORTIZ, V. (1986). Doctoral Thesis. Politechnic University of Madrid

OUHAYOUN, J. (1985). *Proceedings de la Association pour la promotion Industrie-Agriculture, Toulouse,* pp. 117–142

PAIRET, M., BOUYSSOU, TH., AUVERGNE, A., CANDAU, M and RUCKEBUSCH, Y. (1986). *Reproduction, Nutrition, Development,* **26**, 85–95

PARIGI-BINI, R. and CHIERICATO, G.M. (1974). *Rivista de Zootecnia e Veterinaria,* **3**, 202–212

PARIGI-BINI, R. and RIVE, V.D. (1978). *Rivista de Zootecnia e Veterinaria,* **4**, 242

PARIGI-BINI, R. and XICCATO, G. (1986). *Conigliocoltura,* **23**, 54–56

PARTRIDGE, G.G. (1986). *Proceedings of the III World Congress on Animal Nutrition,* pp. 271–277. Madrid

PARTRIDGE, G.G. and ALLAN, S. J. (1982). *Animal Production,* **35**, 145–155

PARTRIDGE, G.G., FULLER, M.F. and PULLAR, J.D. (1983). *British Journal of Nutrition,* **49**, 507–515

PARTRIDGE, G.G., FINDLAY, M. and FORDYCE, R.A. (1986). *Animal Feed Science and Technology,* **16**, 109–117

PARTRIDGE, G.G., LOBLEY, G.E. and FORDYCE, R.A. (1986). *British Journal of Nutrition,* **56**, 199–207

PEREZ DE AYALA, P. and FRAGA, M.J. (unpublished)

PORTSMOUTH, J. (1977). *The Nutrition of Rabbits.* In *Nutrition Conference for Feed Manufacturers,* pp. 93–111. Butterworths, London

PROHÁSZKA, L. (1980). *Zentralblatt für Veterinarmedizin, Reihe B,* **27**, 631–639

PROTO, V. (1976). *Conigliocoltura,* **7**, 15–33

PROTO, V., GARGANO, D. and GIANANI, L. (1968). *Produzione Animale,* **7**, 157–171

ROLFE, R.D. (1984). *Infective Immunity,* **45**, 185–191

SALZE, A. (1983). *Cuni Sciences,* **1**, 28–45

SÁNCHEZ, W.K., CHEEKE, P.R. and PATTON, N.M. (1985). *Journal of Animal Science,* **60**, 1029–1039

SANTOMÁ, G., CARABAÑO, R., DE BLAS, J.C. and FRAGA, M.J. (1985). *Anales del Instituto Nacional de Investigaciones Agrarias*, **22**, 75–82

SANTOMÁ, G., DE BLAS, J.C., CARABAÑO, R. and FRAGA, M.J. (1987). *Animal Production*, **45**, 291–300

SATTER, L.D. and SLYTER, L.L. (1974). *British Journal of Nutrition*, **32**, 199–208

SCHULTZE, W.H., SMITH, W.C. and MOUGHAM, P.J. (1988). *Animal Production*, **47**, 303–310

SEROUX, M. (1986). *4emes Journes de la Recherche Cunicole*, Communication no. 10. Paris

SPREADBURY, D. (1978). *British Journal of Nutrition*, **39**, 601–613

SPREADBURY, D. and DAVIDSON, J. (1978). *Journal of the Science of Food and Agriculture*, **29**, 640–648

STEPHENS, A.G. (1977). *Proceedings of the Nutrition Society*, **36**, 4A

SURDEAU, PH. and HENAFF, R. (1983). *La production de Lapin*, pp. 201–202. J.B. Bailliere, Paris. (Spanish edition. Mundi-Prensa), Madrid

TOOFANIAN, F. and HAMMEN, D.W. (1986). *American Journal of Veterinary Research*, **47**, 2423–2425

VILLAMIDE, M.J. and DE BLAS, J.C. (unpublished data)

WARNER, A.C.I. (1981). *Nutrition Abstracts and Reviews. Series B*, **51**, 789–820

WOLTER, R., NOUWAKPO, F. and DURIX, A. (1980). *Reproduction, Nutrition, Development*, **20**, 1723–1730

ZINA, E. (1988). *Conigliocoltura*, 21–23

III

Ruminant Nutrition

7

PREDICTING THE NUTRITIVE VALUE OF SILAGE

G. D. BARBER and N. W. OFFER
West of Scotland College, Auchincruive, Ayr, UK

and

D. I. GIVENS
ADAS Feed Evaluation Unit, Alcester Road, Stratford-upon-Avon, UK

Introduction

The economic value of a silage depends upon the level of animal production it will support. This depends on both nutritive value (the quantities of nutrients the animal obtains from a unit weight of silage) and fermentation quality (the type and extent of fermentation and its consequences for both preservation and voluntary feed intake).

This chapter is concerned with nutritive value, aspects of fermentation quality having been reviewed recently (AFRC, 1987a and b).

The nutritive value of silage is defined by its contribution to the energy and protein needs of the animal and the emphasis of the research effort to date has been directed at characterizing silage as an energy feed. In consequence, the main thrust of this chapter is concerned with the prediction of the energy value of silage, some consideration also being given to the implications of recent developments in the concept of nutrient response based systems of rationing. Aspects of silage protein evaluation have been reviewed elsewhere (AFRC, 1987b).

Is silage evaluation necessary?

Critics of the 'scientific approach' to livestock rationing often advance the view that animals provide the only meaningful assessment of a silage and that investment of time and money in silage evaluation and ration formulation is unjustified. This argument cannot be sustained. Feed costs make up a high proportion of the variable costs of intensive ruminant production and there is ample evidence that production levels, the need for purchased concentrates, and the economic performance of a farm are all very sensitive to silage quality. Silage is the key (and most seasonally variable) resource associated with winter animal production. Every farm business which relies on grass silage can benefit substantially from advance knowledge of its nutritive value. Silage evaluation, undertaken before winter feeding begins, is the first essential step in obtaining answers to questions such as:

- Is there sufficient silage in store?
- What quantities of concentrates are required, and when?
- What quality of concentrates is needed?
- How can 'straights' and alternative feeds be used to advantage?
- What daily rations should be set for different stock?
- Is the silage quality appropriate for the desired type and level of production or should grassland management and ensilage technique be reviewed?
- What are the relative feed costs of alternative strategies for achieving target output?
- What is the likely profitability of the enterprise?

Figure 7.1 shows the effects which silages of different metabolizable energy (ME) value have on calculated winter feed requirements for a 100-cow autumn-calving dairy herd yielding an average 5500 litres per lactation. The calculation used ARC (1980) nutrient requirements and the intake prediction equation of Lewis (1981) in Scottish Agricultural Colleges (SAC) rationing software. Over the range of ME values shown (9.5–12.5 MJ/kg DM), an increase of one unit of silage ME value reduces herd concentrate requirements over winter, by approximately 30 tonnes at the expense of about 60 extra tonnes of silage. At current prices, this represents a

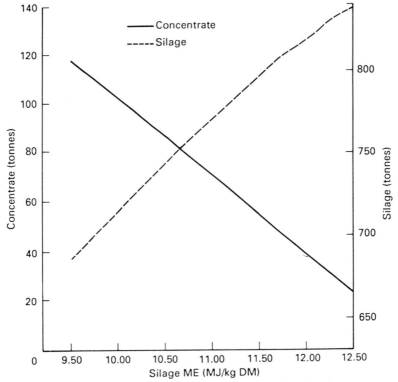

Figure 7.1 Effects of silage ME content on calculated winter feed requirements for a 100-cow autumn dairy herd

saving of about £4500 per MJ increase in silage ME value. Thus differences of silage ME as small as 0.25 MJ/kg DM may have significant economic consequences for the farm business.

Animals are the ultimate judges of silage quality and 'fine-tuning' of rations in response to observed performance is both inevitable and desirable. But failure to undertake detailed feed planning based on accurate silage evaluation may be a very expensive omission, leading to wasted outlay on feeds or to lost production.

SAMPLING OF SILAGE CLAMPS

Sampling of silage clamps appears to present a formidable problem. How can the final sample of a few grammes of material be representative of perhaps 1000 tonnes of silage, often consisting of several cuts ensiled at different times? Alexander (1960) found coefficients of variation for horizontal and vertical variation of silage DM content in a large farm clamp to be 4.6% and 17.1% respectively. The clamp had been filled with grass, cut at different times, and wilted to different extents. Horizontal variation was assessed by analysis of a series of core samples taken along a diagonal and showed little variation. Most of the variation was due to cores taken at the front of the clamp where grass which had formed ramp during filling had been placed. However, vertical variation, measured at 20-cm deep intervals at the mid-point of a half-diagonal was much greater, reflecting the presence of layers of different grass within the clamp. Alexander (1960) recommended that silage clamps should be sampled by coring in at least two positions and preferably four. Using small scale silos, he showed that sampling in this way gave an unbiased sample even when variation within the clamp, as assessed by multiple grab samples, was considerable. A clear conclusion from this work is that grab samples obtained from the silage face are worthless, yet properly taken core samples can yield a truly representative sample of the silage in the clamp.

Intake prediction

Intake is the key biological constraint in the calculation of rations for livestock. Its prediction is the second essential step in answering the practical questions posed above. Table 7.1 shows the extreme sensitivity of the rationing outcome to assumed intake. Yet much less research effort has been applied to predicting voluntary feed intake than to establishing nutrient requirements and responses to nutrients.

Table 7.1   THE EFFECT OF ASSUMED DM INTAKE OF A DAIRY COW ON RATION FORMULATION ESTIMATED ON A FRESH WEIGHT (FW) BASIS

Total DM intake (kg/day)	Silage FW (kg/day)	Compound feed FW (kg/day)
16.0	17	13.5
16.5	27	11.3
17.0	37	9.0
17.5	46	6.8
18.0	56	4.6

Assumptions:
  600 kg Friesian cow yielding 30 kg/day milk
  silage DM 250 g/kg, ME 10.5 MJ/kg DM compound feed DM 870 g/kg, ME 13.2 MJ/kg DM

Various approaches to intake prediction have been reviewed by Forbes (1988). Most common is the use of empirically-derived multivariate regression equations containing a number of animal and feed factors such as live weight, stage of lactation, milk yield, concentrate intake and silage dry matter (DM), ammonia-nitrogen and digestibility. This approach is hampered by lack of adequate data sets containing sufficient detail of animal and diet and by interdependence of many of the variables. Neal, Thomas and Cobby (1984) compared a number of regression equations as predictors of intake using, as a reference, measured intakes of cows at the Grassland Research Institute. The equations of Vadiveloo and Holmes (1979) and Lewis (1981) proved most accurate but their respective mean square prediction errors of 2.2 and 2.8 kg DM per day suggest that the accuracy needed for reliable ration formulation has yet to be achieved. Forbes (1988) discussed the development of dynamic prediction models which, as they attempt to simulate the fundamental mechanisms controlling feed intake, should be less dependent on animal or diet type. However, he concluded that, at the current state of knowledge, such models are unlikely to be as accurate as those based on regression equations. Further work is urgently needed as intake is currently the greatest uncertainty in ration calculation.

History of silage energy evaluation

Assessment of the energy value of grass silage, as with other forages, has concentrated on measurement of dry matter or organic matter (OM) digestibility (DMD and OMD) *in vivo*, together with development of laboratory methods which allow their prediction. Use of laboratory measurements to predict silage digestibility has a long history and crude fibre (CF) has played a major role. Early studies in the Netherlands (Dijkstra, 1949) related CF content to starch equivalent value, although at the same time, Hallsworth (1949) reported that the relationship between CF and OMD in nine silages was not significant. Hallsworth (1949) did however show a very significant relationship between these two parameters in another population of 36 silages.

The use of CF for silage OMD prediction has survived to the present time (Korva and Tuori, 1986) and it is currently used for extension purposes in the Netherlands (CVVB, 1977). More recent attempts at developing prediction relationships for DMD and OMD from various cell wall fractions include the use of lignin, neutral detergent fibre, acid detergent fibre (Bosman, 1970; Joshi, 1972; Aerts *et al.*, 1977), cellulose, hemicellulose and the use of summative equations (Aerts *et al.*, 1977).

In attempts to improve prediction power beyond that of cell wall fractions, various methods of estimating silage digestibility *in vitro* have been developed. Barnet (1957) used a rumen liquor *in vitro* system to estimate cellulose digestibility and related these measurements to CF digestibility *in vivo*. Tilley and Terry (1963) developed a similar *in vitro* system for herbages using rumen liquor and pepsin, and this was modified and used for predicting silage digestibility *in vivo* by several workers including Alexander and McGowan (1969), O'Shea, Wilson and Sheenan (1972), Larson (1974) and Aerts *et al.* (1977). More recently, *in vitro* digestion systems using cell-free cellulose-type enzymes have been introduced (Jones and Hayward, 1975; Dowman and Collins, 1982) and their application to silage evaluation has been studied by Barber, Adamson and Altman (1984), Givens,

Everington and Adamson (1989) and Barber *et al.* (1989). The faecal liquor *in vitro* method of El Shaer *et al.* (1987) has not yet been validated against *in vivo* digestibility on a large population of silages. To develop prediction equations for use in extension work in the UK, measurements of digestible organic matter in the dry matter (DOMD) were made on some 80 silages on MAFF Experimental Husbandry Farms (EHFs) in England and Wales during the 1960s. The data were related initially to CF content (Alderman, Collins and Dougall, 1971) and later to modified acid detergent fibre (MADF) using the method of Clancy and Wilson (1966). Following the establishment of feedstuff evaluation units (FEUs), at the Rowett Research Institute (RRI) in 1973 and by ADAS near Stratford-upon-Avon, in 1975, a programme of *in vivo* evaluation of silages was begun with the intention of providing a bank of data for this purpose. Additional *in vivo* data have been accumulated in the UK at the North of Scotland College of Agriculture (NOSCA) and by the Department of Agriculture for Northern Ireland (DANI). The current UK collection of *in vivo* data is summarized in Table 7.2. These data have been collected using wether sheep fed at maintenance. There are other data produced in Northern Ireland using dairy cows.

Givens (1986) described a study with 70 silages from the ADAS FEU in which relationships based on MADF were developed to predict DOMD and ME. These equations were based on *in vivo* measurements made using fresh silage and so included OM and ME in volatile compounds lost during oven drying. However, use of the equations for extension purposes required conversion of MADF measured on oven dried silage to a 'true' dry matter base. In the absence of a routine laboratory method for measurement of true dry matter, this was achieved by assuming that the true dry matter content was higher than the measured oven dry matter by a constant 19 g/kg, and thus calculating a 'corrected' dry matter.

Table 7.2   UK *IN VIVO* DATA ON GRASS SILAGES

Silage type	Year of study	n	Digestibility of:			Energy values		References
			DM	OM	Proximates	DE	ME	
Clamp	1975	16	Y	Y	N	Y	Y	(b)
Clamp	1978	16	Y	Y	N	Y	Y	(c)
Clamp	1984	16	Y	Y	N	Y	Y	(d)
Clamp	1984/86	27	Y	Y	N	N	N	(e)
Clamp	1978/86	122	Y	Y	Y	Y	Y(a)	(f)
Clamp		42	Y	Y		Y	Y	(g)
Big bale	1983/87	28	Y	Y	Y	Y	Y(a)	(h)

[a] ME calculated using predicted methane energy losses
Y Yes
N No
[b] RRI (1975)
[c] Wainman *et al.* (1978)
[d] Wainman *et al.* (1984)
[e] NOSCA (unpublished)
[f] Givens *et al.* (1989)
[g] Unsworth (unpublished)
[h] Everington and Givens (1989)

The equations were adopted by the Government Advisory Services in England, Wales and Northern Ireland and used during the period 1986/1988. In Scotland, SAC continued to use predictions based on oven dry matter values for advisory purposes because of the lack of a method for routine measurement or prediction of the amount of volatiles in individual silages and the observed absence of bias in the relationship between predicted and measured dairy cow performance (Offer, Castle and Barber, unpublished, Figure 7.3).

Energy prediction and energy prediction relationships

While work on silage evaluation has been concentrated on predicting digestibility, it has been recognized for some time that silage, unlike other forages, can exhibit a rather poor relationship between digestibility and digestible energy (DE) or ME content. This largely stems from the variability in the gross energy (GE) content of silages (Givens and Brunnen, 1987). The need to include energy related terms, such as ether extract, crude protein and GE, as bivariates with estimates of digestibility, to allow a more accurate prediction of DE, has been described by Alderman *et al.* (1971), ARC (1976) and Terry and Osbourn (1980).

It is important to recognize that prediction of ME is only the first part of the rationing process. It should not be considered in isolation. Animal performance depends on the net energy (NE) value of the silage and therefore the ultimate criterion for choosing an evaluation method is the accuracy with which NE values can be obtained.

OPTIONAL ROUTES FOR NE PREDICTION

Figure 7.2 represents the route options available for prediction of silage NE. The green route, represented by dotted lines, has been used since 1975 by UK advisory services. This route was dictated by the absence of measured ME values on which

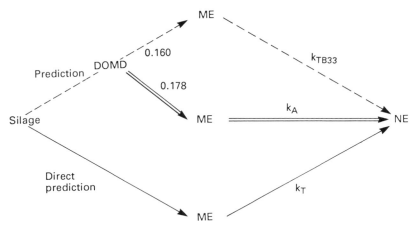

Figure 7.2 Optional routes for NE prediction. k_{TB33}, efficiencies of utilization of ME calculated according to MAFF *et al.* (1975). k_A, mean published values for efficiencies of utilization of ME. k_T, predicted *in vivo* values for efficiencies of utilization of ME

to base predictive relationships. The red route, represented by a solid line in Figure 7.2 is the only route which is scientifically sound, since it is based entirely upon measured parameters. Both the green route and the intermediate route (double lines) assign constant value to parameters which have been shown to vary. For example: the routes through DOMD assume a constant GE of DOM. As greater numbers of silages with measured ME values become available from FEUs, so consideration must be given to adoption of the red route for advisory purposes.

The problems to be resolved before the red route can be adopted include: measurement of GE consumed by the animal; insufficient data *in vivo* on methane energy losses and the difficulty of predicting efficiencies of utilization of silage ME.

Measurement of GE consumed by the animal

Two difficulties arise in this: firstly, satisfactory subsampling of heterogeneous materials which contain energy-dense volatile substances and secondly, quantification of the amount of the volatiles actually consumed by the animal.

Givens and Brunnen (1987) showed that when ME of silage is predicted directly, R^2 is lower and standard error of calibration (SEC) is higher than when OMD is predicted. Inclusion of GE as a bivariate with another predictive parameter in the prediction of ME improves both R^2 and SEC. However inclusion of GE in ME prediction relationships involves an element of autocorrelation as well as the possibility of transfer of error. The heterogeneous nature of the material provides problems in the accurate measurement of silage GE, as does possible loss of volatile, energy-dense substances prior to measurement. Such errors will appear both as random error and as bias in predictive relationships.

Norcross, Weir and Offer (unpublished) have shown that the GE of silage samples from the exposed face of self-feed clamps is lower than that of the same material sampled by coring from the same volume of the clamp before the face was exposed. Thus energy-dense volatile substances may be lost before consumption by the animal. If this occurred during the measurement *in vivo* of ME, then ME would be overestimated by the magnitude of the energy lost by volatilization.

These problems cast some doubt on the accuracy of currently measured ME values.

Shortage of measured methane energy losses in vivo

Losses of energy as methane gas have been measured *in vivo* in only a limited number of cases in the UK (48 silages at RRI and 50 silages by DANI). These numbers are insufficient to form an adequate base for derivation and validation of prediction equations. The ADAS FEU have completed measurements *in vivo* on some 122 silages, but, because of lack of availability of respiration chambers, have been unable to measure methane losses until very recently.

Until the FEUs have produced sufficient silage evaluations with measured methane losses, it is necessary to assume or predict the losses of methane in most of the data used to develop predictive relationships for ME. Two options are available: either to predict methane loss using, for example, the equation developed by Blaxter and Clapperton (1965) or to assume a fixed value based upon existing data. Edwards (1986) compared measured methane energy losses with those predicted by the Blaxter and Clapperton equation. For 48 silages evaluated at the RRI (RRI, 1975; Wainman, Dewey and Boyne, 1978; Wainman, Dewey and Brewer, 1984). The mean bias (observed-calculated) was $-0.48\% \pm 0.71\%$ for an observed mean loss of gross energy of 7.7%. The regression of predicted energy loss as methane on measured loss gives an R^2 of 0.00 (Kridis, personal communication), and this, together with the bias towards overprediction, suggests no basis for the use of the Blaxter and Clapperton equation in this context. This is not surprising since the equation was not developed using silage-based diets. These comparisons do, however, suggest that the use of a mean value for methane losses from silage diets is unlikely to introduce either significant error or bias.

Calculation of ME from DOMD relies on mean values both for the ME/DE ratio and for the GE content of digestible organic matter (DOM). Such calculations ignore variations in both GE of silage dry matter and urine and methane energy losses. Table 7.3 presents mean values for the GE of DOM and for ME/DE ratio measured in nine silage populations together with the mean overall conversion factor from DOMD to ME.

Table 7.3 MEAN GE OF DOM, ME/DE RATIO AND OVERALL CONVERSION FACTOR FROM DOMD (g/kg) TO ME (MJ/kg DM) FOR NINE SILAGE POPULATIONS

n	Species	GE of DOM (MJ/kg)	ME/DE	Overall factor	References
16	Sheep	20.6	0.82	0.0169	RRI (1975)
12[a]	Sheep	20.6	NR		Terry and Osbourn (1980)
16	Sheep	20.0	0.84	0.0168	Wainman *et al.* (1978)
16	Sheep	19.8	0.82	0.0162	Wainman *et al.* (1984)
2[c]	Lactating cows	NR	0.85		Unsworth *et al.* (1984)
4[c]	Lactating cows	NR	0.86		Unsworth *et al.* (1987)
122	Sheep	19.9	0.81[b]	0.0161	Givens *et al.* (1989)
42[c]	Sheep	21.0	0.85	0.0178	Unsworth (pers. comm.)
3	Sheep	NR	0.84		Unsworth *et al.* (1989)

NR, not reported
[a] Included three lucerne silages
[b] Energy losses as methane predicted
[c] Harvested in Northern Ireland

The silages evaluated in Northern Ireland had mean GE contents higher than those measured elsewhere. Excluding those silages, the overall mean of 170 measurements of GE of DOM was 19.94 ± 1.78 MJ/kg. The overestimation of methane energy loss, which occurs when the equation of Blaxter and Clapperton (1965) is used, largely accounts for the ME/DE ratio of 0.81 reported by Givens *et al.* (1989) which is lower than all studies in which methane losses have been measured (mean value of all data, 0.84). Combining an ME/DE ratio of 0.84 and a GE of DOM of 19.94 MJ/kg produces a DOMD to ME conversion factor of 0.0167. For the population of 42 silages studied in Northern Ireland, the conversion factor is 0.0178. These values are high compared with the factor of 0.0160 which is currently used by ADAS and by SAC in evaluating advisory silage samples.

Prediction of values for efficiency of utilization of silage ME

With the first publication of measured *in vivo* ME values (RRI, 1975), it was recognized that they were substantially higher than those predicted via the green route in Figure 7.2. Barber, Offer and Barber (1980) showed that use of red route ME values in rationing led to overestimation of milk yields, whereas the green route showed no bias between predicted and actual performance of dairy cows. One explanation of this observation may be that efficiencies of utilization of ME were lower than those assumed by MAFF *et al.* (1975). The subsequent accumulation of data supports this thesis (Table 7.4).

Table 7.4   EFFICIENCIES OF UTILIZATION OF SILAGE ME

	n	*Determined ± SD*	*Calculated*[a] *± SD*	*SE of the difference*	*Significant at p =*	*Reference*
k_m	8	0.68 ± 0.023	0.72 ± 0.022	0.011	0.007	b
k_f	14	0.43 ± 0.112	0.52 ± 0.069	0.031	0.011	c
k_l	4	0.54 ± 0.035	0.65 ± 0.006	0.018	0.010	d

[a] ARC (1980)
[b] Thomas and Thomas (1985)
[c] McDonald (1983)
[d] Unsworth and Gordon (1985)

The measured efficiencies contrast with the fixed values for k_m and k_l adopted by MAFF *et al.* (1975, 1984) of 0.72 and 0.62 respectively. If the 5% safety margin adopted by MAFF *et al.* (1975, 1984) is taken into account, these efficiency values are effectively 0.68 and 0.59 respectively.

PRACTICALITIES

Using the red route to predict NE of silages would require the ability to:

(1) predict ME directly and
(2) predict efficiencies of utilization of ME for individual feedstuffs or diets.

At present, neither of these requirements can be met with sufficient accuracy and precision for advisory purposes. It is therefore necessary, either to continue to use the green route, or to choose an intermediate path which predicts, from DOMD, ME values typical of those measured *in vivo* (by using a mean measured value for GE of DOM and ME/DE ratio). The latter course would then impose the use of mean measured efficiency values to calculate NE (Figure 7.2 and Table 7.5).

Although the ME values arrived at by these routes differ widely, the value for NE are similar. Both approaches suffer from the disadvantage that they use mean values in the calculations and so fail to describe real variations in GE, urine and methane energy losses, and k_{mp} among silages.

The approach of MAFF *et al.* (1975, 1984) uses values for both conversion steps which have now been shown to be incorrect. However, the overall factor for converting DOMD to NE in Table 7.5, has a value of 0.0099 which is close to the

Table 7.5 ME AND NE FOR MAINTENANCE AND PRODUCTION CALCULATED IN DIFFERING WAYS FOR A 600 kg COW YIELDING 30 kg MILK

Silage DOMD (g/kg DM)	$ME = DOMD \times 0.0160$ $NE = ME \times 0.62^a$		$ME = DOMD \times 0.0178$ $NE = ME \times 0.58^b$	
	ME	NE	ME	NE
500	8.0	5.0	8.9	5.2
600	9.6	6.0	10.7	6.2
700	11.2	6.9	12.5	7.2
800	12.8	7.9	14.2	8.2

$^a\ k_m = 0.68,\ k_1 = 0.59$
$^b\ k_m = 0.68,\ k_1 = 0.54$

mean measured value of 0.0103 (though that is based upon only four silages). The former route has the advantage that the resultant ME values are approximately comparable to those of other feedstuffs, can therefore be used additively in the ME rationing system currently operated by the advisory services for lactating ruminants and are readily appreciated by the farming community. The considerably higher ME values yielded by the alternative approach present a need to adjust current rationing systems and fail to describe silage feeding value in a way which conforms to the farmer's experience of animal responses. Similar arguments apply to the use of silage ME values for growing ruminants, although in this case, a variable NE system is already used in rationing.

It is important to recognize that, in following the green route in Figure 7.2, although some precision is unavoidably sacrificed, no bias is involved. Offer, Castle and Barber (unpublished) demonstrated this by comparing predicted against recorded yields for 63 individual dairy cows, fed silage only in trials at the Hannah Research Institute (Figure 7.3). Predicted yields were calculated using ME values predicted by the green route and efficiencies calculated from MAFF *et al.* (1975).

Until GE of DOM can be measured or predicted satisfactorily and adequate predictions can be made of efficiency values, there is no advantage to be gained by adopting ME values in line with those measured *in vivo*, since no gain in precision can be demonstrated. Also, such a modification to current practice implies losing the benefits of additivity currently enjoyed by users of the MAFF *et al.* (1975, 1984) ME system for lactating ruminants. Certainly, with the limited data currently available, the introduction of feedstuff-specific efficiency values would be an unjustified step.

Recent advances in the prediction of OMD

Table 7.6 shows that, prior to 1986, the UK government advisory laboratories used a variety of analytical methods to predict ME. In all cases, ME was calculated by assuming a fixed relationship with DOMD, and the silage populations used to develop the prediction equations were relatively small and varied between laboratories.

After 1986, both ADAS and SAC introduced the use of near infrared reflectance spectroscopy (NIR) for silage analysis. However, with relatively small *in vivo*

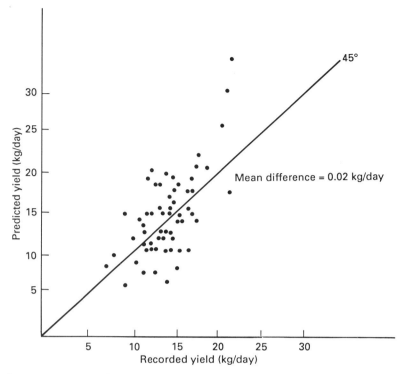

Figure 7.3 A comparison of predicted with recorded yields of individual dairy cow

populations available, it was not possible to predict either ME or DOMD directly and NIR was used to predict MADF and LIGA prior to use of the existing regressions to DOMD and subsequent calculation of ME. While the use of NIR in this way enabled these organizations to provide a more rapid and economic service for farmers, some accuracy had to be sacrificed, since the errors of NIR prediction of MADF and LIGA were added to the existing errors of DOMD prediction from these parameters.

Table 7.6 METHODS USED TO PREDICT ME (1975–86)

	ADAS	*DANI*	*ESCA*	*NOSCA*	*WSC*	*References*
MADF	x	x	x			Clancy and Wilson (1966)
IVOMD					x	Alexander and McGowan (1969)
LIGA				x		Morrison (1972)

ESCA   East of Scotland College of Agriculture
WSC   West of Scotland College
IVOMD   *in vitro* organic matter digestibility
LIGA   Acetyl bromide lignin

In 1985, a collaborative programme of research was begun, involving laboratories in both SAC and ADAS (Barber *et al.*, 1989). This work was aimed at collecting together all of the available samples of grass silage for which *in vivo* data were available and using these data sets to evaluate existing chemical, biological and physical OMD and ME prediction methods. Investigation of the NIR technique for direct prediction of *in vivo* OMD and ME was a major objective.

Barber *et al.* (1984) showed that the relationships between MADF and *in vivo* DOMD differed substantially between the data sets of the ADAS FEU and the RRI FEU. Such differences might be explained in one or more of three ways:

(1) between-laboratory differences in analytical results;
(2) between-laboratory differences in *in vivo* results;
(3) real between-population differences in the relationship between the analytical parameter and the *in vivo* data.

The first of these options is easily tested and, with proper between-laboratory control, is of little consequence. The second option has extremely serious implications for the evaluation of feedstuffs since, unless it can be shown that there are no significant between-centre differences in results obtained *in vivo,* the basis of both feedstuff evaluation (Schneider and Flatt, 1975) and animal rationing is discredited. The third option may be held to indicate that the particular predictive parameter is of little value, since even if the relationship obtained for a particular population (of limited size) is excellent, it will generally prove to be impossible to identify which particular relationship will apply to a particular advisory silage sample. Such relationships may have a limited application if they apply to populations which may be defined unambiguously, for example to all first cut silages.

To eliminate differences between populations in predictive relationships caused by between-laboratory analytical differences, all samples were analysed for each parameter by a single laboratory.

The results obtained were used in univariate and multivariate regression analyses to derive predictive relationships with OMD and ME. All of the samples from one population and 20 samples, selected at random, from another, were reserved to provide an independent test set for validation of regression relationships. The remaining 122 samples were used as a calibration set.

The results confirmed the earlier work of Givens and Brunnen (1987), showing that the direct prediction of ME was in all cases poorer, having lower R^2 and higher SEC values than those obtained for the direct prediction of OMD. In view of the arguments outlined above, which establish the difficulties implicit in adopting direct prediction of ME at this time, the approach used concentrated on the prediction of OMD.

For each predictive approach, analysis of variance was used to detect differences between populations in intercept and slope of the regression lines obtained. Where no significant differences were revealed, it was possible to use a single regression relationship to describe all of the data.

Chemical measurements of fibre components (MADF and LIGA) gave regression lines which differed significantly between silage populations in intercept while the pepsin-cellulase (PCOMD) technique of Jones and Hayward (1975) produced regression lines which differed significantly between silage populations in both intercept and slope. IVOMD (Alexander and McGowan, 1969) regression lines for the silage populations did not differ significantly in either intercept or slope. The same situation applied to NIR analysis using the approach of Norris *et al.* (1976) , provided that more than five spectral segments were used to derive the multivariate relationship.

The MADF results compared closely with the data presented by Barber *et al.* (1984). However on this occasion, all of the laboratory analyses were performed at a single laboratory, confirming that the differences in intercept between populations cannot be explained by between-laboratory analytical differences. Between-centre differences in *in vivo* measurements are an unlikely explanation, as both NIR and IVOMD showed no significant between-population differences. This suggests that, where between-population differences were observed (MADF, LIGA, PCOMD), they reflected real population-dependent differences in the relationship between predictive parameter and *in vivo* OMD. It is possible that this may be a reflection of the heterogeneous nature of the fibre components in forage plants, particularly where, as in many silages, there are different proportions of grass species and varieties and the presence or absence of varying proportions of clover. Whatever the explanations of this phenomenon, it is now clear that those analytical methods which demonstrate it are entirely unsuitable as tools for use in the evaluation of farm silages for advisory purposes.

It is essential to test any predictive relationship against an independent test set of samples which were not included in the population used in the initial calibration. This approach has not usually been adopted for the traditional methods of silage OMD prediction because of shortage of *in vivo* data. The validation test made by Barber *et al.* (1989) confirmed the inadequacy of predictions based upon MADF and LIGA (R^2 0.20 and 0.14, SEP 5.1 and 5.3, respectively) and clearly indicated the superiority of direct prediction of *in vivo* OMD using NIR (R^2 0.76, SEP 2.6).

As a consequence of this work, the government advisory services in the UK have, from 1988, agreed to adopt a common method of silage evaluation using NIR to predict OMD directly, based upon a large *in vivo* population combined from all centres.

Nutrient response based systems of rationing

For many years, agricultural advisers have made use of rationing systems based upon descriptions of requirements for individual nutrients. However, during the 1970s and 1980s, the emphasis of research has changed in an attempt to elucidate underlying mechanisms and so devise rationing systems based on mathematical models which are mechanistic rather than empirical in nature. The underlying philosophy and the specialized terminology of this development have been described by France, Thornley and Beever (1984). A critical feature of such models is integration of genetic potential with utilization of absorbed nutrients. The key advantage of this approach is that it should allow prediction of production responses to changes in quantity and/or quality of the ration. Existing requirement-based models define a requirement for each nutrient, in isolation, for a given target production level. Such models fail to predict adequately the consequences of providing too little or too much of a particular nutrient or the results of nutrient interactions.

Even so, AFRC (1987a) considered that: '. . . despite the inherent drawbacks to the ME/DE systems of evaluation, practical use will be made of these systems for the forseeable future and further work aimed at their development and improvement is warranted.'

This apparent conflict is a reflection of the inevitable and perfectly proper difference of viewpoint between the researcher (whose interests lie in achieving advances in scientific knowledge) and the adviser (who must place his reputation at risk every time he gives advice to a farmer). The adviser's advice may result in financial disaster for his client; understandably, he therefore relies to a large extent upon hard won experience, using scientific systems only as a base. The introduction of a radically different scientific system presents the established adviser with a major problem since it removes him from the firm ground of his experience.

The introduction of nutrient response based systems is likely to proceed slowly, either alongside the continued use of existing rationing systems or as gradual developments from them.

Before considering the demands upon feedstuff characterization which are likely to be made by nutrient response based rationing systems, it is worth contemplating the lessons which might be learned from the history of development of existing rationing systems.

While feed intake is the key biological constraint in the calculation of rations for livestock, it is also the area of greatest uncertainty. The current paucity of knowledge in this vitally important area reflects the lack of research attention given to it. Similarly, inadequate attention has been given in the past to the evaluation of feedstuffs. This is evidenced by the fact that, although the ME system was recommended for advisory use by ARC (1965) and adopted by the advisory services in 1975, a firm basis for the prediction of the ME of the primary conserved forage has become available only in 1988. Examination of research expenditure would reveal disproportionally huge investments in investigations into the metabolic basis of animal nutrient requirements and nutrient responses with comparatively little attention to feed intake and feedstuff characterization. Yet, conceptual advances in aspects of metabolism have limited value to the agricultural industry until they can be applied through knowledge of these factors.

It is clear from the literature on a nutrient response systems that their implementation will require a more complex analytical base than does the current amalgam of nutrient requirement systems. For example, it is a fundamental tennet of the new thinking that energy is not a single nutrient, and that consideration must be given to the variety of forms in which energy is absorbed. Thus, a single analysis from which ME can be calculated must give way to an analytical system which allows the prediction of the absorption of individual energy-yielding nutrients from the gut. This may require detailed chemical analysis of the feed, but it will also be necessary to provide information about the physiological dynamics of digestion to allow for the effects on nutrient absorption of such interacting variables as physical form of the diet, rate of eating, rate of passage through different segments of the gut and rate of digestion.

For forages, prediction of the rate of cell wall degradation will be a fundamental requirement. The model proposed by Dewhurst *et al.* (1986) suggested the use of a relationship involving lignin and β-glycans for this purpose. The problems associated with this approach are remarkably similar to those previously discussed in relation to the prediction of OMD from fibre parameters.

The practical implementation of this increase in complexity, which is an order of magnitude greater than any previously contemplated, has become potentially feasible only because of the widespread availability of cheap computing power. However, the decision to change from existing rationing systems to newly devised ones will depend not only upon the ability of the scientists to convince advisers of the conceptual superiority of the new systems but upon the cost to the farmer of the whole advisory package. The necessary ingredients for success are:

(1) an acceptable conceptual metabolic framework;
(2) satisfactory prediction of feed intake;
(3) an adequate analytical data base for feedstuffs which exhibit only limited variation in composition;
(4) the ability to provide a rapid and economic analytical service for feedstuffs which vary widely in composition;
(5) control of feeding systems on farms consistent with the degree of sophistication used in rationing.

While considerable progress is being made towards achieving the first of these requirements, the problems which abound in respect of the remainder are considerable.

It is necessary to recognize that analytical limitations may reduce the applicability of a proposed model. Indeed, the model itself is likely to be based upon rather limited data; Thomas and Thomas (1985) commented on the poor quality and paucity of analytical data reported in the animal experimentation literature and, by definition, new concepts in modelling metabolism are unlikely to be supported by the existing analytical base.

Traditional chemical analytical techniques are unlikely to form a satisfactory base for implementing a new rationing system as many of them are slow and expensive. It is therefore essential to look to 'high-tech' analytical systems capable of rapid, multi-component analysis at low cost. NIR is an example of a technique which may meet these requirements. In addition to providing estimates of nutrient content of feedstuffs, the successful use of NIR to predict *in vivo* OMD (Barber *et al.*, 1989) suggests that prediction of potential rates and end-products of organic matter digestion in the rumen might be feasible.

Satisfactory use of NIR is dependent on the existence of adequate calibration and validation data. Abrams *et al.* (1987) showed that NIR calibration sets need to contain at least 100 samples to minimize SEP. Thus it may take many years to accumulate the necessary reference data for the prediction of the complex dynamic variables demanded by the new nutrient response based approach.

It is an open question whether the response-based rationing systems proposed for the future will be economically viable if it proves necessary to provide such large data bases to support them, particularly considering the difficult, time consuming and expensive *in vivo* measurements which may have to be made for a wide range of feedstuffs. Although there is no doubt that the scientific base of current rationing systems is incomplete, it is a matter of opinion whether or not they are adequate for farm advisory work. In the current political climate in the UK, the opinion that will count in the long term is that of the research customer.

Acknowledgements

The contribution of Dr E. F. Unsworth in discussion of aspects of energy utilization and related calculations, and his permission, together with that of Mr J. Weir, Dr J. K. Thompson and Dr M. E. Castle to use and refer to unpublished data is gratefully acknowledged.

We wish to thank Professor P. C. Thomas, Dr A. Macpherson and Mr J. Weir for their advice on preparation of the manuscript.

References

ABRAMS, S.M., SHENK, J.S., WESTERHAUS, M.O. and BARTON, F.E. (1987). *Journal of Dairy Science,* **70**, 806–813

AERTS, J.V., DE BRABANDER, D.L., COTTYN, B.G. and BUYSSE, F.X. (1977). *Animal Feed Science and Technology,* **2**, 337–349

AGRICULTURAL AND FOOD RESEARCH COUNCIL (1987a). Technical Committee on Responses to Nutrients, Report No. 1, Characterisation of Feedstuffs: Energy. *Nutrition Abstracts and Reviews, Series B: Livestock Feeds and Feeding,* Vol. 57, pp. 507–523

AGRICULTURAL AND FOOD RESEARCH COUNCIL (1987b). Technical Committee on Responses to Nutrients, Report No. 2, *Nutrition Abstracts and Reviews, Series B: Livestock Feeds and Feeding,* Vol. 57, pp. 713–736

AGRICULTURAL RESEARCH COUNCIL (1965). *The Nutrient Requirements of Farm Livestock. No.2, Ruminants.* HMSO, London

AGRICULTURAL RESEARCH COUNCIL (1976). *The Nutrient Requirements of Farm Livestock. No.4, Composition of British Feedingstuffs.*

AGRICULTURAL RESEARCH COUNCIL (1980). *The Nutrient Requirements of Ruminant Livestock.* Commonwealth Agricultural Bureaux, Farnham Royal, Slough

ALDERMAN, G., COLLINS, F.C. and DOUGALL, H.W. (1971). *Journal of the British Grassland Society,* **26**, 109–111

ALEXANDER, R.H. (1960). *Journal of Agricultural Engineering Research,* **5**, 118–121

ALEXANDER, R.H. and McGOWAN, M. (1969). *Journal of the British Grassland Society,* **24**, 195–198

BARBER, W.P., ADAMSON, A.H. and ALTMAN, J.F.B. (1984). In *Recent Advances in Animal Nutrition – 1984,* pp. 161–176. Ed. Haresign, W. and Cole, D.J.A. Butterworths, London

BARBER, G.D., GIVENS, D.I., KRIDIS, M.S., OFFER, N.W. and MURRAY, I. (1989). *Animal Feed Science and Technology,* (in press)

BARBER, W.P., OFFER, N. W. and BARBER, G.D. (1980). *Proceedings of the 4th COSAC Study Conference,* Perth

BARNETT, A.J.G. (1957). *Journal of Agricultural Science,* **49**, 467–474

BLAXTER, K.L. and CLAPPERTON, J.L. (1965). *British Journal of Nutrition,* **19**, 511–522

BOSMAN, M.S.M. (1970). Mededelingen, Instituut voer Biologischeen Scheikundige Onderzoek Landbouwgewassen, Wageningen, No.413, p.15

CLANCY, M.J. and WILSON, R.K. (1966). *Proceedings of the 10th International Grassland Congress,* pp. 445–453. Helsinki, Finland

CVVB (1977). *Manual for the calculation of the nutritive value of roughages.* CVVB in Nederland, Lelystad

DEWHURST, R.J., WEBSTER, A.J.F., WAINMAN, F.W. and DEWEY, P.J.S. (1986). *Animal Production,* **43**, 183–194

DIJKSTRA, N.D. (1949). *Verslagen der Rijkslandbouwproefstations.*'s Gravenhage, **55**, 15

DOWMAN, M.G. and COLLINS, F.C. (1982). *Journal of the Science of Food and Agriculture,* **33**, 689–696

EDWARDS, R.A. (1986). In *Ruminant Feed Evaluation,* pp. 35–58. Ed. Stark, B.A., Givens, D.I. and Wilkinson, J.M. Chalcombe Publications, Marlow

EL SHAER, H.M., OMED, H.M., CHAMBERLAIN, A.G. and AXFORD, R.F.E. (1987). *Journal of Agricultural Science,* **109**, 257–259

EVERINGTON, J.M. and GIVENS, D.I. (1989). (in preparation)

FORBES, J.M. (1988). In *Nutrition and Lactation in the Dairy Cow*, pp. 294–312. Ed. Garnsworthy, P.C. Butterworths, London

FRANCE, J., THORNLEY, J.H.M. and BEEVER, D.E. (1984). *Research and Development in Agriculturre*, **2**, 65–71

GIVENS, D.I. (1986). In *Developments in Silage 1986*. Ed. Stark, B.A. and Wilkinson, J.M. Chalcombe Publications, Marlow

GIVENS, D.I. and BRUNNEN, J.M. (1987). In *Proceedings of the 8th Silage Conference*, pp. 55–56. Hurley

GIVENS, D.I., EVERINGTON, J.M. and ADAMSON, A.H. (1989). *Animal Feed Science and Technology*, (in press)

HALLSWORTH, E.G. (1949). *Journal of Agricultural Science*, **39**, 254–258

JONES, D.I.H. and HAYWARD, M.V. (1975). *Journal of the Science of Food and Agriculture*, **26**, 711–718

JOSHI, D.C. (1972). *Acta Agriculturae Scandinavia*, **22**, 243–247

KORVA, J. and TUORI, M. (1986). *Journal of Agricultural Science in Finland*, **58**, 175–183

LARSEN, R.E. (1974). *Ghana Journal of Agricultural Science*, **7**, 195–202

LEWIS, M. (1981). *Proceedings of the VI Silage Conference, Edinburgh*, pp. 35–36

McDONALD, P. (1983). *Hannah Research Institute Report*, pp. 56–67

MINISTRY OF AGRICULTURE, FISHERIES AND FOOD (1976). *ADAS Advisory Paper No. 11*, 2nd Edition

MINISTRY OF AGRICULTURE, FISHERIES AND FOOD, DEPARTMENT OF AGRICULTURE AND FISHERIES FOR SCOTLAND AND DEPARTMENT OF AGRICULTURE FOR NORTHERN IRELAND (1975). *Technical Bulletin No. 33*

MINISTRY OF AGRICULTURE, FISHERIES AND FOOD, DEPARTMENT OF AGRICULTURE AND FISHERIES FOR SCOTLAND AND DEPARTMENT OF AGRICULTURE FOR NORTHERN IRELAND (1984). *Reference Book 433*

MORRISON, I.M. (1972). *Journal of the Science of Food and Agriculture*, **23**, 455–463

NEAL, H.D.S., THOMAS, C. and COBBY, J.M. (1984). *Journal of Agricultural Science, Cambridge*, **103**, 1–10

NORRIS, K.H., BARNES, R.F., MOORE, J.E. and SHENK, J.S. (1976). *Journal of Animal Science*, **43**, 889–897

OLDHAM, J.D. and WEBSTER, A.J.F. (1986). Paper N1.8, *Commission of Animal Nutrition, 37th Annual Meeting of the EAAP*, Budapest

O'SHEA, J., WILSON, R.K. and SHEENAN, W. (1972). *Irish Journal of Agricultural Research*, **11**, 175–179

ROWETT RESEARCH INSTITUTE (1975). *Feedstuffs Evaluation Unit First Report*. Department of Agricultural and Fisheries for Scotland, Edinburgh

SCHNEIDER, B.H. and FLATT, W.P. (1975). *The Evaluation of Feeds through Digestibility Experiments*. The University of Georgia Press, Athens, GA, USA

TERRY, R.A. and OSBOURN, D.F. (1980). In *Forage Conservation in the 80s*, pp. 315–381. Ed. Thomas, C. British Grassland Society Occational Publication No.11

THOMAS, C. and THOMAS, P.C. (1985). In *Recent Advances in Animal Nutrition – 1985*, pp. 223–256. Ed. Haresign, W. and Cole, D.J.A. Butterworths, London

TILLEY, J.M.A. and TERRY, R.A. (1963). *Journal of the British Grassland Society*, **18**, 104–111

UNSWORTH, E.F. and GORDON, F. (1985). *Agricultural Research Institute of Northern Ireland 58th Annual Report*, pp. 13–20

UNSWORTH, E.F., WYLIE, A.R.G. and ANDERSON, R. (1984). In *Proceedings of the 7th Silage Conference*, pp. 35–36. Belfast, Queen's Unive;rsity

UNSWORTH, E.F., WYLIE, A.R.G. and ANDERSON, R. (1987). In *Proceedings of the 8th Silage Conference*, pp. 57–58. Hurley

VADIVELOO, J. and HOLMES, W. (1979). *Journal of Agricultural Science, Cambridge,* **93**, 553–562

WAINMAN, F.W., DEWEY, P.J.S. and BOYNE, A.W. (1978). *Rowett Research Institute, Feedstuffs Evaluation Unit Second Report.* Department of Agriculture and Fisheries for Scotland, Edinburgh

WAINMAN, F.W., DEWEY, P.J.S. and BREWER, A.C. (1984). *Rowett Research Institute, Feedstuffs Evaluation Unit Fourth Report.* Department of Agriculture and Fisheries for Scotland, Edinburgh

8

EFFECT OF SILAGE ADDITIVES AND WILTING ON ANIMAL PERFORMANCE

F. J. GORDON

Agricultural Research Institute of Northern Ireland, Hillsborough, Co. Down, UK

Introduction

The on-farm profitability of ruminant livestock enterprises in Europe is increasingly being limited by the restrictions on the volume of product which can be produced. With milk, these restrictions arise through externally imposed quotas while with beef output is limited by the declining number of calves available. In such circumstances the industry is forced towards reducing the costs of producing each unit of output, and in many parts of Europe this implies a greater reliance on grass, either grazed or conserved. Ensiling is now accepted as the major method of forage conservation and much research has been undertaken into the factors which influence its quality (see reviews by McDonald, 1983 and Thomas and Thomas, 1985). In spite of this there continues to be confusion with regard to many aspects of silage production, particularly in terms of their influence on animal performance. The multiplicity of silage additives available, and lack of decisiveness with regard to advice on either their optimum composition or use highlights one such example. Similarly, the implications of prewilting of herbage prior to ensiling has, over many years, been a major source of debate. The present chapter is aimed at examining some of the aspects within both of these areas, particularly in relation to recent information. However, within this chapter it is not possible to provide a comprehensive review of the full data in these total areas.

The use of silage additives

SILAGE FERMENTATION

The major objective in silage fermentation is to achieve a stable pH at which biological activity virtually ceases. In this way, preservation is obtained while minimizing nutrient losses and avoiding adverse changes in the chemical composition of the material. This is achieved by discouraging the activities of undesirable microorganisms and encouraging the development of bacteria which produce lactic acid. The undesirable microorganisms are mainly clostridia, coliforms and yeasts and if these proliferate they will compete with lactic acid bacteria for sugars. In addition they will metabolize lactic acid to produce end-products such as acetate, butyrate, propionate, ethanol and butanol (Thomas and Morrison, 1982). Clostridia and coliforms will also metabolize amino acids to produce a variety of products including higher volatile fatty acids, amines and ammonia (McDonald, 1981). Since butyric acid is a weaker acid than lactic acid, and many of the products of amino acid breakdown are bases, clostridial activity slows, or reverses, the normal reduction in pH which occurs in the silo.

The role of additives is to assist in achieving a stable pH within the silo. While there are presently a wide range of additive types available for ensiling McDonald (1981) has classified them into four categories. These are:

(1) fermentation stimulants, which encourage a lactic acid fermentation;
(2) fermentation inhibitors, which either partially or completely inhibit microbial growth;
(3) aerobic deterioration inhibitors, which are primarily aimed at controlling the deterioration of silage on exposure to air, and
(4) nutrients, which are added to crops at the time of ensiling in order to improve the nutritional value of the silage.

However, it is recognized that these categories are not mutually exclusive and there may be considerable overlap. This is particularly so where some materials may be used to both assist fermentation and act as sources of nutrients in their own right. In addition it could be considered that a new class of additive is now emerging, effluent absorbents, although in some instances these may also have characteristics which would place them in other groupings.

Within this chapter it is impossible to provide a review of all additive types, and hence discussion will be restricted to those which can be considered as either fermentation stimulants or inhibitors, and even within these categories only to those that are of either major importance or for which new and relevant information has become available.

FERMENTATION INHIBITORS

Formic acid

Although the use of formic acid as a silage additive was suggested many years ago it was not until the introduction of the forage harvester, and the development of simple methods of applying the additive, that its use became widespread. Its addition to crops, at moderate levels, rapidly reduces the pH thereby inhibiting the activities of coliforms and clostridia. At higher levels of application the fermentation by lactic acid bacteria is also inhibited and the water soluble carbohydrate contents in the silage may even be greater than those of the original herbage. However, formic acid does not inhibit the growth of yeasts and this may often result in silages with high ethanol contents, a condition which may lead to increased dry matter (DM) losses (McDonald, 1983).

Extensive reviews of the effects of formic acid on fermentation have been undertaken by Waldo (1978) and Thomas and Thomas (1985). These have shown that formic acid results in a reduction in fermentation, with lower proportions of acetic and butyric acids, and reduced proteolysis in the resulting silage. In addition, Parker and Crawshaw (1982) have summarized the data from 22 comparisons undertaken at ADAS Experimental Husbandry Farms and demonstrated the ability of formic acid to improve the fermentation of low DM crops.

Table 8.1   EFFECTS OF ADDITIVES WHICH INHIBIT FERMENTATION ON DIGESTIBILITY, FEED INTAKE AND ANIMAL PERFORMANCE

	Direct cut				Wilted	
	Untreated	Formic	Formic and formaldehyde	Formaldehyde	Untreated	Formic
Digestibility of energy (%)	63.8	64.0	63.4	63.5	63.2	61.9
Growing cattle						
DE intake (k cal/kg$^{0.75}$)	217	261	246	267	305	324
Liveweight gain – cattle (g/day)	431	735	718	750	547	691
Lactating cattle						
DM intake (kg DM/day)	8.3	9.3	9.4	–	9.3	10.1
Milk yield (kg/day)	16.9	17.7	17.8	19.1 ($n = 1$)	16.4	16.7

(Waldo, 1978)

Animal performance   The effects of any additive on animal production are derived from its combined effects on digestibility, feed intake and nutrient utilization. The data on these aspects from the review by Waldo (1978) are given in Table 8.1. It was concluded that the fermentation quality of silage has minimal effects on digestibility and that those effects which have been recorded in some studies are artefacts of the higher losses of volatiles which occur during sample preparation with highly fermented, untreated silages. However, Parker and Crawshaw (1982), from their review, concluded that where the control silage was well preserved formic acid did not influence digestibility but when it prevented the development of a clostridial fermentation digestibility was increased.

The data of Waldo (1978) indicate that formic acid increased silage DM intake of growing cattle by 20% and dairy cows by 12%, with the respective increases in animal performance being 71% and 5%. Parker and Crawshaw (1982) reported mean increases in silage DM intake and liveweight gain by growing cattle of 5% and 25% respectively. Thomas and Thomas (1985) in a review of 41 studies with growing cattle reported a mean response in livewweight gain of 32%. Data from a series of 24 comparisons carried out at Greenmount College of Agriculture and Horticulture in Northern Ireland, and involving formic acid, have shown mean responses in intake and liveweight gain of 9% and 17% respectively (Table 8.2, S. Kennedy, personal communication). Over this data set, formic acid resulted in significant improvements in silage fermentation, silage digestibility, silage intake and liveweight gain with the mean responses of the latter two variables being 9% and 17% respectively. However, it is widely recognized that the response in animal performance to additive is reduced as level of supplementation increases and this is also demonstrated by the data in Table 8.2. For example using the data relating only to the lowest level of supplementation in each experiment indicates a mean response in liveweight gain of 24% rather than the 17% recorded when using all supplementation levels. However, the calculation of means in each of these reviews hides very major differences in responses between experiments. Those studies in which the control silage was well preserved have generally shown no beneficial effects from formic acid yet where the fermentation quality of the control was poor there were large beneficial effects (Parker and Crawshaw, 1982).

Table 8.2 COMPARISONS OF UNTREATED AND FORMIC ACID TREATED SILAGES FOR FINISHING CATTLE AT GREENMOUNT COLLEGE OF AGRICULTURE AND HORTICULTURE

	Untreated	Formic acid treated	SEM		n
Fermentation					
pH	4.40	3.97	0.05	☆☆☆	21
NH₃N (g/kg total N)	95	60	4.3	☆☆☆	21
Butyric acid (g/kg DM)	5.0	2.3	0.41	☆☆☆	15
Animal data					
In vivo DOM (g/kg DM)	661	679	5.5	☆	12
Silage intake (kg/day)					
Lowest conc. inputs	6.29	6.94	0.09	☆☆☆	21
All comparisons	6.36	6.93	0.075	☆☆☆	24
Liveweight gain (kg/day)					
Lowest conc. inputs	0.58	0.72	0.016	☆☆☆	21
All comparisons	0.66	0.77	0.015	☆☆☆	24

(S. Kennedy, personal communication)

The production responses from the use of formic acid with lactating cows have generally been below those reported for growing cattle. Waldo (1978) reported mean responses in silage intake and milk yield of 12% and 5% respectively and more recent evidence would support these small responses. For example, Chamberlain, Thomas and Robertson (1987) reported two studies in which formic acid resulted in increases in silage intake of 9% and 6% but in neither study was there any increase in the total outputs of milk solids (fat, protein and lactose). Similarly in a series of three studies at this Institute (Gordon, 1989a and b and Mayne, unpublished), in which control silages were well fermented, trends towards negative responses to formic acid in milk yield, and only marginal positive responses in milk energy output were obtained (Table 8.3). Such effects, when considered in conjunction with those already published would suggest that, in relation to silage for dairy cattle, there is a much greater need for clearer definition of when it may be beneficial to use formic acid.

Table 8.3 EFFECTS OF UNTREATED, FORMIC ACID-TREATED AND INOCULANT-TREATED SILAGES ON THE PERFORMANCE OF LACTATING COWS (MEAN OF THREE STUDIES)

	Untreated	Formic acid treated	Inoculant treated	SEM	
Composition of silages					
Dry matter (g/kg)	184	191	183	1.9	NS
Ammonia N (g/kg total N)	81	58	64	3.5	☆
pH	3.9	3.8	3.8	0.05	NS
Butyrate (g/kg DM)	0.34	0.10	0.30	0.115	NS
Feed intake and performance					
Silage DM intake (kg/day)	9.4	9.7	10.4	0.23	NS
Milk yield (kg/day)	22.9	22.2	24.4	0.16	☆☆
Milk energy output (MJ/day)	69.2	70.1	74.1	0.62	☆

(Gordon 1989a and b and Mayne, unpublished)

Sulphuric acid

The early work on inhibition of silage fermentation involved the use of high levels of mineral acids (Virtanen, 1933). However, the high rate of acid use in this system with resulting detrimental effects on silage intake and concerns over operator safety, resulted in this approach not being widely adopted in the UK. However the recent work of Flynn and his colleagues in the Republic of Ireland, using low levels of sulphuric acid, has resulted in an upsurge in its popularity. O'Kiely, Flynn and Poole (1989a) have demonstrated that sulphuric acid (45% w/w) is as effective as formic acid in terms of both controlling fermentation and enhancing silage intake by growing cattle (Table 8.4). However, the effects on animal performance have

Table 8.4 COMPARISON OF FORMIC AND SULPHURIC ACIDS AS SILAGE PRESERVATIVES (MEAN OF TWO EXPERIMENTS)

	No additive	*Formic acid*	*Sulphuric acid*
Dry matter (g/kg)	168	190	188
pH	5.1	4.2	4.2
NH$_3$-N (g/kg N)	233	74	71
Crude protein (g/kg DM)	185	178	187
In vitro DMD (g/kg DM)	544	658	675
DM intake[a] (g/kg liveweight)	16.2	22.6	23.5

[a] Relates to one experiment only
(O'Kiely, Flynn and Poole 1989a)

not been so clear. Weddell and Yackiminie (1988) compared an additive based on sulphuric acid with either untreated or formic acid treated silages and concluded that animal performance from the former was only marginally below that of formic acid. O'Kiely, Flynn and Poole (1989a and b) and O'Kiely *et al.* (1989) have reported a series of comparisons between formic and sulphuric acids and concluded that there were no differences in animal performance. However, in this latter series of studies there were no negative control treatments. Recently, Kennedy (1989a) has reported a series of five studies, with finishing cattle, embracing untreated, formic acid-treated and sulphuric acid-treated silages produced under a range of ensiling conditions. From these data it is concluded that, on average, animal performance with the sulphuric acid treatment was no better than that of untreated silage. There is little information available on the use of sulphuric acid as an additive when used with lactating cattle although Murphy (1986) has compared formic and sulphuric acids and reported no differences in animal performance. The data presented in Table 8.5 summarize the available animal production information with sulphuric acid and would suggest that, in terms of animal performance, sulphuric acid is not as effective as formic acid.

These discrepancies between the effectiveness of sulphuric acid in animal production studies are surprising in view of the fact that in general sulphuric acid has tended to result in improvements in silage preservation. For example, Table 8.6 is derived from the data of Kennedy (1989a) and indicates that, over the five trials within the series, although sulphuric acid generally improved both silage fermentation and silage DM intake and had no detrimental effects on silage digestibility, it did not improve carcass gain. The reasons for such differences are unclear. Sulphuric acid is known to have detrimental effects on copper status of the animal, particularly when used in conjunction with low levels of concentrate supplementation, and this could impair animal performance (O'Kiely *et al.*, 1989). However, the commercial additive used by Kennedy (1989a) was fortified with copper although it may be possible that the inclusion level was not adequate to sustain animal performance. For example in other work (S. Kennedy, personal

Table 8.5 RELATIVE ANIMAL PERFORMANCES FROM FORMIC ACID AND SULPHURIC ACID-TREATED SILAGES (COMPARED TO AN INDEX VALUE OF 100 FOR THE FORMIC ACID TREATED SILAGE)

		Control	Formic acid	Sulphuric acid
Beef cattle				
Weddell and Yackiminie (1988)	(1)	71	100	98
	(2)	87	100	96
O'Kiely, Flynn and Poole (1989a)	(1)	–	100	102
O'Kiely *et al.* (1989)	(2)	–	100	93
	(3)	–	100	111
	(4)	–	100	98
Kennedy (1989a)	(1)	102	100	104
	(2)	75	100	83
	(3)	104	100	99
	(4)	91	100	83
	(5)	93	100	79
Dairy cattle				
Murphy (1986) (Fat and protein yeild)		–	100	99
	Means	90.9[a]	100	95.4 SEM 2.3

[a] Includes five missing observations

Table 8.6 COMPARISON OF FORMIC AND SULPHURIC ACIDS AS ADDITIVES FOR SILAGE OFFERED TO FINISHING CATTLE (DATA DERIVED AS A MEAN OF FIVE STUDIES)

	Untreated	Formic acid	Sulphuric acid	SEM
Fermentation characteristics				
Dry matter (g/kg)	192	200	19.4	2.9
pH	4.2	3.9	4.0	0.04
NH_3-N (g/kg N)	94	55	71	4.7
Butyrate (g/kg DM)	8.6	4.9	5.7	0.67
Animal data				
In vivo DOM (g/kg DM)	649	657	650	8.3
Silage DM intake (kg/day)	6.39	6.88	6.64	0.19
Liveweight gain (kg/day)	0.66	0.77	0.65	0.027
Carcass gain (kg/day)	0.49	0.53	0.47	0.017

(Kennedy, 1989a)

communication) copper deficiency symptoms have been recorded when using another proprietary sulphuric acid additive which contained low levels of added copper.

In the light of these contrasting effects of sulphuric acid on animal performance caution should be exercised in terms of its use. Nevertheless, in view of its undoubted benefits in terms of cost effectiveness as a preservative it is extremely important that further animal production studies, preferably with dairy cattle, are undertaken at the earliest date.

Formaldehyde and acid/formaldehyde mixtures

While formaldehyde has primarily been used as a means of restricting fermentation it has also the ability to bind with plant proteins and reduce the rate of breakdown of protein both in the silo and in the rumen. However, controlling these processes is extremely difficult, with low levels of formaldehyde tending to stimulate the growth of undesirable organisms, such as clostridia, while high levels may result in a irreversible bonding with protein. This latter aspect reduces protein digestibility and results in decreased protein utilization by the animal. For these reasons, formaldehyde is generally considered to be most effective when used at low levels in a mixture with a normal rate of acid (McDonald, 1983).

Waldo (1978) has reviewed the data relating to feed intake and animal performance using formaldehyde and formic/formaldehyde mixtures for both growing cattle and lactating cows (Table 8.1). While the number of studies have been small in some groupings, the data would suggest formic/formaldehyde mixtures to result in similar levels of intake and animal performance to that achieved with formic acid. Thomas and Thomas (1985) have also concluded that formaldehyde will produce positive animal responses when compared with control silages without additive, but little response when an acid additive has been used as the control.

In more recent work O'Kiely and Flynn (1988) have reported a series of three studies to compare formic acid and formic/formaldehyde mixtures for finishing cattle. In general, both additives were equally effective at controlling silage preservation and in none of the studies were there any differences between additives in carcass gain. These studies also explored the interaction between additive type and protein content of the supplement and the data suggest that the inclusion of formaldehyde did not enhance protein utilization from the silages. McCullough (1984) also reported a series of five studies using mixtures of formaldehyde and formic acid and formaldehyde and sulphuric acid and found neither of these mixtures to improve performance over that achieved with formic acid alone.

While in most animal production studies the use of formaldehyde in mixtures with acids has been as effective as formic acid alone no advantage has been demonstrated. Concern is also expressed in relation to the inclusion of high levels of formaldehyde with Kaiser *et al.* (1981) showing high inclusion rates to result in major reductions in DM digestibility, liveweight gain and nitrogen retention.

FERMENTATION STIMULANTS

These additives fall into two main categories: carbohydrate rich materials and bacterial inoculants or enzymes. The major additives in the first category are molasses and sugars although others such as cereals or sugar beet pulp may also, at least partially, influence fermentation by providing energy yielding substrates for the growth of lactic acid bacteria. A large number of studies have been carried out on the effectiveness of additives such as molasses and it is generally agreed that, provided application rates are adequate, they can provide adequate control of silage fermentation and enhance animal performance (McDonald, 1981; Murdoch, 1961; Fox, Brown and McCullough, 1972). However, a problem with many of these bulky additives is ensuring that the material is intimately mixed with the grass at ensiling (O'Kiely, 1988).

It is not intended here to review the additives within this group.

Inoculants

The possibility of using inoculants as silage additives has been recognized for many years, but in the major proportion of the early studies the encouraging results obtained in laboratory experiments often proved disappointing when taken into farm-scale operations. For example Bolsen (1978) concluded that additional research was required if inoculants were to make a significant contribution to silage production in the future. More recently Done (1986) has reviewed the effects of inoculants on the fermentation of grass silages produced under Western European conditions. From this she concluded that, although some inoculants had demonstrated their ability to improve fermentation in laboratory scale studies, their effectiveness as silage preservatives under farm conditions remained unproven. However, while silage fermentation is undoubtedly important it should not be regarded as an end in itself and the main issue must be the effect of inoculants on animal performance. This conflict between the effects on fermentation and on animal performance is highlighted in Table 8.7, derived from the data of Steen *et al.* (1989). In this study the inoculant did not improve fermentation over that of the untreated material yet both DM intake and animal performance were considerably increased. In fact the intake and performance with the 'poorly fermented' inoculant-treated silage was similar to that achieved with the well fermented formic acid-treated silage.

A recent series of three studies with lactating cows at this Institute have also shown considerable benefits from the use of an inoculant without any apparent

Table 8.7 EFFECTS OF UNTREATED, FORMIC ACID-TREATED AND INOCULANT-TREATED SILAGES FOR CALVES

	Untreated	*Formic acid*	*Inoculant*	*SEM*	
Fermentation					
pH	4.5	3.9	4.5	0.05	***
NH_3-N (g/kg total N)	162	91	176	6.0	***
Butyric acid (g/kg DM)	12.1	1.3	16.6	1.37	***
Animal data					
Silage DM intake (kg/day)	2.28	2.49	2.44	0.037	***
Liveweight gain (kg/day)	0.88	0.94	0.95	0.022	*

(Steen *et al.*, 1989)

benefit in terms of either the fermentation or the chemical compositions of the silages (Table 8.3, Gordon 1989a and b; Mayne, unpublished). The inoculant resulted in increases in silage intake of 10% and milk energy output of 7% and these responses were significantly greater than those recorded with the formic acid treated silages. Using the data of Gordon (1984), relating the response in milk yield to level of concentrate supplementation, would indicate that the response from the inoculant could enable a reduction in supplementary feed of 1.8 kg/day and yet maintain a similar level of milk output to that on the control silage. Recent data from Moorepark, in the Republic of Ireland (J. Murphy, personal communication), have also shown a significant milk production response (+6%) to the use of the same inoculant as that used at this Institute when compared with an acid additive.

There are other recent studies which would indicate that beneficial effects from inoculation can be achieved (Anderson *et al.,* 1989; Rooke *et al.,* 1988; Henderson *et al.,* 1987; Appleton and Done, 1987). However, other studies have shown no benefits in terms of animal performance (Kennedy *et al.,* 1989; Kennedy, 1987a; Kennedy, 1989b; Haigh, Appleton and Clench, 1987). It is also recognized that recently Chamberlain, Thomas and Robertson (1987) were unable to show any benefits from the use of an inoculant with lactating cattle.

Enzymes

It is often considered that under many ensiling conditions the availability of energy yielding substrate is a more limiting factor in achieving a satisfactory fermentation than the quantity of bacteria available. Recently enzymes, either on their own or in combination with bacterial inoculants, have been examined as a means of releasing suitable energy substrates. Some small scale studies have indicated beneficial effects on rate of silage fermentation when cell wall degrading enzymes have been added either with or without an inoculant (Merry and Braithwaite, 1987; Henderson, McGinn and Kerr, 1987). Animal production studies, using enzymes alone, have not clearly demonstrated significant responses in animal performance (Kennedy, 1987b; Chamberlain, Thomas and Robertson, 1987). For example, over two dairy cow studies Chamberlain, Thomas and Robertson (1987) reported the yield of milk solids (fat, protein and lactose) to be on average only 2% greater with an enzyme-treated than an untreated silage. However, more recent work by Chamberlain and Robertson (1989) has shown significant milk yield responses to an enzyme as an additive.

Wilting

Wilting of grass prior to ensiling has been recognized for many years as an effective means of improving the fermentation quality of silage, particularly where an effective additive has not been used. However the effects on animal performance have often been less clear although it has often been regarded as a prerequisite for producing silage with the potential to produce high levels of animal performance.

GROWING CATTLE

Wilkins (1984), prior to the commencement of the series of 'Eurowilt' studies, undertook a review of data from 110 production studies and concluded that wilting increased silage DM intake, although the effects were marginal in those studies in which an additive had been used with the unwilted silages. In summarizing the results of the subsequent Eurowilt studies Rhor and Thomas (1984) concluded that wilting increased silage DM intake of growing cattle by 9% but reduced liveweight gain by 5%. More recently Steen (1984) has reviewed the data from 40 comparisons throughout the UK and the Republic of Ireland and concluded that wilting resulted in a mean increase in silage DM intake of 18% and liveweight gain of 4%. It was suggested that the anomaly between the responses in intake and those in animal performance was accentuated if the latter was assessed in terms of carcass gain, as cattle offered wilted silage had lower dressing proportions than those offered unwilted material (O'Kiely and Flynn, 1980). Steen (1984) also reported the results from four recent comparisons of wilted and unwilted silages at this Institute and the mean effects are given in Table 8.8. These data are in line with other data in the literature and show wilting to depress the digestibility of the total diet, increase silage DM intake and depress both liveweight gain and carcass gain by 3% and 7% respectively. Within these studies the effects on carcass composition were also assessed and Steen (1984) concluded that, although animals offered the wilted silages had a marginally lower fat content, it was likely that this reflected the lower energy intake on these diets rather than wilting *per se*.

Very few studies have provided sufficient data to enable the effects of wilting on the efficiency of energy utilization for carcass gain to be assessed. Charmley and Thomas (1987) reported a slaughter study with growing cattle and concluded that, although wilting increased silage DM intake by 6%, and liveweight gain by 55 g/day, because of both the lower energy digestibility of the wilted silage and the lower energy content of the empty body gain of the animals receiving the wilted material, there were no differences between silages in terms of efficiency of utilization of digestible energy for energy retention. These data are in agreement with the calorimetric data of Anderson (1984) using sheep. Steen (1984) also compared the relative digestible energy intakes and metabolizable energy requirements on wilted and unwilted silage treatments, and suggested that there

Table 8.8 COMPARISON OF UNWILTED AND WILTED SILAGES FOR GROWING CATTLE (MEAN OF FOUR STUDIES AT HILLSBOROUGH

	Unwilted	*Wilted*	*SE*	*Response to wilting*
DM at ensiling	163	263	3.33	
Fermentation				
NH$_3$-N (g/kg total N)	62.5	92.8	6.95	
Butyric acid (g/kg DM)	1.7	4.4	1.11	
Animal data				
OMD (total (diets) (g/kg)	780	758	4.1	
Silage DM intake (kg/day)	5.72	6.00	0.066	+5%
Liveweight gain (kg/day)	0.90	0.87	0.024	−3%
Dressing proportion (g/kg)	551	547	1.5	
Carcass gain (kg/day)	0.60	0.56	0.013	−7%

(Steen, 1984)

was no effect of wilting on efficiency of digestible energy utilization for liveweight gain.

LACTATING CATTLE

At the conclusion of the 'Eurowilt' series of studies Rhor and Thomas (1984) concluded that wilting of silage resulted in 4% higher intake by lactating cattle and 2% lower milk output. A series of eight similar studies has been carried out at this Institute to compare wilted and unwilted silages, with both silages receiving an additive at ensiling. Summaries of the effects on silage intake and milk yield recording in these studies are given in Table 8.9. This shows wilting to result in mean increases in silage DM intake of 6% and depressions in milk yield of 3%. Superficially these data would suggest that the DM from wilted silage was utilized less efficiently for milk production than that from unwilted material. However, in the major proportion of these studies DM intakes were not adjusted for differences in alcohol content and, in view of the high alcohol content of some formic acid-treated, unwilted silages, this would have resulted in an overestimation of the differences in DM intakes. In a proportion of these experiments, total ration digestibility and feed utilization studies were also undertaken and these have shown that although silage DM intake was increased following wilting the increase in digestible energy intake was very small, or in some studies negative. This mainly arose because of the higher gross energy contents and digestibilities of the unwilted silages (Unsworth and Gordon, 1985). A series of three energy balance studies with lactating cows were undertaken concurrently with these production studies by Unsworth and colleagues and the data have been reported by Unsworth and Gordon (1985) and Unsworth, Wylie and Anderson (1987). The animal performance data from these studies were directly in line with the previous studies; wilting increased DM intake by 5% but reduced milk output by 5%. In spite of these differences, no effect of wilting on the efficiency of utilization of metabolizable energy for milk production was recorded, the mean efficiencies being 0.56 and 0.55 for the wilted and unwilted silages respectively.

Table 8.9 INFLUENCE OF WILTING PRIOR TO ENSILING ON SILAGE INTAKE AND MILK YIELD OF LACTATING COWS (DATA FROM A SERIES OF STUDIES AT HILLSBOROUGH)

	Silage DM intake (kg/day)			Milk yield (kg/day)		
	Unwilted	Wilted	Response (%)	Unwilted	Wilted	Response (%)
Gordon (1980a)	9.3	9.5	+2	23.6	23.9	+1
Gordon (1980b)	10.7	11.7	+9	28.3	27.0	−5
	10.0	9.6	−4	28.3	26.8	−7
	8.1	9.3	+15	26.4	26.0	−2
Gordon (1981)	9.2	9.8	+6	23.6	21.7	−8
Steen and Gordon (1980)	9.4	9.5	+1	26.1	24.7	−5
Gordon and Peoples (1986)	8.1	8.3	+3	21.3	20.7	−3
Peoples and Gordon (1989)	6.2	7.3	+18	19.0	19.3	+2
		Mean	+6.3			−3.4

TOTAL HARVESTING SYSTEMS

The selection of a particular system of silage production for use on an individual farm must take many factors into account. These include capital and running costs of machinery, power and labour requirements, potential output per hour, degree of chop required for ancillary equipment for ensiling and feeding and the ability to handle silage effluent. However, it must also embrace some assessment of the effect of the system on the output of marketable animal product, both per animal and per hectare of grassland.

Series of dairy cows (Gordon, 1988) and growing cattle (Steen, 1984) studies have been carried out at this Institute to assess the effects of wilting on animal output per hectare. In each of these series multi-harvest systems of silage production, two or three harvests per season from the same area of land have been compared and the systems examined ranged from simple flail-harvesting systems through to precision-chop systems coupled with wilting. The mean effects of the systems on output of animal product per hectare are presented in Table 8.10. These have shown considerable improvements in output of both carcass gain and milk per hectare by the adoption of an unwilting approach (mean response +13%), with a further improvement (+7%) being obtained when a simple flail harvesting system was adopted. These values do not include the animal production which could be produced from the effluent resulting from unwilted silages. In this latter context, Steen (1984) has suggested that, in view of the high nutritive value of effluent DM, this could increase the output of animal product per hectare from unwilting systems by a further 5%.

Table 8.10   EFFECTS OF SYSTEM OF HARVESTING GRASS FOR SILAGE ON THE OUTPUT OF ANIMAL PRODUCT PER HECTARE (EXPRESSED RELATIVE TO AN INDEX VALUE OF 100 FOR PRECISION-CHOPPED, WILTED SILAGE)

	Flail system Direct	*Precision chop/ Unwilted*	*Precision chop/ Wilted*
Finishing cattle (relative carcass gains/ha)			
Expt. 1	–	125	100
2	–	112	100
3	113	112	100
4	99	102	100
Lactating cows (relative milk outputs/ha)			
Expt. 1	118	–	100
2	121	107	100
3	124	111	100
Means	120.3	113.2	100
		SEM 1.58 ***	

(Steen, 1984 and Gordon, 1988)

Conclusions

While there are a wide range of additives available for use in silage production the information available on their effectiveness, in terms of inducing improvements in animal performance, is often extremely limited. It is also clear that where additives rely on producing their beneficial effects through improvements in silage fermentation alone then in many situations they may not be producing positive effects on animal performance. Indeed numerous studies have indicated that when control silages are well preserved these additives will not improve animal performance and, in view of a lack of other positive benefits, would be of negative financial value. This aspect is particularly important as present ensiling techniques have become increasingly effective and can now in many instances achieve good fermentation without additives. There is therefore a greater need for improved methods of making decisions on when such additives should be used.

While the evidence on the beneficial effects of inoculants as aids to preservation of grass is equivocal, recent information suggests that animal production responses may be achieved without any apparent improvement in silage fermentation characteristics. This raises considerable questions with regard to the laboratory evaluations of such silages.

It is now clearly recognized that wilting of silage prior to ensiling, while enhancing DM intake, is more likely to decrease, rather than increase, animal performance. However, there is no evidence to suggest that wilting has any effect on the efficiency of energy utilization by either growing or lactating cattle. Nevertheless, wilting can result in considerable detrimental effects on the output of animal product per hectare and recent studies with beef and dairy cattle would indicate depressions in output of animal product per hectare of around 13% following wilting. Further improvements are also possible through the use of a simple harvesting system.

References

ANDERSON, R. (1984). *Animal Feed Science and Technology,* **12**, 109–118

ANDERSON, R., GRACEY, H.I., KENNEDY, S.J., UNSWORTH, E.F. and STEEN, R.W.J. (1989). *Grass and Forage Science,* (in press)

APPLETON, M. and DONE, D. (1987). In *Proceedings of the 8th Silage Conference, Institute for Grassland and Animal Production, Hurley,* pp. 15–16

BOLSEN, K. (1978). In *Fermentation of Silage – a Review,* pp. 181–200. Ed. McCullough, M.E. National Feed Ingredients Association, Iowa

CHAMBERLAIN, D.G. and ROBERTSON, S. (1989). In *Silage for Milk Production,* pp. 187–195. Ed. Mayne, C.S. Occasional Symposium No. 23, British Grassland Society

CHAMBERLAIN, D.G., THOMAS, P.C. and ROBERTSON, S. (1987). In *Proceedings of the 8th Silage Conference, Institute for Grassland and Animal Production, Hurley,* pp. 31–32

CHARMLEY, D. and THOMAS, C. (1987). *Animal Production,* **45**, 191–203

DONE, D.L. (1986). *Research and Development in Agriculture,* **3**, 83–88

FOX, J.B., BROWN, S.M. and McCULLOUGH, I.I. (1972). *Record of Agricultural Research N. Ireland,* **20**, 45–51

GORDON, F.J. (1980a). *Animal Production,* **30**, 29–37

GORDON, F.J. (1980b). *Animal Production*, **31**, 35–41

GORDON, F.J. (1981). *Animal Production*, **32**, 171–178

GORDON, F.J. (1984). *Journal of Agricultural Science, (Cambridge)*, **102**, 163–179

GORDON, F.J. (1988). In *Nutrition and Lactation in the Dairy Cow*, pp. 355–377. Ed. Garnsworthy, P.C. Butterworths, London

GORDON, F.J. (1989a). *Grass and Forage Science*, (in press)

GORDON, F.J. (1989b). *Grass and Forage Science*, (in press)

GORDON, F.J. and PEOPLES, A.C. (1986). *Animal Production*, **43**, 355–366

HAIGH, P.M., APPLETON, M. and CLENCH, S.F. (1987). *Grass and Forage Science*, **42**, 405–410

HENDERSON, A.R., ANDERSON, D.H., NEILSON, D.R. and HUNTER, E.A. (1987). In *Proceedings of the 8th Silage Conference, Institute for Grassland and Animal Production, Hurley*, pp. 13–14

HENDERSON, A.R., McGINN, R. and KERR, W.D. (1987). In *Proceedings of the 8th Silage Conference, Institute for Grassland and Animal Production, Hurley*, pp. 29–30

KAISER, A.G., TAYLER, J.C., GIBBS, B.G. and ENGLAND, P. (1981). *Journal of Agricultural Science, (Cambridge)*, **97**, 1–11

KENNEDY, S.J. (1987a). *Irish Grassland and Animal Production Association Conference – March 1987*

KENNEDY, S.J. (1987b). In *Proceedings of the 8th Silage Conference, Institute for Grassland and Animal Production, Hurley*, pp. 25–26

KENNEDY, S.J. (1989a). *Grass and Forage Science*, (in press)

KENNEDY, S.J. (1989b). *Grass and Forage Science*, (in press)

KENNEDY, S.J., GRACEY, H.I., UNSWORTH, E.F., STEEN, R.W.J. and ANDERSON, R. (1989). *Grass and Forage Science*, (in press)

McCULLOUGH, I.I. (1984). *Record of Agricultural Research N. Ireland*, **32**, 90–98

McDONALD, P. (1981). In *The Biochemistry of Silage*. John Wiley and Sons, Chichester

McDONALD, P. (1983). *Hannah Research Institute Annual Report*, pp. 59–67

MERRY, R.J. and BRAITHWAITE, G.D. (1987). In *Proceedings of the 8th Silage Conference, Institute for Grassland and Animal Production, Hurley*, pp. 27–28

MURDOCH, J.C. (1961). *Journal of the British Grassland Society*, **16**, 253–259

MURPHY, J.J. (1986). *Irish Grassland and Animal Production Association Journal*, **18**, 50–58

O'KIELY, P. (1988). *Farm and Food Research*, **19**(3), 8–10

O'KIELY, P. and FLYNN, A.V. (1980). *An Foras Taluntais, Animal Production Research Report*, pp. 25–26

O'KIELY, P. and FLYNN, A.V. (1988). *Irish Journal of Agricultural Research*, **27**(2), (in press)

O'KIELY, P., FLYNN, A.V. and POOLE, D.B.R. (1989a). *Irish Journal of Agricultural Research*, (in press)

O'KIELY, P., FLYNN, A.V. and POOLE, D.B.R. (1989b). *Irish Journal of Agricultural Research*, (in press)

O'KIELY, P., FLYNN, A.V., POOLE, D.B.R. and ROGERS, P.A.M. (1989). *Irish Journal of Agricultural Research*, (in press)

PARKER, J.W.G. and CRAWSHAW, R. (1982). *Grass and Forage Science*, **37**, 53–58

PEOPLES, A.C. and GORDON, F.J. (1989). *Animal Production*, **48**, 305–317

RHOR, K. and THOMAS, C. (1984). *Landbauforschung Volkenrode, Sonderheft*, **69**, 64–70

ROOKE, A.J., MAYA, F.M., ARNOLD, J.A. and ARMSTRONG, D.G. (1988). *Grass and Forage Science,* **43**, 87–95

STEEN, R.W.J. (1984). *57th Annual Report, Agricultural Research Institute of N. Ireland*, pp. 21–32

STEEN, R.W.J. and GORDON, F.J. (1980). *Animal Production,* **30**, 341–354

STEEN, R.W.J., UNSWORTH, E.F., GRACEY, H.I., KENNEDY, S.J. and ANDERSON, R. (1989). *Grass and Forage Science*, (in press)

THOMAS, P.C. and MORRISON, I.M. (1982). In *Silage for Milk Production*, pp. 13–38. Ed. Rook, J.A.F. and Thomas, P.C., *Technical Bulletin 2*. National Institute for Research in Dairying and Hannah Research Institute

THOMAS, C. and THOMAS, P.C. (1985). In *Recent Advances in Animal Nutrition – 1985*, pp. 223–256. Ed. Haresign, W. and Cole, D.J.A. Butterworths, London

UNSWORTH, E.F. and GORDON, F.J. (1985). *58th Annual Report, Agricultural Research Institute of N. Ireland*, pp. 13–20

UNSWORTH, E.F., WYLIE, A.R.G. and ANDERSON, R. (1987). In *Proceedings of the 8th Silage Conference, Institute for Grassland and Animal Production, Hurley*, pp. 57–58

VIRTANEN, A.I. (1933). *Journal of Experimental Agriculture,* **1**, 143–145

WALDO, D.R. (1978). In *Fermentation of Silage – a Review*, pp. 120–180. National Feed Ingredients Association, Iowa

WEDDELL, J.R. and YACKIMINIE, D.S. (1988). In *Efficient Beef Production from Grass*, pp. 251–253. Ed. Frame, J., Occasional Symposium No. 22, British Grassland Society

WILKINS, R.J. (1984). *Landbauforschung Volkenrode, Sonderheft,* **69**, 5–12

9

OPTIMIZING COMPOUND FEED USE IN DAIRY COWS WITH HIGH INTAKES OF SILAGE

D. G. CHAMBERLAIN, P. A. MARTIN and S. ROBERTSON
The Hannah Research Institute, Ayr KA6 5HL, UK

Introduction

High-quality grass silage can be given as the sole feed to dairy cows; cows in the first 13 weeks of lactation yielded an average of 21 kg milk/day (Rae *et al.*, 1987). However, constraints on the voluntary intake of silage (see Thomas and Rae, 1988) ensure that the production level is likely to be well below the potential of the cow. It therefore seems unlikely that all-silage diets would be optimal in financial terms. However, since the cost of metabolizable energy (ME) from silage is approximately half that from purchased concentrate feeds (Leaver, 1981), there must be good economic reasons for maximizing the proportion of silage in the diet of dairy cows. Constraints on voluntary intake due to a low energy density can still apply when the diet contains a high proportion of silage. Maximizing the proportion of silage in the diet therefore demands that the silage be of high ME concentration. This is well illustrated in experiments at Crichton Royal Farm (Leaver, 1981). With high-quality forage (11.0 MJ ME/kg DM), maximum ME intake was achieved when forage constituted about 55% of total DM but when the forage was of moderate quality (10.0 MJ ME/kg DM), the corresponding contribution of forage was only 30% of total DM.

The subject of concentrate supplementation of silage for dairy cows has been reviewed recently (Thomas and Rae, 1988). It is proposed here to take a more restricted view and to focus attention on diets containing a minimum of around 60% of the total DM as silage. Consequently, this chapter will be concerned, almost exclusively, with the use of high digestibility (D-value > 67), well-preserved grass silages.

Nutritional limitations of silage-based diets

Thomas and Rae (1988) have reviewed these in some detail. Briefly, they relate to limitations of voluntary intake, amino acid supply and ME supply and utilization. These will be considered in the context of the use of supplements in silage diets containing at least 60% silage in the total DM.

THE EFFECTS OF SUPPLEMENTS ON THE VOLUNTARY INTAKE OF SILAGE DIETS

Usually, when silages offered *ad libitum* are supplemented with concentrates the intake of silage is reduced. The reduction (kg) in the intake of silage DM for each kg increase in concentrate DM is termed the substitution rate. Factors affecting substitution rates in dairy cows given diets based on grass silage have been reviewed by Thomas (1987) who concluded that, in general, substitution rate is greater for high-digestibility silages with high-intake characteristics. Substitution rate is dependent also on the composition of the supplement.

Carbohydrate

Starch can reduce cellulolytic activity in the rumen (Terry, Tilley and Outen, 1969; Mould, Ørskov and Mann, 1983) which, in turn, can reduce the intake of forage. Reductions in the number and activity of rumen cellulolytic bacteria when forage diets are supplemented with starchy concentrates occur as a result of the reduced rumen pH induced by the rapid fermentation of starch, and also because the ready availability of the carbohydrate allows amylolytic species to compete successfully for an increased proportion of the non-carbohydrate nutrients that are essential for growth (Mould and Ørskov, 1983; Mould, Ørskov and Mann, 1983).

Replacing starch-based supplements with those containing digestible fibre, such as sugar beet pulp or highly digestible dried grass, can lead to substantially higher intakes of silage when the level of supplementation is low (Table 9.1) although, for sugar beet pulp, the results are not consistent. Attention has also been given to treatment of cereals to reduce the rate of ruminal degradation of starch in an

Table 9.1 THE EFFECT OF THE CARBOHYDRATE COMPOSITION OF THE SUPPLEMENT ON SILAGE INTAKE

Supplement type	Intake (kg DM/day)		Source
	Supplement	Silage	
Barley	5.4	9.3	Chamberlain *et al.* (1984)
Beet pulp	5.5	10.3	
Barley	5.8	9.4	
Beet pulp	5.1	9.0	
Barley	5.8	7.5	
Beet pulp	5.7	7.8	
Barley	6.1	7.4	Thomas *et al.* (1986)
Beet pulp/rice bran	6.0	8.4	
Barley	3.3	9.7	Castle and Watson (1975)
Dried grass	3.4	10.1	
Barley	4.6	9.7	
Dried grass	4.6	10.3	
Barley	6.0	8.1	
Dried grass	6.3	9.4	
Barley	6.1	7.0	Taylor and Aston (1976)
Dried grass	6.0	8.6	

attempt to reduce the substitution rate. In some experiments, treatment of rolled barley with acid-formaldehyde produced a substantial reduction in silage substitution rate (Kassem *et al.*, 1987). Although information is lacking for silage diets, sodium hydroxide treatment of barley (Ørskov and Fraser, 1975) and oat grain (P. A. Martin, unpublished) reduced the substitution rate of hay when compared with the untreated, rolled cereals.

Lipid

In general, forage intake is depressed when lipid is included in supplements for silage diets (Table 9.2). Unfortunately, few relevant studies have been reported and interpretation of the data is difficult as, within these studies, the basal supplements were altered in differing ways during lipid inclusion. Where lipid has been added directly to a basal supplement, substitution rate for the lipid itself can be calculated; values ranging from 0.5 to 4.1 have been found (Clapperton and Steele, 1983; Steele, 1985). However, the effect of lipid inclusion on the substitution rate of the complete supplement – the figure of more practical importance – cannot be calculated unless silage diets containing the basal supplement and the lipid-enriched supplement have been compared with a corresponding all-silage diet. In experiments of this type, Clapperton and Steele (1983) found substitution rates ranging from 0.36 to 0.90 for supplements containing varying amounts of tallow, but as the composition of the basal concentrate was modified during fat inclusion the significance of these values is unclear.

It is likely that the effects of lipid on silage intake derive from adverse effects on digestion in the rumen; free fats, and especially unsaturated oils, reduce ruminal fibre digestion in sheep (Devendra and Lewis, 1974; McAllan, Knight and Sutton, 1983), although this effect may be diminished in lactating cows (see Palmquist, 1984). It is clear from the data available that responses in silage intake to lipid

Table 9.2 THE EFFECT OF INCLUSIONS OF FREE FAT OR OIL IN SUPPLEMENTARY CONCENTRATES ON SILAGE INTAKE

Lipid inclusion		*Response in silage intake* (kg DM/day)	*Reference*
Type	*Amount* (g/day)		
Tallow	400 (0.11)[a]	−0.2	Clapperton and Steele (1985)
	420 (0.11)[a]	−0.7	
	450 (0.08)[b]	+1.2	Murphy and Gleeson (1979)
	490 (0.08)[c]	+0.5	
	832 (0.12)[c]	−0.8	Murphy and Morgan (1983)
Soyabean oil	320 (0.15)[b]	−1.10	Steele (1985)
	400 (0.11)[a]	−0.73	Clapperton and Steele (1985)
	420 (0.11)[a]	−1.72	
	530 (0.17)[a]	−1.80	Steele (1985)

[a] Given in addition to basal supplement
[b] Substituted for an isoenergetic amount of basal supplement
[c] Substituted for an equivalent weight of basal supplement
The values in parentheses show the proportion of the supplement DM that the added lipid represents

inclusions are influenced by the fatty acid composition of the lipid, the composition of the basal supplement and the form in which the lipid is given.

In line with the observed effects of lipids of differing composition on rumen metabolism, while isoenergetic substitution of 450 g/day of tallow for part of a supplement increased silage intake by an amount so large as to suggest a reduction in substitution rate (Murphy and Gleeson, 1979), isoenergetic substitution of 320 g soyabean oil/day was found to reduce silage consumption despite the reduction in the amount of supplement offered (Steele, 1985). Similarly, although both tallow and soyabean oil reduced silage intake when given in addition to a basal supplement, the effect was greater for soyabean oil which is more unsaturated (Clapperton and Steele, 1985). The results of this study also demonstrate an effect of the composition of the basal supplement; within lipid sources the depression in silage intake was greater when the basal supplement was soyabean meal rather than barley. Interestingly, Steele (1985) also found that reductions in silage intake associated with soyabean oil inclusion could not be prevented by increasing the protein content of the supplement. However, alteration of the carbohydrate portion of the supplement may prove beneficial; by progressively replacing barley with sugar beet pulp and soyabean meal in isoenergetic supplements containing increasing amounts of tallow, Clapperton and Steele (1985) were able to include 749 g tallow/day (12% of the concentrate DM) without adversely affecting silage intake. By contrast, silage intake was depressed when 832 g tallow/day was substituted for an equal weight of a supplement given with a silage of lower digestibility, 61D-value (Murphy and Morgan, 1983).

Modifications to the form in which lipid is presented can improve responses to its inclusion in hay-based diets. For example, when equivalent amounts of soyabean oil are given as either crushed soyabeans or free oil, adverse effects on rumen digestion appear to be smaller than when the soyabeans are given (Banks *et al.*, 1980). However, where silage was offered *ad libitum*, intake was lower with crushed soyabeans than with soyabean oil (Steele, 1985). With respect to the use of 'protected' forms of fat, which are more resistant to degradation in the rumen, adverse effects of tallow on silage intake can be reduced by encapsulating the fat in formaldehyde-treated casein (Murphy and Morgan, 1983). However, this option is currently of limited practical importance given the cost and variability of 'protected' lipid products. Lipid inclusions in the form of calcium soaps or fat prills (Jenkins and Palmquist, 1984; Gummer, 1988) appear to have relatively small effects on rumen digestion but their use in supplements for silage-based diets has not been studied.

Where free lipid is to be included to increase the energy density of supplements for silage diets, it appears that a relatively saturated fat such as tallow is preferable. With the proviso that the composition of the basal supplement is suitable, it appears that tallow can be included as approximately 12% of the supplement DM without depressing silage intake. Where the supplement constitutes 40% of the total DM, this level of tallow inclusion represents 5% of the total ration. This is the highest level of fat inclusion suggested to be practicable by Palmquist and Jenkins (1980) in their review of the use of fats in lactation rations. With respect to the composition of the basal supplement, there may be advantages in terms of silage intake associated with the inclusion of an increased proportion of a fibrous source of carbohydrate, though this may well partially offset the increase in energy density due to lipid inclusion. Increasing the protein content of supplements containing lipids does not appear to improve silage consumption.

Protein

In contrast to carbohydrate and lipid supplements, the feeding of protein concentrates such as groundnut or soya (Castle, 1982) or mixtures of fishmeal and soya (Rae *et al.*, 1986) as the sole supplement, does not reduce silage intake and may even increase it. Furthermore, the inclusion of protein in cereal concentrates reduces substitution rate by an amount that is greater than would be expected simply from the reduction in starch content (Castle, 1982), suggesting that supplementary protein has a stimulatory effect on silage intake. There is evidence that, even for highly digestible grass silages, physical 'rumen fill' mechanisms are important in the control of intake (Thomas and Chamberlain, 1982). Increased intakes of silage in response to protein supplementation could arise from rumen-based effects, whereby a stimulation of microbial activity produces a faster rate of digestion, and/or the absorbed products of protein digestion could lead to a 'resetting' of the digesta load at which 'fill' becomes a limiting factor (see Gill, Rook and Thiago, 1988).

Taking dairy cow diets as a whole, the intake responses to increasing the CP content of the diet are extremely variable (Oldham and Alderman, 1982). Even if consideration is limited to diets based on grass silages, the responses vary widely; differences in digestibility and fermentation quality have been implicated in the cause of the variability (Thomas, 1987). However, even if consideration is further restricted to those diets based on high digestibility (>67D-value), well-preserved grass silages produced at this Institute and offered *ad libitum* with fixed rates of barley-based concentrates such that silage accounts for 60%–70% of the total ration DM, the responses remain variable, albeit less so than the data compiled by Oldham and Alderman (1982) (Table 9.3). Since the Institute studies referred to in Table 9.3 were carried out with protein supplements of different types (groundnut, soya, fishmeal), it might be argued that the variability of response was associated with differences of protein source. There are insufficient data to provide a clear-cut answer, but there is no indication in the data of consistent differences between the protein sources.

Table 9.3 RESPONSES OF INTAKE TO INCREASES IN THE CONTENT OF CRUDE PROTEIN IN THE DIET

	Increase in DM intake (kg)/*unit increase in CP% in the diet*
All rations (Oldham and Alderman, 1982)	0.34 ± 0.07 ($n = 23$) range: -0.24 to 1.35
High D-value silage plus barley-based concentrates (HRI)[a]	$0.29 + 0.04$ ($n = 10$) range: 0.12 to 0.51

[a] Source of data: Castle and Watson (1969); Castle and Watson (1976); Thomas *et al.* (1984); Thomas and Chamberlain, unpublished observations

The mechanism of the protein-linked increase in food intake is being investigated in experiments with dairy cows in which responses to dietary additions of protein are compared with the responses when protein is infused into the abomasum. The results of a preliminary experiment are shown in Table 9.4. These results show clearly that the intake response can be mediated via an increased supply of protein to the abomasum. However, that intake responses can also occur as a result of effects at a ruminal level is suggested by the well-established effects of protein concentration in the diet on the digestibility of the ration (Oldham, 1984).

Table 9.4 THE EFFECT OF PROTEIN SUPPLEMENTATION
VIA ADDITION TO THE DIET OR VIA INFUSION INTO THE
ABOMASUM ON SILAGE INTAKE (FROM A PRELIMINARY
3 × 3 LATIN SQUARE EXPERIMENT)

Treatment	DM intake (kg/day)	
	Silage	Concentrate
Basal diet (140 g CP/kg DM)	7.5	6.0
Basal + 1.2 kg P[a]	8.5	6.0
Basal + 0.6 kg P[b] via abomasum	8.7	6.0

[a] A mixture containing (g/kg DM): 500 fishmeal, 300 bloodmeal and 200
meat and bone meal
[b] Calculated to supply an equivalent amount of protein at the
abomasum, assuming P has a rumen degradability of 0.50
Girdler, Thomas and Chamberlain (unpublished observations)

AMINO ACID SUPPLY FROM SILAGE-BASED DIETS

Silage is a food that has already been exposed to extensive microbial activity and although the extent of the resultant microbial modifications depends on the ensiling conditions (see Thomas and Thomas, 1985), events within the silo have important consequences for the pattern of digestion of silage diets.

After the grass has been cut, and subsequently during storage in the silo, proteolysis due to plant and microbial proteases increases the non-protein nitrogen (NPN) content such that it eventually constitutes 40%–70% of the total nitrogen; only if additives containing formaldehyde are used, will the content of NPN be lower than this. Provided the silage is well preserved, protein breakdown does not proceed beyond the level of free amino acids. During fermentation, sugars are replaced by fermentation products, notably lactic and acetic acids with variable, but usually small, amounts of ethanol.

As might be expected, the 'prefermented' nature of silage has immediate consequences for the microbes in the rumen. For protein synthesis, rumen bacteria require a supply of nitrogen in the form of ammonia, amino acids or peptides, the preferred source varying with species, and a supply of fermentable substrate to provide ATP. Silage NPN and soluble protein are rapidly degraded to ammonia in the rumen (see Thomas and Thomas, 1985) and this may mean that the availability of amino acids and peptides limits protein synthesis. What is more, the rapid release of ammonia means that, for its efficient fixation into bacterial protein, a readily-fermentable source of energy is needed. However, readily-fermentable sugars in the grass have been replaced by fermentation end-products that are of low energy yield to the rumen organisms (see Chamberlain, 1987). With diets containing a high proportion of silage, the combination of high ruminal concentrations of ammonia and a low availability of readily-fermentable substrates results in low rates of microbial growth and substantial loses of nitrogen before the small intestine (Thomas and Thomas, 1985; Chamberlain, Thomas and Quig, 1986; Rooke and Armstrong, 1987).

These events in the rumen have consequences for nutrient supply to the host animal. Not only is there a low supply of total amino acids to the small intestine, in addition, the reduced synthesis of bacterial protein results in low concentrations in duodenal digesta of particular amino acids, notably methionine and lysine (Chamberlain and Thomas, 1980; Thomas *et al.*, 1980; Chamberlain, Thomas and Quig, 1986).

It would therefore be expected that cows given silage diets would be responsive to protein supplementation and Thomas and Rae (1988) concluded that increasing the proportion of protein in the supplement increased milk and milk protein yield, in some cases beyond that expected with dried diets.

There are three strategies that can be employed to overcome the limitations of amino acid supply: (a) protein supplements can be used, with the intention of supplying UDP to the small intestine; (b) supplements of specific amino acids can be used to remove deficiencies in supply of particular amino acids; and (c) steps can be taken to improve microbial utilization of nitrogen in the rumen. Each of these aspects will be discussed in turn.

Protein supplementation

Responses of milk production to protein supplementation are usually accompanied by increases in ME supply arising from increases in silage intake and/or increases in ration digestibility (Oldham, 1984), so complicating the interpretation of the responses. However, that substantial milk production responses can occur in the absence of effects on voluntary intake is illustrated in Table 9.5.

Thomas and Rae (1988) examined the question of whether responses to supplementation varied with the source of protein and concluded that reports in the literature were conflicting. A number of experiments have been carried out in recent years at the Hannah Research Institute in which responses to soyabean meal and to fishmeal or fishmeal-based products have been measured. In an attempt to eliminate variability of responses due to differences in silage characteristics, proportion of silage in the diet and concentrate type, results have been compiled for diets of high-digestibility grass silage and barley-based concentrates where silage made up at least 60% of the total DM (Table 9.6). In view of the variability of the intake response to protein supplements with diets of this type (Table 9.3), it

Table 9.5 MILK PRODUCTION OF COWS GIVEN *AD LIBITUM* ACCESS TO A COMPLETE MIX DIET OF GRASS SILAGE, BARLEY AND SOYABEAN MEAL (680; 290; 30 g/kg DM RESPECTIVELY), CONTAINING 150 g CP/kg DM, AND SUPPLEMENTS OF 1.1 kg DM/day AS BARLEY (B) OR A MIXTURE OF FISHMEAL, BLOODMEAL AND MEAT AND BONEMEAL (500; 300; 200 g/kg DM RESPECTIVELY) (P)

	Supplement	
	B	*P*
DM intake (kg/day)	17.0	17.2
Milk (kg/day)	19.8	23.0
Fat (g/day)	853	949
Protein (g/day)	653	758

Girdler, Thomas and Chamberlain (1988b)

Table 9.6 MILK PRODUCTION RESPONSES OF COWS GIVEN DIETS OF GRASS SILAGE
AND BARLEY-BASED CONCENTRATES CONTAINING SOYABEAN MEAL OR FISHMEAL

Expt	CP (g/kg DM total diet)		Protein source	Increase in yield/increase in CP intake (g/g)		
	Basal	Supplemented		Milk	Milk protein	Milk fat
1	129	154	S	3.8	0.11	0.10
	129	154	F	6.0	0.24	0.05
	129	186	S	1.5	0.08	−0.05
	129	186	F	3.2	0.12	0.11
2	150	186	F[a]	4.4	0.15	0.13
3	145	186	S	3.5	0.14	0.14
	145	186	F[a]	4.5	0.15	0.15
4	170	190	S	3.4	0.15	0.13
	138	161	S	3.3	0.12	0.09
5	121	155	F	2.5	0.08	0.07
	Overall mean		S	3.1 ±0.4	0.12 ±0.01	0.08 ±0.03
			F	4.1 ±0.6	0.15 ±0.03	0.10 ±0.02

[a] a 'fishmeal blend' containing (g/kg DM): 500 fishmeal, 300 bloodmeal and 200 meat and bone meal
1 from Thomas and Chamberlain (unpublished observations)
2 and 3 from Girdler, Thomas and Chamberlain (1988)
4 from Thomas *et al.* (1984)
5 from Kassem *et al.* (1987)

is not surprising that the milk production responses are variable. The responses to
soya are similar to those reported by Gordon (1979) for diets containing about 50%
of the DM as silage. The data in Table 9.6 show a general tendency for bigger
responses to fishmeal than to soya.

The pattern of response of milk production to level of protein supplementation
has been examined in three experiments in which cows were given *ad libitum* access
to complete-mix diets supplemented with increasing amounts of protein (Table
9.7). Substantial milk production responses were seen with all three basal diets.
The pattern of response of milk fat output was particularly variable. However, the
responses of milk protein output to increases in CP intake were reasonably close to
linear if the value for the highest level of supplementation with diet A (for which
DM intake was depressed) is excluded (Figure 9.1); furthermore, the average
response, at about 0.13 g milk CP/g increase in CP intake, is very similar to the
values in Table 9.6.

Amino acid supplements

Whether the bigger responses to fishmeal than to soya that are seen in some
experiments relate to differences in their UDP contents or to differences in their
amino acid composition is not known. Interpretation of the results of production

Table 9.7 MILK PRODUCTION RESPONSES TO INCREASING LEVELS OF PROTEIN SUPPLEMENTATION IN COWS GIVEN *AD LIBITUM* ACCESS TO COMPLETE-MIX DIETS

Expt	Amount of protein supplement[a] (kg/day)					
	0	*0.8*	*1.2*	*1.6*	*2.4*	*SED*
(1) Basal diet A						
DM intake (kg/day)	16.5	17.4	16.7	18.1	17.5	0.5
Milk yield (kg/day)	22.4	25.2	24.7	26.1	25.3	0.7
Fat (g/day)	876	964	995	926	951	54
Protein (g/day)	681	788	780	823	813	28
(2) Basal diet B						
DM intake (kg/day)	15.0	16.1	15.4	16.5	17.0	0.4
Milk yield (kg/day)	21.8	23.8	24.0	24.9	25.9	1.0
Fat (g/day)	808	882	824	968	949	61
Protein (g/day)	604	681	698	725	779	26
	0	0.3	0.6	0.9	1.2	
(3) Basal diet C						
DM intake (kg/day)	12.2	12.9	13.1	13.3	13.6	0.2
Milk yield (kg/day)	18.0	19.1	19.5	19.9	20.1	0.3
Fat (g/day)	661	663	663	697	720	20
Protein (g/day)	517	542	564	609	621	14

Basal diet A: Silage and barley-soya (60:40 on a DM basis) containing 140 g CP/kg DM
Basal diet B: Silage and dried grass (60:40 on a DM basis) containing 140 g CP/kg DM
Basal diet A: Silage and barley-soya (80:20 on a DM basis) containing 140 g CP/kg DM
Expts 1 and 2 from Thomas *et al*. (1987)
Expt 3 from Chamberlain, Robertson and Choung (unpublished observations)
[a] a protein mix containing 500 g fishmeal, 300 g bloodmeal and 200 g meat and bone meal/kg DM

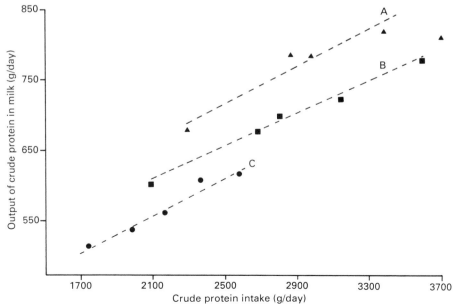

Figure 9.1 Relation of crude protein output in milk to crude protein intake in three complete-mix diets supplemented with protein (see Table 9.7 for details of diets A, B and C)

trials in terms of protein supply to the small intestine is hindered by the knowledge that, for both fishmeal and soya, rumen degradability varies, not only with source but apparently, also with the level of inclusion in the diet (Gill and Beever, 1982; Rooke, Alvarez and Armstrong, 1986; Rooke and Armstrong, 1987).

Fishmeal supplementation of grass silage diets has been associated with a more efficient transfer of duodenal non-ammonia CP into milk protein than was the case for supplementation with soya (Oldham *et al.*, 1985) suggesting an effect linked to a superior amino acid balance of fishmeal. Soya and fishmeal differ in their content of essential amino acids, fishmeal being richer in lysine and especially methionine. However, where a fishmeal supplement induced a bigger milk production response than an equivalent amount of protein as soya, adding rumen-protected methionine and lysine to the soya-supplemented diet did not increase milk yield (Girdler, Thomas and Chamberlain, 1988a).

In cows in late lactation, receiving a diet of grass silage alone and yielding about 5 kg milk/day, Rogers, Bryant and McLeay (1979) reported an increase in milk protein yield of 13% in response to an abomasal infusion of methionine. A series of experiments has been undertaken at the Hannah Research Institute to examine the responses to supplementary methionine and combinations of methionine and lysine. In these studies, the amino acids have been added to the diet in a rumen-protected form or infused, intravenously or intra-abomasally. In summary, milk production responses have been limited to inconsistent responses in milk fat content sometimes, but not always, accompanied by increases in fat yield. Except in one experiment (Girdler *et al.*, 1986), supplementation of silage diets with methionine and lysine has been singularly unsuccessful in promoting increases in either milk yield or milk protein yield (Chamberlain and Thomas, 1982; Shamoon, 1984; Girdler, Thomas and Chamberlain, 1988a and b; Thomas and Chamberlain, unpublished observations). In keeping with these observations, cows given similar diets did not respond to supplements of methionine hydroxy analogue (Gordon and Unsworth, 1986).

Improving microbial utilization of nitrogen in the rumen

As described earlier, the efficient incorporation of silage nitrogen into microbial protein requires a supplement supplying readily fermentable substrate. The traditional choice of 'energy' supplement for silage is barley. However, it is clear that the responses to barley supplements in terms of avoiding losses of nitrogen across the rumen can be disappointing. Even with modest concentrations of CP in the silage, losses of nitrogen across the rumen are not prevented unless barley accounts for around 50% of the total diet DM (Table 9.8).

From a series of experiments, Chamberlain *et al.* (1985) concluded that there were marked differences between carbohydrate sources in their effects on nitrogen metabolism in the rumen. Starch was relatively ineffective in reducing the high ruminal concentrations of ammonia typical of silage diets when compared with isocaloric amounts of sugars. Furthermore, the suggestion from these experiments was that the differences between starch and sugars were related to the different effects of carbohydrates on the numbers of rumen protozoa; starch, by increasing protozoa numbers markedly, probably stimulated wasteful intraruminal recycling of nitrogen, so limiting its effect on the efficiency of microbial synthesis.

The effectiveness of sugar as a supplement to silage has been further

Table 9.8 RATIO OF DUODENAL NON-AMMONIA NITROGEN (NAN) NITROGEN (NI) INTAKE IN COWS GIVEN DIETS OF GRASS SILAGE SUPPLEMENTED WITH BARLEY

Ratio silage:barley (DM basis)	NAN/NI (g/g)	Source
65:25[a]	0.79	Chamberlain, Thomas and Quig (1986)
60:40[b]	0.90	Rooke and Armstrong (1987)
50:50[a]	1.07	Brett *et al.* (1979)
50:60[a]	1.06	Rooke *et al.* (1987)

[a] Silages contained approx. 150 g CP/kg DM
[b] Silage contained 128 g CP/kg DM

demonstrated by the study of Rooke, Lee and Armstrong (1987) in which silage was supplemented with glucose syrup, urea, casein or casein plus glucose syrup (Table 9.9). It is worth noting that the marked beneficial effects of glucose syrup on ruminal metabolism of nitrogen (NAN supply to the duodenum increased by 19%) were achieved with a sugar supplement accounting for only 16% of the total diet DM. The apparent synergism between glucose syrup and casein supplements (Table 9.9) may reflect a shortage of preformed amino acids and/or peptides when silage is the only source of dietary protein. However, reports of the effects of protein supplementation of silage diets on rumen microbial protein synthesis are contradictory. The efficiency of microbial protein synthesis can be improved by soyabean inclusion (Brett *et al.*, 1979; Rooke *et al.*, 1985) but the effects may depend on the level of inclusion of soya (Rooke, Alvarez and Armstrong, 1986). On the other hand, fishmeal inclusion did not increase the efficiency of microbial protein synthesis with diets of silage alone (Bever *et al.*, 1987) or silage and barley (Rooke and Armstrong, 1987).

A potential drawback to the use of sugar supplements is that their rapid rate of fermentation can induce a low rumen pH which, in turn, can restrict microbial growth (Russell, Sharp and Baldwin, 1979). Sodium bicarbonate in combination with sucrose prevented the low rumen pH induced by sucrose alone and markedly improved the effectiveness of the sugar supplement in reducing ammonia concentrations (Chamberlain *et al.*, 1985). Subsequently, beneficial responses of

Table 9.9 THE EFFECT OF INTRARUMINAL INFUSIONS OF CASEIN, UREA AND GLUCOSE SYRUP ON RUMINAL DIGESTION IN COWS GIVEN A DIET OF GRASS SILAGE

	Infusate				
	Water	Casein	Urea	Glucose	Casein + glucose
CP in diet[a] (g/kg DM)	115	137	139	98	114
Duodenal NAN/N intake (g/g)	0.95	0.81	0.76	1.23	1.35
Microbial N at the duodenum (g/day)	63	75	68	81	109

[a] Including infused nutrients
(from Rooke, Lee and Armstrong, 1987)

Table 9.10   THE EFFECTS OF DIETARY ADDITIONS OF SODIUM
BICARBONATE ON RUMEN DIGESTION IN SHEEP GIVEN A DIET
OF GRASS SILAGE AND A SUPPLEMENT OF BARLEY AND
SUCROSE (60:20:20 IN THE DM)

	Sodium bicarbonate (g/day)			
	0	25	50	75
Nitrogen intake (g/day)	20.1	20.1	20.1	20.1
Rumen NH_3-N (mg/l)	170	152	143	125
Rumen liquid clearance rate (proportion of volume/h)	0.08	0.09	0.09	0.11
Rumen pH	6.1	6.2	6.4	6.4
Duodenal NAN (g/day)	19.8	19.7	22.4	21.2

Newbold, Thomas and Chamberlain (1988)

duodenal nitrogen flow to dietary additions of bicarbonate were demonstrated by
Newbold, Thomas and Chamberlain (1988) (Table 9.10).

The use of sugars and bicarbonate as a supplement to silage has been examined
in a dairy cow production trial (Table 9.11). The response to the molasses/
bicarbonate treatment appears disappointing. However, it should be noted that
such a relatively small inclusion of sugars could have only a limited effect on the
utilization of nitrogen in the rumen. Clearly, the amount of sugar required will
depend on the nitrogen content of the silage. The substantial responses to glucose
supplementation obtained by Rooke, Lee and Armstrong (1987), with a silage of
relatively low protein content, were achieved when sugar accounted for 16% of the
dietary DM. In the study of Rae *et al.* (1987) (Table 9.11), with a high-protein
silage, sugar accounted for less than 10% of the diet DM.

Table 9.11   THE EFFECT OF SUGAR, SODIUM BICARBONATE
AND PROTEIN SUPPLEMENTS ON MILK PRODUCTION OF COWS
OFFERED GRASS SILAGE *AD LIBITUM*

Supplement (kg DM/day)	*Silage intake* (kg DM/day)	*Milk yield* (kg/day)	*Yield* (g/day)	
			Fat	*Protein*
Cereal concentrate (4.3)	11.1	21.6	901	670
Fishmeal (0.4) Soyabean meal (1.2)	12.2	20.4	759	667
Molasses (1.6) Bicarbonate (0.4)	12.0	17.8	728	566
Fishmeal (0.4) Soyabean meal (1.2) Molasses (1.6) Bicarbonate (0.4)	12.4	22.5	860	729

Rae, Thomas and Reeve (unpublished observations; cited by Thomas and Rae,
1988)

Further milk production trials are required to establish the potential of sugar supplements. If molasses is the sugar source to be used, it can be calculated that an intake of about 4 kg/day of molasses would probably be necessary to obtain measurable responses of milk production from an improved supply of microbial protein to the small intestine. Dairy cows receiving high quality grass silage readily consumed in excess of this amount of molasses with no apparent ill effects (Chamberlain and Robertson, unpublished observations).

SUPPLY AND UTILIZATION OF ME FROM SILAGE-BASED DIETS

Energy supply can limit the efficiency with which milk is produced from diets containing a high proportion of silage in a number of interrelated ways. The energy density of the diet may restrict ME intake, and hence milk production, to levels below the cow's maximum, such that maintenance requirements represent a relatively large proportion of energy expenditure and potential gross efficiency is not achieved. Also, the ME available above maintenance may be partitioned towards body tissue gains rather than milk production and/or the energy that is partitioned towards milk production may be used inefficiently. Given these limitations, supplements for high-silage diets should be designed to increase dietary energy supply and to promote efficient utilization of energy for milk production.

Increasing the energy supply

One approach to increasing energy supply is to increase the energy density of the diet directly by addition of high-energy supplements containing lipid. As the energy value of lipid is approximately 2.25 times that of carbohydrate or protein, the inclusion of fats or oils increases the energy density of supplements, though whether this translates to an increase in energy intake depends on the effects of the lipid-rich supplement on silage intake (Table 9.2) and any effects of the added lipid on the digestibility of the remaining dietary components. In the absence of determined ME values for a range of fat-supplemented diets based on silage, it is difficult to assess the importance of the associative effects of different types and levels of lipid inclusion on dietary ME supply. However, on the basis of calculated ME values, energy intake is consistently increased when tallow is given, despite reductions in silage consumption (Table 9.12). Conversely, energy intake is generally reduced when soyabean oil is included as this is associated with relatively large reductions in silage intake.

Alternatively, energy density can be increased indirectly by the use of supplements that increase the digestibility of the diet. The most well-documented examples of effects of supplements on digestibility are those of supplementary protein (Oldham, 1984). Indeed, it can be calculated (see Oldham, 1984) that a substantial part, if not all, of the reported milk production responses to protein supplements can be explained by increases in ME supply arising from increases in ration digestibility and voluntary intake.

Utilization of ME for milk production

Calorimetric determinations of the efficiency of utilization of silage ME for maintenance (k_m) and for fattening (k_f) have shown that the values tend to be lower

Table 9.12   THE EFFECT OF LIPID INCLUSIONS ON CALCULATED
ME INTAKE FROM SILAGE-BASED DIETS

Lipid inclusion[a]		Change in ME intake (MJ/day)
Type	*Amount* (g/day)	
Tallow	400	+13.0
	420	+9.2
	450	+12.2
	490	+12.0
	749[b]	+0.6
	832	+7.6
	832[c]	+10.5
Soyabean oil	320	−13.4
	400	+7.0
	420	−3.5
	530	−3.5

[a] See Table 9.2 for references
[b] Clapperton and Steel (1983)
[c] Added in a protected form. Murphy and Morgan (1983)

than those predicted from equations of ARC (1980) (see Thomas and Thomas, 1985). There are very few data from calorimetric studies with dairy cows given silage diets but for diets containing a high proportion of grass silage, the efficiency of utilization of ME for lactation (k_l) was about 85% of that predicted from ARC (1980) (Unsworth and Gordon, 1985). Calculations of k_{l0} (efficiency of ME utilization at zero bodyweight change) from feeding trials are open to considerable error but calculations for diets containing high proportions of silage produced values of 39% to 64% with higher values being found when supplementary protein was given (Thomas and Castle, 1978). Similar calculations can be performed on the data from the three production trials shown in Table 9.7. Again, the calculations show consistent increases in k_{l0} in response to protein supplementation (Table 9.13). These apparent effects on k_{l0} need to be confirmed in detailed calorimetric studies but they are a further indication that milk production responses to protein supplementation derive from effects of protein/amino acid supply on metabolism as a whole (Oldham, 1984).

There are theoretical reasons, supported by studies where 'protected' lipids have been given, to suppose that dietary lipid inclusions can improve the efficiency with which energy is utilized for milk production (a) by improving the balance of nutrients available and (b) by increasing the amount of dietary long-chain fatty acids transferred into milk fat (see Palmquist, 1984; Smith, 1988). However, the increased yields of milk and milk fat associated with tallow inclusion in silage-based diets (Table 9.14) appear to be attributable simply to increases in energy supply (Table 9.12). On the basis of the limited number of k_{l0} values calculated from these studies it appears that, in general, efficiency of energy utilization was virtually unchanged by tallow inclusion and was, in some cases, reduced when supplementary soyabean oil was given. There is clearly a need for more information regarding the effects of lipids on energy supply and energy utilization if the energy-dense nature of lipids is to be exploited to the best advantage in silage-based diets.

Table 9.13 THE EFFECT OF PROTEIN SUPPLEMENTATION OF COMPLETE MIX DIETS[a] ON THE CALCULATED EFFICIENCY OF UTILIZATION OF ME FOR LACTATION (k_{10})

	Amount of protein supplement (kg/day)				
	0	*0.8*	*1.2*	*1.6*	*2.4*
Basal diet A[b]	0.47	0.53	0.55	0.53	0.56
k_{10}		(0.51)	(0.53)	(0.51)	(0.52)
Basal diet B	0.55	0.65	0.62	0.61	0.66
k_{10}		(0.63)	(0.60)	(0.58)	(0.61)
	0	0.3	0.6	0.9	1.2
Basal diet C	0.52	0.57	0.57	0.62	0.57
k_{10}		(0.56)	(0.55)	(0.60)	(0.54)

[a] See Table 9.7 for details of the diets
[b] ME contents of silages calculated from DOMD $\times$ 0.16
The energy value of milk and the ME for gain or loss of body weight were taken from MAFF (1984)
The values in parentheses show the k_{10} values after allowing for increases in ME supply from increases in digestibility of the diet (values from Gordon, Unsworth and Peoples, 1981)

Table 9.14 THE EFFECT OF LIPID INCLUSIONS ON CALCULATED EFFICIENCY OF UTILIZATION OF ME FOR LACTATION (k_{10})

Lipid inclusion[a]		*Response*			
Type	*Amount* (g/day)	*Milk yield* (kg/day)	*Fat yield* (g/day)	*Protein yield* (g/day)	k_{10}
Tallow	400	+2.0	+109	+45	0.55→0.50
	420	+0.5	+61	−10	0.84→0.64
	450	−0.3	−22	−13	0.57→0.51
	490	+1.1	+55	+28	0.57→0.50
	749	+1.1	+117	+7	0.45→0.48
	832	+0.7	+19	−15	0.62→0.59
	832[b]	+1.6	+159	+25	0.62→0.65
Soyabean oil	320	−2.3	−88	−84	0.76→0.71
	400	+1.3	+10	+19	0.55→0.57
	420	+0.7	−23	−8	0.84→0.64
	530	−1.3	−99	−41	0.76→0.62

[a] see Table 9.12 for references
[b] Added in a protected form

Conclusions

High-quality silage is a prerequisite if limitations of energy intake are not to restrict milk yields to unacceptable levels. Indeed, the primary objective with diets containing a high proportion of silage must be to maximize dry matter intake since with these diets, energy intake is likely to be constrained below maximum.

The formulation of the supplement should aim to minimize the substitution rate of silage whilst providing substrates that encourage efficient microbial utilization of silage nitrogen in the rumen. From the standpoint of ruminal nitrogen metabolism, this argues for a low starch content and a high sugar content; depending on the level of sugar inclusion, bicarbonate (or other suitable buffer) may be beneficial. However, more information is needed on the effects of sugar inclusion on substitution rate of silage.

Supplements containing added fat can be expected to increase milk production. As well as increasing the energy density of the diet, and hence ME intake, lipid inclusion can also inhibit rumen protozoa (Newbold and Chamberlain, 1988) and thereby may improve microbial protein synthesis. Successful exploitation of this dual effect of lipid inclusion needs further research since clearly, there will be a need to optimize the effects of the level of inclusion and the fatty acid composition of the lipid on silage intake and microbial metabolism.

Production trials show the protein level in the supplement to be particularly important, although the precise details of the way in which it exerts its effects have yet to be uncovered. The extent of involvement of effects on voluntary intake and ration digestibility, i.e. energy supply, and effects due to improved amino acid supply *per se* varies with dietary circumstances but, for whatever reason, milk production responses can be expected up to concentrations of 180–190 g CP/kg DM in the total diet. The economics of these responses depend on the changes in the yields of milk fat and protein and on the cost of the supplementary protein. For diets of this type, the data available suggest that, at equivalent levels of CP, responses to soya are about 75% of those obtained with high-UDP sources containing fishmeal. What remains to be demonstrated is to what extent the responses to protein supplementation can be reduced by enhanced flows of microbial protein from the rumen when a better-balanced supply of carbohydrates is provided in the supplement.

In reviewing the information available, the approach taken has been one of confining consideration to a particular diet type in the hope that this would improve the predictability of responses to changes in composition of the supplement. In some respects, this has proved to be the case but it is also only too clear that, in many instances, despite operating within this relatively well-defined diet type, responses of milk production in terms of individual milk constituents, especially milk fat, fall short of an acceptable degree of predictability.

References

AGRICULTURAL RESEARCH COUNCIL (1980). *The Nutrient Requirements of Farm Livestock*. Commonwealth Agricultural Bureaux, Farnham Royal, Slough
BANKS, W., CLAPPERTON, J.L., KELLY, M.E., WILSON, A.G. and CRAWFORD, R.J.M. (1980). *Journal of the Science of Food and Agriculture*, **31**, 368–374

D. G. Chamberlain, P. A. Martin and S. Robertson 191

BEEVER, D.E., GILL, E.M., EVANS, R.T., GALE, D.L. and WILTON, J.C. (1987). *Proceedings of the Nutrition Society*, **46**, 38A

BRETT, P.A., ALMOND, M., HARRISON, D.G., ROWLINSON, P., ROOKE, J.A. and ARMSTRONG, D.G. (1979). *Proceedings of the Nutrition Society*, **38**, 148A

CASTLE, M.E. (1982). In *Silage for Milk Production*, pp. 127–150. Ed. Rook, J.A.F. and Thomas, P.C. Technical Bulletin No. 2, NIRD, Reading

CASTLE, M.E. and WATSON, J.N. (1969). *Journal of the British Grassland Society*, **24**, 187–192

CASTLE, M.E. and WATSON, J.N. (1975). *Journal of the British Grassland Society*, **30**, 217–222

CASTLE, M.E. and WATSON, J.N. (1976). *Journal of the British Grassland Society*, **31**, 191–195

CHAMBERLAIN, D.G. (1987). *Process Biochemistry*, **22**, 60–63

CHAMBERLAIN, D.G. and THOMAS, P.C. (1980). In *Proceedings of the 3rd Symposium on Protein Metabolism and Nutrition*, pp. 422–431. Ed. Oslage, H.J. and Rohr, K. Information Centre of Bundesförschungsanstalt für Landwirtschaft, Braunschweig

CHAMBERLAIN, D.G. and THOMAS, P.C. (1982). *Journal of Dairy Science Research*, **49**, 25–28

CHAMBERLAIN, D.G., THOMAS, P.C. and QUIG, J. (1986). *Grass and Forage Science*, **41**, 31–38

CHAMBERLAIN, D.G., THOMAS, P.C., WILSON, W., NEWBOLD, C.J. and MacDONALD, J.C. (1985). *Journal of Agricultural Science, Cambridge*, **104**, 331–340

CLAPPERTON, J.L. and STEELE, W. (1983). *Journal of Dairy Science*, **66**, 1032–1038

CLAPPERTON, J.L. and STEELE, W. (1985). *Journal of Dairy Science*, **68**, 2908–2913

DEVENDRA, C. and LEWIS, D. (1974). *Animal Production*, **19**, 67–76

GILL, M. and BEEVER, D.E. (1982). *British Journal of Nutrition*, **48**, 37–48

GILL, M., ROOK, A.J. and THIAGO, L.R.S. (1988). In *Nutrition and Lactation in the Dairy Cow*, pp. 262–279. Ed. Garnsworthy, P.C. Butterworths, London

GIRDLER, C.P., THOMAS, P.C. and CHAMBERLAIN, D.G. (1988a). *Proceedings of the Nutrition Society*, **47**, 50A

GIRDLER, C.P., THOMAS, P.C. and CHAMBERLAIN, D.G. (1988b). *Proceedings of the Nutrition Society*, **47**, 82A

GORDON, F.J. (1979). *Animal Production*, **28**, 183–189

GORDON, F.J. and UNSWORTH, E.F. (1986). *Grass and Forage Science*, **41**, 1–8

GORDON, F.J., UNSWORTH, E.F. and PEOPLES, A.C. (1981). *54th Annual Report of the Agricultural Research Institute of Northern Ireland*, pp. 13–23. HMSO, Belfast

GRUMMER, R.R. (1988). *Journal of Dairy Science*, **71**, 117–123

JENKINS, T.C. and PALMQUIST, D.L. (1984). *Journal of Dairy Science*, **67**, 978–986

KASSEM, M.M., THOMAS, P.C., CHAMBERLAIN, D.G. and ROBERTSON, S. (1987). *Grass and Forage Science*, **42**, 175–183

LEAVER, J.D. (1981). In *Recent Advances in Animal Nutrition – 1981*, pp. 71–80. Ed. Haresign, W. Butterworths, London

McALLAN, A.B., KNIGHT, R. and SUTTON, J.D. (1983). *British Journal of Nutrition*, **49**, 433–440

MINISTRY OF AGRICULTURE, FISHERIES AND FOOD (1984). *Energy Allowances and Feeding Systems for Ruminants*. Technical Bulletin 433. HMSO, London

MOULD, F.L. and ØRSKOV, E.R. (1983). *Animal Feed Science and Technology*, **10**, 1–14

MOULD, F.L., ØRSKOV, E.R. and MANN, S.O. (1983). *Animal Feed Science and Technology*, **10**, 15–30

MURPHY, J.J. and GLEESON, P.A. (1979). *Irish Journal of Agricultural Research*, **18**, 245–251

MURPHY, J.J. and MORGAN, D.J. (1983). *Animal Production*, **37**, 203–210

NEWBOLD, C.J. and CHAMBERLAIN, D.G. (1989). *Proceedings of the Nutrition Society* (in press)

NEWBOLD, C.J., THOMAS, P.C. and CHAMBERLAIN, D.G. (1988). *Journal of Agricultural Science, Cambridge*, **110**, 383–386

OLDHAM, J.D. (1984). *Journal of Dairy Science*, **67**, 1090–1114

OLDHAM, J.D. and ALDERMAN, G. (1982). In *Protein and Energy Supply for High Production of Milk and Meat*, pp. 33–63. Pergamon Press, Oxford

OLDHAM, J.D., PHIPPS, R.H., FULFORD, R.J., NAPPER, D.J., THOMAS, J. and WELLER, R.F. (1985). *Animal Production*, **40**, 519

ØRSKOV, E.R. and FRASER, C. (1975). *British Journal of Nutrition*, **34**, 493–500

PALMQUIST, D.L. (1984). In *Fats in Animal Nutrition*, pp. 357–381. Ed. Wiseman, J. Butterworths, London

PALMQUIST, D.L. and JENKINS, T.C. (1980). *Journal of Dairy Science*, **63**, 1–14

RAE, R.C., GOLIGHTLY, A.J., MARSHALL, D.R. and THOMAS, C. (1986). *Animal Production*, **42**, 435 (abstract)

RAE, R.C., THOMAS, C., REEVE, A., GOLIGHTLY, A.J., HODSON, R.G. and BAKER, R.D. (1987). *Grass and Forage Science*, **42**, 249–257

ROGERS, G.L., BRYANT, A.M. and McLEAY, L.M. (1979). *New Zealand Journal of Agricultural Research*, **22**, 533–541

ROOKE, J.A. and ARMSTRONG, D.G. (1987). *Journal of Agricultural Science, Cambridge*, **109**, 261–272

ROOKE, J.A., ALVAREZ, P. and ARMSTRONG, D.G. (1986). *Journal of Agricultural Science, Cambridge*, **107**, 263–272

ROOKE, J.A., LEE, N.H. and ARMSTRONG, D.G. (1987). *British Journal of Nutrition*, **57**, 89–98

ROOKE, J.A., BRETT, P., OVEREND, M.A. and ARMSTRONG, D.G. (1985). *Animal Feed Science and Technology*, **13**, 255–267

RUSSELL, J.B., SHARP, W.M. and BALDWIN, R.L. (1979). *Journal of Animal Science*, **48**, 251–255

SHAMOON, S.A. (1984). Amino acid supplements for ruminant farm livestock with special reference to methionine. PhD Thesis, University of Glasgow, UK

SMITH, N.E. (1988). In *Nutrition and Lactation in the Dairy Cow*, pp. 216–231. Ed. Garnsworthy, P.C. Butterworths, London

STEELE, W. (1985). *Journal of Dairy Science*, **68**, 1409–1415

TAYLOR, J.C. and ASTON, K. (1976). *Animal Production*, **23**, 197–209

TERRY, R.A., TILLEY, J.M.A. and OUTEN, G.E. (1969). *Journal of the Science of Food and Agriculture*, **20**, 317–320

THOMAS, C. (1987). In *Recent Advances in Animal Nutrition – 1987*, pp. 205–218. Ed. Haresign, W. Butterworths, London

THOMAS, C. and RAE, R.C. (1988). In *Nutrition and Lactation in the Dairy Cow*, pp. 327–354. Ed. Garnsworthy, P.C. Butterworths, London

THOMAS, C. and THOMAS, P.C. (1985). In *Advances in Animal Nutrition*, pp. 223–256. Ed. Haresign, W. Butterworths, London

THOMAS, C., ASTON, K., DALEY, S.R. and BASS, J. (1986). *Animal Production*, **42**, 315–325

THOMAS, P.C. and CASTLE, M.E. (1978). Annual Report of the Hannah Research Institute, pp. 108–117

THOMAS, P.C. and CHAMBERLAIN, D.G. (1982). In *Silage for Milk Production*, pp. 63–102. Ed. Rook, J.A.F. and Thomas, P.C. NIRD, Technical Bulletin No. 2, Reading

THOMAS, P.C., CHAMBERLAIN, D.G., CHOUNG, J.J. and ROBERTSON, S. (1987). *Proceedings of the 8th Silage Conference*, Hurley, pp. 169–170. Institute for Grassland and Animal Production, Hurley

THOMAS, P.C., CHAMBERLAIN, D.G., KELLY, N.C. and WAIT, M.K. (1980). *British Journal of Nutrition*, **43**, 469–479

THOMAS, P.C., CHAMBERLAIN, D.G., ROBERTSON, S., SHAMOON, S.A. and WATSON, J.N. (1984). *Proceedings of the 7th Silage Conference*, Belfast, pp. 45–46. Ed. Gordon, F.J. and Unsworth, E.F. Queen's University, Belfast

UNSWORTH, E.F. and GORDON, F.J. (1985). *58th Annual Report of the Agricultural Research Institute of Northern Ireland*, pp. 13–20. HMSO, Belfast

10

NUTRITION OF INTENSIVELY REARED LAMBS

R. JONES
NAC Sheep Unit, Stoneleigh, Warks

R. KNIGHT
BP Nutrition (UK) Ltd, Wincham, Cheshire

A. WHITE
Avondale Veterinary Group, Warwick, UK

Introduction

The seasonal breeding pattern of native ewe breeds in the UK has led traditionally to a glut of lambs reaching the market place during the summer and early autumn. This has been reflected in the pattern of market prices during the season, with prices being lowest during the period July to October gradually increasing to reach a peak during March and April. In order to exploit these price differences many producers have been looking at systems of producing lambs to reach the market at this time of shortage. The most common breed used to date has been the Dorset Horn or its crosses, particularly the Finn-Dorset, which have a much longer breeding season than most of our other native breeds. But recent advances in oestrus manipulation has led to many other breeds being induced to lamb out of their normal season.

Early lambing (November–January) systems differ from the more traditional grass-based summer production in three important aspects – lower lambing percentage, higher feed costs and high stocking rates. Gross margins per ewe from the two systems are normally similar and profitability is dictated by achieving high stocking rates of dry ewes and high growth rates of lambs to reach slaughter weights during the targeted marketing period of January to May.

Apart from some areas of the south-west where grass-based systems are adopted, out of season lamb production has involved some form of housing for both ewes and lambs. Lambs are creep fed and weaned at an early age, usually 6 weeks, before being finished indoors on rations which are both palatable and give high growth rates.

The rations used for intensively fed lambs need to be correctly balanced for energy, protein and minerals and composition of the diet needs to stimulate the development of the rumen.

Energy nutrition

ENERGY DIGESTION AND EFFICIENCY OF UTILIZATION

The molar proportions of the major volatile fatty acids (VFA) in rumen liquor – acetic, propionic and butyric acids – produced by fermentation, vary with the composition of the diet. In general terms the acetate:propionate ratio tends to increase as dietary fibre content increases and vice versa. The efficiency of utilization of metabolizable energy for fattening (k_f) has been traditionally regarded as being negatively related to acetate proportion in rumen liquor.

In recent studies with lambs sustained by intragastric infusion of VFAs, Ørskov *et al.* (1979) measured k_f values for a range of VFA mixtures (Table 10.1).

Table 10.1 MOLAR PROPORTIONS OF VOLATILE FATTY ACIDS (%) IN LAMBS SUSTAINED BY INTRAGASTRIC INFUSION AND THE EFFICIENCY OF ENERGY UTILIZATION FOR GROWTH

	Molar proportion					
Acetic acid	35	45	55	65	75	85
Propionic acid	55	45	35	25	15	5
Butyric acid	10	10	10	10	10	10
k_f	0.78	0.64	0.57	0.61	0.61	0.59

Only at what in practical terms might be considered an unusually high level of propionate was k_f significantly higher than values observed with other mixtures of VFAs. For mixtures which are likely to be encountered with practical rations, differences in k_f values were small and of no significance.

Energy requirements for growth

Energy requirements for growth will reflect the proportions of fat, protein and water deposited. Net energy requirements per kg empty body gain range from 5 to 34 MJ/kg (NRC, 1985).

Figure 10.1 gives a comparison of energy requirements from NRC (1985) and ARC (1980).

The increasing energy requirement can be compared with changes in body composition with increasing live weight. Theriez *et al.* (1981) measured the chemical composition of lambs at different live weights (Table 10.2). Crude protein and ash contents are relatively static while fat content increases as that of water diminishes. Concomitantly, energy content increases.

Protein nutrition

PROTEIN REQUIREMENTS

Early studies with lambs fed concentrate rations indicated significant responses to dietary crude protein content. Andrews and Ørskov (1970) fed rations based on cereal and soyabean meal, ranging from 100 to 200 g/kg protein, to lambs from 16 to 40 kg live weight. Growth rate responses curvilinearly to dietary protein content

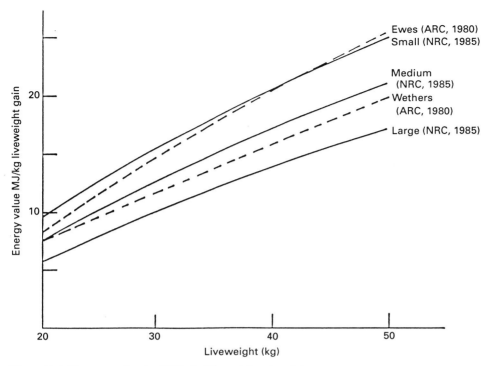

Figure 10.1 The energy density (MJ/kg) of liveweight gains of sheep

Table 10.2   THE PROPORTION (g/kg) OF WATER, FAT, CRUDE
PROTEIN, AND ASH AND THE GROSS ENERGY CONTENT (MJ/kg) IN
THE EMPTY BODY OF CROSS BRED LAMBS AT DIFFERENT
WEIGHTS

Empty body weight (kg)	25	30	35
Body composition (g/kg)			
Water	646	618	595
Fat	152	179	207
Protein	166	165	164
Ash	33	33	32
Gross energy MJ/kg	6.69	10.79	11.82

(Figure 10.2), and for fast growing lambs gaining around 300 g/day one could
conclude an optimal dietary crude protein content of 160–170 g/kg.

In a subsequent trial, lambs were fed rations based on barley and fishmeal
containing 110, 157 and 194 g/kg crude protein *ad libitum* (Ørskov *et al.*, 1971).
Growth rate, feed intake and feed conversion efficiency were all improved with
increasing dietary protein content (Table 10.3). Differences in feed intake were
small at the higher protein levels.

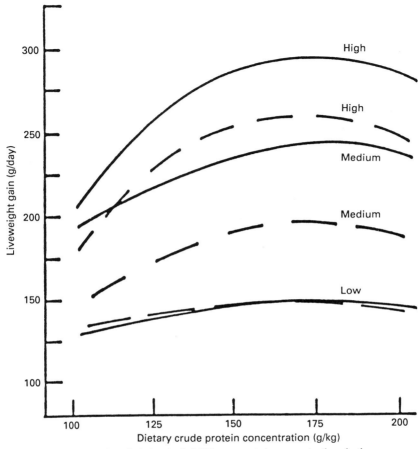

Figure 10.2 Liveweight gain in lambs fed different protein concentrations in the diet at three levels of feeding (high, medium and low)

Table 10.3 RATE OF GROWTH (kg/day), FEED INTAKE (kg/day) AND FEED CONVERSION EFFICIENCY IN LAMBS FED *AD LIBITUM* DIETS DIFFERING IN CRUDE PROTEIN CONTENT

	Dietary protein content (g/kg)		
	110	157	194
Feed intake (kg/day at 20 kg live weight)	0.658	0.746	0.711
Feed intake (kg/day at 30 kg live weight)	0.960	1.192	1.116
Feed conversion efficiency (kg feed/kg gain) (15–30 kg liveweight)			
Males	3.70	2.76	2.37
Females	3.75	3.46	2.76
Daily gain (kg)			
Males	0.191	0.270	0.330
Females	0.177	0.225	0.301

More recently the conceptual approach to describing protein utilization from a standpoint of the way it is utilized in the rumen (ARC, 1980) has indicated the importance of dietary protein quality for fast growing lambs.

Storm, Ørskov and Stuart (1983) indicated efficient utilization of rumen microbial protein by growing lambs sustained by intragastric nutrition. However protein furnished by rumen microbial synthesis is inadequate to sustain the high growth rates demanded by modern intensive production systems. For instance, Storm and Ørskov (1984) identified rumen microbial protein to be limiting in four essential amino acids, namely methionine, lysine, arginine and histidine.

It is not surprising therefore that the ARC (1980) recommendations for protein requirements of fast growing limbs indicate substantial undegradable dietary protein (UDP) requirements. For example a 30 kg lamb fed a high energy diet and gaining 0.3 kg/day was calculated to require 100 g rumen degradable protein and 25 g UDP daily.

However, more recently Hovell *et al.* (1983) have suggested that ARC (1980) estimates of protein requirements for basal metabolism were significantly underestimated by 3- to 4-fold. Certainly practical experience would associate maximum growth rates in intensively fed lambs with higher inputs of protein and particularly UDP than indicated by ARC (1980) for instance. Perhaps one can draw a parallel with intensively fed cattle. Newbold (1987) observed improvements in daily liveweight gain of over 0.1 kg/day in fast growing implanted steers when dietary UDP content was increased from 20 to 40 g/kg DM by incorporation of protected soyabean meal.

Ration composition and rumen development

Intake of dry feed is a better guide to weaning than live weight. Lambs suckling ewes milking heavily would meet liveweight criteria but might have a less developed rumen owing to minimal consumption of dry feed.

Rumen development is stimulated by the intake of solid feed and its fermentation to volatile fatty acids (Ørskov, 1983). The provision of fresh dry feed to young ruminants prior to weaning has been shown to promote early intake, and ease of consumption (i.e. pelleted feed) is a factor affecting intake (Hodgson, 1971). Additionally, the incorporation of fibrous feeds into rations for calves has been associated with early rumen development and early stability of rumen conditions as judged by pH of rumen fluid (Williams *et al.*, 1985). Calves were offered rations from 14 days of age and weaned at 56 days of age. Straw (chopped, 20 mm length) either untreated or treated with ammonia was incorporated into a cereal based ration at 150, 200 and 250 g/kg. Straw inclusion was associated with improved intakes of dry feed, higher liveweight gains, earlier stability of rumen pH, greater rumen volume and enhanced fibre digestion (Table 10.4).

The implication in this study for promoting intake of dry feed and the enhancement of stable rumen conditions are clear. However it is notable that in the post-weaning period high levels of straw inclusion limited intake, presumably owing to a physiological restriction.

While similar information is lacking for young lambs it is widely accepted that the incorporation of a 'fibrous' raw material at a level of 100 to 150 g/kg is desirable to promote early and greater feed intake and the early development of a stable rumen. Higher levels, however, are likely to be associated with reduced performance post-weaning.

Table 10.4 THE EFFECTS OF DIFFERING DIETARY INCLUSIONS
OF STRAW ON FEED INTAKE AND LIVEWEIGHT GAIN IN CALVES
AND THE EFFECT ON RUMEN VOLUME, RUMEN pH AND FIBRE
(ADF) DIGESTION

	Straw inclusion (g/kg)			
	0	*150*	*200*	*250*
Daily feed intake (kg) (14–56 days)	0.62	0.97	1.04	1.02
Daily liveweight gain (kg)	0.30	0.49	0.48	0.46
Rumen volume (litres) (42 days)	2.64	5.19	6.17	5.55
Digestibility of ADF	0.02	0.16	0.20	0.20
pH of rumen liquor:				
33 days	5.4	6.3	6.7	6.6
63 days	5.4	6.3	6.7	6.8
77 days	6.2	6.3	6.8	7.0

Feeding and carcass composition

Effects of diet on chemical composition of the carcass have been found to be small
compared with genotype differences reported in the literature (Theriez *et al.*,
1981). In surveying carcass characteristics of the main types of British lamb (421
carcasses – seven breed types) Kempster and Cuthbertson (1977) observed
significant breed differences in fat distribution and lean:bone ratio. Similarly the
percentage lean differed at constant subcutaneous fat levels.

In early studies, Morgan and Owen (1972a) restricted the nutrient intake of
lambs by diluting a high energy 140 g/kg protein ration with 400 g/kg oat husks.
Effects on carcass quality as judged by dissected components were small, although
restriction (to 70% of *ad libitum* intake) in the final stage of the production cycle in
a further trial, tended to reduce fat content and increase protein content. However,
in contrast, feed restriction in a further trial led to a higher fat content and a
reduced protein content in the carcass (Morgan and Owen, 1973).

Murray and Slezacek (1976) compared different growth rates in lambs, as
achieved by *ad libitum* and restricted feeding of a 140 g/kg protein, lucerne/cereal
based ration of 0.23 kg/day and 0.09 kg/day on carcass composition of lambs
slaughtered at 40 kg live weight.

Carcass dissection indicated similar proportions of muscle, bone, connective
tissue and total fat. However, carcasses from animals fed *ad libitum* contained more
subcutaneous fat and less intramuscular fat than those from animals fed restricted
quantities of feed.

Searle, Graham and Donnelly (1982) found lambs gaining at 0.1 kg/day
contained more fat in the carcass than lambs gaining at 0.2 kg/day. Protein content
of the carcass was similar at a given empty body weight for either growth rate. In
this particular trial lambs were fed a high protein ration (236 g/kg DM) and it is
unlikely that those lambs fed for the lower growth rate were deprived of protein.

Table 10.5 CARCASS CHARACTERISTICS FOR
LAMBS SLAUGHTERED AT THE SAME WEIGHT AND
DIFFERENT AGES (H AND LL) AND AT THE SAME
WEIGHT AND AGE (LH AND HL) HAVING
FOLLOWED DIFFERENT GROWTH PATTERNS

	Rearing treatment			
	H	LL	LH	HL
Carcass weight (kg)	15.2	16.6	24.7	23.1
Killing out %	45.7	48.5	50.4	48.3
Dissected tissues (kg)				
Lean	8.29	8.24	9.65	10.12
Bone	2.42	2.36	2.76	3.00
Total fat	4.82	6.04	13.08	10.58
Intermuscular fat	2.02	2.58	4.66	4.04
Subcutaneous fat	1.56	2.06	5.10	3.87
KKCF	0.40	0.47	1.30	0.93
Omental fat	0.39	0.54	1.19	1.06

Clearly the evidence for the effect of feeding level on carcass composition is conflicting, although there is perhaps a consensus of opinion which suggests that fast growing animals will be fatter at a given weight (see Butler-Hogg and Johnson, 1986).

More recently Butler-Hogg and Johnson (1986) have examined the effects of growth pattern on carcass composition of crossbred ewe lambs. From 4 to 20 weeks of age intakes of compound feed were controlled to produce gains of 0.21 kg/day (H) or 0.11 kg/day (L). From weeks 20 to 36 growth patterns, achieved by control of food intake, were LL, HL, or LH (Table 10.5).

Growth pattern had little effect on the content of lean and bone in the carcass. On the other hand there were considerable differences in the patterns of fat deposition. There were differences in total body fat content and in fat distribution. Slow growing lambs (LL) contained 1.2 kg more total fat compared with fast growing lambs (H). Changing the pattern of growth from 20 weeks of age led to greater differences in fat content with LH lambs containing 2.5 kg more total fat than HL lambs. These differences in fatness were thought to be the result of changes in protein:energy ratio.

The rates of tissue gain have important implications for site of fat deposition in relation to growth pattern (Table 10.6).

These results, therefore, are indicative that there is some potential manipulation of carcass characteristics, particularly fatness, through dietary manipulation. However one should qualify this by pointing out that daily growth rates achieved in this study, even at the highest level, fall some way short of those currently achieved in commercial intensive lamb fattening systems. Thus one can only extrapolate principles to present systems.

Table 10.6 TISSUE GROWTH RATES (g/day) IN LAMBS
FOLLOWING DIFFERENT GROWTH PATTERNS

| | *Rearing treatment and time period* | | | | |
| | *Weeks 4–20* | | *Weeks 20–36* | | |
	H	L	HL	LH	LL
Daily gain (kg)	0.216	0.118	0.105	0.209	0.112
Tissue growth rate (g/day)					
Lean	97.2	52.2	32.8	66.4	46.2
Bone	13.8	7.7	5.1	8.3	5.7
Total fat	13.3	4.4	17.5	29.8	14.1
Subcutaneous fat	11.5	2.0	20.4	38.0	14.5
KKCF	1.8	–0.2	5.4	10.0	3.3
Omental fat	2.9	0.5	5.9	9.2	3.9

Dietary mineral related disorders

COPPER TOXICITY

In intensively fed sheep, copper toxicity poses a real problem. This is in direct contrast to extensively fed sheep in which copper deficiency is generally the greater problem. Copper is likely to be more available in the intensive situation owing to lower dietary concentrations of molybdenum and sulphur than those encountered in the extensive situation.

Assuming 9% availability of dietary copper, ARC (1980) suggest a requirement of 5.1 mg/kg DM for a 40 kg lamb gaining 300 g per day. This is low relative to levels of 10–15 mg/kg DM to be found in compounds from background levels. Similarly it is low relative to the current maximum of 20 mg/kg DM prescribed by the Feedingstuffs Regulations.

Copper intoxication is the result of an accumulation of copper in the liver. Levels of hepatic copper considered to be normal in sheep are in the range of 500 µg/kg, while haemolytic crisis is associated with levels of 1000–3000 µg/kg (Bostwick, 1982). Jaundice (readily detected in the eyes) and haemoglobinuria are characteristic signs of toxicity and ultimately fatality ensues.

Copper toxicity may be controlled by incorporation of sources of molybdenum and sulphur in the diet. For example, Suttle (1977) found that the addition of 4 mg molybdenum and 2 g sulphate/kg feed protected growing lambs from copper toxicosis. Ross (1964; 1966) found supplementation with 50 mg molybdenum (as ammonium molybdate) and 1 g sulphate decreased hepatic copper accumulation and associated liver damage with prevention of clinical toxicosis. However as a generalization copper toxicosis has proved difficult to control once haemolytic crisis has occurred. At a practical level, feed manufacturers are reminded that the maximum content of molybdenum permitted in feeds under the Feedingstuffs Regulations is 2.5 mg/kg.

Table 10.7   THE ESTIMATED RETENTION OF
INGESTED COPPER IN THE LIVER OF CROSSBRED
LAMBS IN A 13-WEEK PERIOD

Sire breed	Proportion of copper retained
Scottish Blackface	0.056
East Friesland	0.067
Finnish Landrace	0.086
Suffolk	0.073
Texel	0.137

Breed plays a significant part in susceptibility to copper toxicity. In summarizing available information at the time, ARC (1980) reported toxicity observed with dietary copper contents ranging from 8 to 29 mg/kg. More recently Woolliams *et al.* (1982) have shown considerable breed differences in susceptibility to copper toxicity. Scottish Blackface lambs retained significantly less copper in the liver than Texel cross lambs (Table 10.7).

Other factors have recently been identified as being implicated in copper metabolism. For example, Ivan, Veira and Kelleher (1986) demonstrated higher absorption and retention of copper in lambs free from ciliate protozoa. In the practical situation, lambs fed 'concentrate' rations intensively are likely to have a relatively low rumen pH and a reduced protozoal population. Ionophores as feed additives have been shown to promote the accumulation of copper in the liver. Van Ryssen and Barrowman (1987) found elevated hepatic copper levels with lambs fed 15 mg/kg monensin sodium in feed containing 36 mg/kg copper for 40 days.

UROLITHIASIS

Urinary calculi or uroliths are mineral deposits which form in the urinary tract (Field, 1969). While both sexes are susceptible, blockage of urine flow as a result of calculi formation tends to occur mainly in entire or castrate males. Blockage of the ureters can lead to rupture of the bladder, hence the name water belly for this condition, which is generally fatal. Straining associated with difficulty of urination and painful urination, slow urination, stamping of the feet and kicking at the area of the penis are indicative of calculi problems (Jensen, 1974).

Urolithiasis is related to the dietary balance and intake of major minerals, and affected animals excrete alkaline urine of high phosphorus content (Crookshank, 1968). Maintenance of the correct dietary balance of major minerals aids considerably in minimizing the incidence of urolithiasis. An excessive intake of phosphorus is to be avoided and a wide calcium to phosphorus ratio (greater than 2:) is preferred (Emerick and Embry, 1963). Similarly an excess of magnesium is undesirable. Generous dietary salt levels will promote water intake and urination accordingly. This will aid in flushing 'the waterworks' and eliminate small forming calculi. Clearly, practical considerations are bedding requirements and water availability.

Recommended dietary targets for the major mineral content of diets for intensively fed lambs are:

Calcium   10–15 g/kg DM
Phosphorus   4.5–5.5 g/kg DM
Magnesium   maximum 2.5 g/kg DM
Salt   15–20 g/kg DM

A further aid in the control of urolithiasis is the dietary incorporation of 'anionic' salts which will reduce the alkalinity of urine, so reducing urolith formation (Crookshank, 1970). Ammonium chloride and ammonium sulphate incorporated into the diet at 5 g/kg DM afforded effective control, with the former being more effective.

Non-dietary disorders

COCCIDIOSIS

Coccidiosis is a problem of intensification of farming and is seen in Britain on an ever-increasing scale. It is seen in the unweaned lamb at grass and particularly in intensive rearing systems where outbreaks can reach astronomic proportions. Coccidiosis, should, therefore be a major concern when planning a health programme for an intensive system of rearing lambs.

Each species of farm animal has its own infective species of coccidiosis which have caused little clinical disease until farming of the domestic animal has become intensified. The first species to have problems were chickens followed by rabbits and cattle. In sheep, the clinical disease has caused increasing concern since the early 1970s.

Diagnosis has been very difficult for several reasons. Firstly there are 11 species of coccidia specific to sheep (Gregory *et al.,* 1980), each with a different pathogenicity. At one time they were thought to be the same as the species in goats but are now known to be different. Differentiation between species, or speciation is a laborious and difficult technique which needs to be performed by a trained technician. However, it is important as two species of coccidia are particularly pathogenic and small numbers of these in the faeces of an infected lamb can be more significant than vast quantities of other species.

Diagnosis is further complicated by apparently healthy lambs having a high coccidial oocyst count in the faeces. Lambs with a negative oocyst count can also be misleading as, in the acute situation, damage to the gut can be so severe that death occurs before any oocysts are produced in the faeces. Thus the simple test in the field of faecal oocyst output cannot be relied upon. It has to be combined with speciation, *postmortem* examination and the history of the flock.

The life-cycle of coccidia is complicated, involving many stages but takes approximately 3 weeks from ingestion of an infective oocyst to the passing of newly formed oocysts in the faeces of the lamb. Thus a 1-week-old lamb can pick up oocysts from the environment voided by ewes and 3 weeks later be producing 10 million oocysts for each one ingested. If 4-week-old lambs start picking up these levels of oocysts, severe damage to the gut will take place 10–14 days later when a similar multiplication occurs in the cells of the gut wall. At the same time lambs are relying on ewes' milk and are losing any maternal immunity they had previously.

Once infection has occurred and been overcome, immunity is rapidly developed and in the adult animal very few oocysts are voided. Occasionally, however, if an adult animal is very stressed acute coccidiosis can occur due to a breakdown in immunity.

It is easy to see that in an intensive situation the bedding has a greater contamination and lambs are more likely to ingest oocysts and develop the disease. But this cannot be the whole story. One farm may get the problem annually, while the next, with similar husbandry methods, never has coccidiosis.

Coccidial oocysts survive best in warm moist conditions and are destroyed by heat, drying and sunlight. Thus lambing pens which are well cleaned out and left dry for a long period over the summer are best. Farms where silage is fed must take more care with bedding due to the increased amounts of urine passed daily. Sufficient straw bedding must be added to reduce contamination and moisture content. The general health status of the flock should be taken into account. In grass-reared lambs acute coccidiosis outbreaks are often apparent after a nematodirus outbreak, as both parasites affect similar areas of gut. But any flock disease causing a lowering of immunity will increase susceptibility to a coccidiosis attack.

Nutrition plays an important role. Single lambs are rarely affected, and this is thought to be due to the ample supply of milk and/or colostrum, resulting in less foraging and possible infection. However, in a severe outbreak well-fed twins suffer a greater setback than poorly fed ones (Gregory, Norton and Catchpole, 1987).

The species of coccidia found on a particular farm are important. Certain combinations may cause more devastating effects than others, particularly if the species present invade the same part of the intestine. Damage to the upper part of the intestine can be compensated for lower down, therefore species invading the lower intestine have greater clinical effects.

The clinical picture is of 4–6-week-old lambs with diarrhoea, often grey/green in colour with a putrid smell (Gregory and Catchpole, 1986). Lambs appear tucked up and dehydrated. Faecal oocyst counts are high and the species involved are mainly the pathogenic *Eimeria ovinoidalis* and *Eimeria crandallis*. In acute infestations damage to the gut can be so great that death can occur before diarrhoea or oocysts appear. On *postmortem* the ileum and caecum are thickened and may show damage to the epithelium and haemorrhage. This picture is common to many diseases of the gut, but scrapings looked at under the microscope show the many infective stages of coccidiosis in the intestinal wall.

Treatment of coccidiosis is restricted by several characteristics of the disease. Firstly, it is theoretically possible for one oocyst ingested to multiply through the various stages in the gut wall to provide 10 million infective oocysts in the faeces in approximately 3 weeks.

Secondly, of the three sources of oocyst available to the lamb, i.e. the environment, the ewe and other lambs, the latter is most important and most difficult to treat in a preventative manner.

During an outbreak, sulphonamides are the drug of choice. Even here the method of administration is important, i.e. oral dosing requiring a minimum of three doses with all the stress of handling, or an injection method which may give carcass damage not many weeks before slaughter. It is uncertain whether dosing earlier will prevent the disease as sulphonamides may not be active against the earlier gut stages of coccidia.

Several methods of prevention of coccidiosis are used in this country although only two products, Deccox and Ovicox, are licensed for sheep. Much research work has been undertaken on the use of monensin sodium (Romensin), which has provided some interesting results, but has its problems. The inclusion rate in feed is lower than that in cattle; feeding for long periods of time at the cattle rate may cause deaths. Palatability is poor, a great drawback in a system where lambs are encouraged to eat as much creep as possible, but some work at the Central Veterinary Laboratory (Catchpole *et al.*, 1986) suggests that incorporation of monensin sodium in creep prevents lambs developing their own immunity to coccidia once they have lost their maternal immunity. Withdrawal of the drug at 8 weeks of age, past the normal risk age for coccidiosis, in a contaminated environment can lead to severe clinical disease 2 weeks later. The answer to this problem may be staged withdrawal of the coccidiostat.

Further research into coccidiosis is needed to understand the disease more fully. Much more needs to be understood about the factors in the intensive lamb rearing systems that predispose to coccidiosis. Once more is known about the role of immunity in both the ewe and the lamb, better methods of prevention and treatment may be introduced.

What is known is that coccidiosis in the intensive lamb rearing systems can strike with devastating results unless the farmer is aware of the potential problem and is ready to treat as soon as coccidiosis occurs.

INTENSIVE LAMBS AT THE NAC

The early lambing flock was set up in 1971, based on Dorset Horn ewes initially lambing three times in 2 years. In 1975 it was decided to concentrate on once a year lambing, in December, and produce finished lambs for the Easter market. In 1980 a crossing programme was started to produce ¼ Finn ¾ Dorset Horn ewes and since 1984 this has been the standard ewe in this flock.

The ewes are lambed in December and remain housed until lambs are weaned at 6 weeks of age. Lambs are introduced to creep from 5 days of age with a target intake of 400–500 g/day at weaning. The need to promote early intake of creep, and hence early rumen development has led to changes in creep area design. Traditionally small creep areas were set up in the corner of the ewe pens (Figure 10.3), but the change to centre creep areas serving two adjacent ewe pens has led to lambs showing much earlier interest in creep feed.

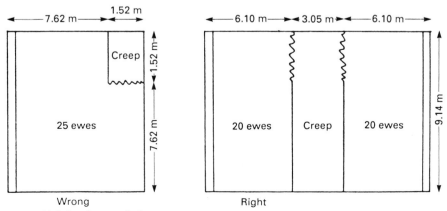

Figure 10.3 Lamb creep design

At weaning lambs are moved to the finishing houses where they are allowed *ad libitum* creep feed of 12.5 MJ ME/kg, 170 g/kg crude protein and 100 g/kg fibre. Growth rates of 400 g/day from weaning to slaughter have been achieved.

Coccidiosis is kept under control by the use of injections of sulphamethoxy-pyridazine at the rate of 250 mg/11 kg live weight. The treatment is administered at weaning. This method of control is only effective where lambs are being moved to clean buildings after treatment. Where lambs are being finished in buildings where ewes and lambs have been previously housed, some form of in-feed control will be required.

Intensive lamb rearing systems are, of necessity, high cost systems and attention to detail, of both feed and general management, is essential if financial margins are to be achieved. There is a growing interest in this form of lamb production, particularly with the increased export opportunities in the European Community. More work needs to be done in developing out of season lambing systems, both in terms of manipulation of the breeding cycle of the ewe and the finishing systems for the weaned lamb.

References

ARC (1980). *Nutrient Requirements of Ruminant Livestock.* Commonwealth Agricultural Bureaux, Slough

ANDREWS, R.P. and ØRSKOV, E.R. (1970). *Journal of Agricultural Science, Cambridge,* **75**, 11–18

BOSTWICK, J.L. (1982). *Journal of the American Veterinary Medical Association,* **180**, 386–387

BUTLER-HOGG, B.W. and JOHNSON, I.D. (1986). *Animal Production,* **42**, 65–72

CATCHPOLE, J., NOLAN, A., GREGORY, M.N. and ARTHUR, M.J. (1986). *Veterinary Record,* **118**, 75–76

CROOKSHANK, H.R. (1968). *Proceedings of a Symposium on Sheep Diseases and Health,* pp. 162–171. University of California, Davis

CROOKSHANK, H.R. (1970). *Journal of Animal Science,* **30**, 1002

EMERICK, R.J. and EMBRY, L.B. (1963). *Journal of Animal Science,* **22**, 510

FIELD, A.C. (1969). *Proceedings of the Nutrition Society,* **28**, 198

GREGORY, M.N. and CATCHPOLE, J. (1986). *International Journal for Parasitology,* **17.6**, 1099–1111

GREGORY, M.N., JOYNER, L.P., CATCHPOLE, J. and NORTON, C.C. (1980). *Veterinary Record,* **106**, 461–462

GREGORY, M.N., NORTON, C.C. and CATCHPOLE, J. (1987). *Le point Vétérinaire,* **19**(103), 29–40

HODGSON, J. (1971). *Animal Production,* **13**, 15–24

HOVELL, F.D.DeB., ØRSKOV, E.R., GRUBB, D.A. and MacLEOD, N.A. (1983). *British Journal of Nutrition,* **50**, 173–187

IVAN, M., VEIRA, D.M. and KELLEHER, C.A. (1986). *British Journal of Nutrition,* **55**, 361–367

JENSEN, R. (1974). *Diseases of Sheep.* Lea and Febiger, Philadelphia

KEMPSTER, A.J. and CUTHBERTSON, A. (1977). *Animal Production,* **25**, 165–179

MORGAN, J.A. and OWEN, J.B. (1972a). *Animal Production,* **15**, 285–292

MORGAN, J.A. and OWEN, J.B. (1972b). *Animal Production,* **15**, 293–300

MORGAN, J.A. and OWEN, J.B. (1973). *Animal Production,* **16**, 49–57
MURRAY, D.M. and SLEZACEK, O. (1976). *Journal of Agricultural Science, Cambridge,* **87**, 171–179
NEWBOLD, J.R. (1987). In *Recent Advances in Animal Nutrition*, pp. 143–172. Ed. Haresign, W. and Cole, D.J.A. Butterworths, London
NRC (1985). *Nutrient Requirements of Sheep.* 6th revised edition, 1985. Sub-Committee on Sheep Nutrition. Committee on Animal Nutrition. Board on Agriculture National Research Council. National Academy Press, Washington DC
ØRSKOV, E.R. (1983). In *Sheep Production*, pp. 155–165. Ed. Haresign, W . Butterworths, London
ØRSKOV, E.R., GRUBB, D.A., SMITH, J.S., WEBSTER, A.J.F. and CORRIGALL, W. (1979). *British Journal of Nutrition,* **41**, 541
ØRSKOV, E.R., McDONALD, I., FRASER, C. and CORSE, E. (1971). *Journal of Agricultural Science, Cambridge,* **77**, 351–361
ROSS, D.B. (1964). *Veterinary Record,* **76**, 875
ROSS, D.B. (1966). *British Veterinary Journal,* **122**, 277
SEARLE, T.W., GRAHAM, N.McC. and DONNELLY, J.B. (1982). *Journal of Agricultural Science, Cambridge,* **98**, 241–245
STORM, E. and ØRSKOV, E.R. (1984). *British Journal of Nutrition,* **52**, 613–620
STORM, E., ØRSKOV, E.R. and SMART, R. (1983). *British Journal of Nutrition,* **50**, 471–478
SUTTLE, N.F. (1977). *Animal Feed Science & Technology,* **2**, 235
THERIEZ, M., TISSIER, M. and ROBELIN, J. (1981). *Animal Production,* **32**, 29–37
VAN RYSSEN, J.B.J. and BARROWMAN, P.R. (1987). *Animal Production,* **44**, 255–261
WILLIAMS, P.E.V., INNES, G.M., BREWER, A. and MAGADI, J.P. (1985). *Animal Production,* **41**, 63–74
WOOLLIAMS, J.A., SUTTLE, N.F., WEINER, R., FIELD, A.C. and WILLIAMS, C. (1982). *Animal Production,* **35**, 299–307

IV

Non-ruminant Nutrition

THE PHYSIOLOGICAL BASIS OF ELECTROLYTES IN ANIMAL NUTRITION

J. F. PATIENCE
Prairie Swine Centre, University of Saskatchewan, Saskatoon, Saskatchewan, Canada

Introduction

From a purely chemical point of view, an electrolyte is any substance that when added to pure water produces a conducting solution (Sienko and Plane, 1979). Thus, any chemical that dissociates into its constituent ions is an electrolyte. A strong electrolyte dissociates completely, or almost completely, while a weak electrolyte dissociates to only a limited extent. Thus, a solution of strong electrolytes will consist mainly of ions and only a few or almost no undissociated molecules in a solution of weak electrolytes, the undissociated molecules will be more common. Table 11.1 lists common substances that are strong or weak electrolytes, or non-electrolytes.

Table 11.1 CLASSIFICATION OF ELECTROLYTES

Electrolytes		Non-electrolytes
Strong	*Weak*	
Hydrogen chloride	Hydrogen fluoride	Glucose
Sodium chloride	Ammonia	Sucrose
Sodium hydroxide	Acetic acid	Ethanol
Potassium fluoride	Mercuric chloride	Oxygen
	Acetone	

In nutrition and physiology, it is common to consider specific electrolytes, i.e. ions found in the various body fluids (plasma, interstitial fluid, intracellular fluid, urine, etc.). For example, sodium, bicarbonate and chloride are the predominant electrolytes in extracellular fluid, with much smaller quantities of potassium, calcium, magnesium, phosphate, organic acids and protein also present. In the intracellular fluid, such electrolytes as potassium, phosphate, magnesium and protein dominate. Sodium, calcium and bicarbonate are less common species. The ionic composition of typical intracellular and extracellular body fluids appears in Table 11.2.

Table 11.2 IONIC COMPOSITION OF BODY FLUIDS
(mEq/l)

Electrolyte	Plasma	Interstitial fluid	Intracellular fluid
Cations			
Sodium	142	145	12
Potassium	4	4	150
Magnesium	1	1	34
Calcium	3	2	4
Anions			
Chloride	104	117	4
Bicarbonate	24	27	12
Phosphate	2	2	40
Protein	14	0	54

Adapted from Rose (1977)

The concentration of electrolytes is often expressed as milliequivalents (mEq) or milliosmoles (mOsm), rather than milligrams (mg), since the electrical and osmotic properties of these ions are of particular interest. Values expressed in milligrams can be converted to milliequivalents by dividing by the molecular weight and multiplying by the valency. Thus, 100 mg of sodium equals 4.3 mEq [(100/23) × 1] and 100 mg of calcium is the same as 5 mEq [(100/40.08) × 2)].

No discussion of electrolytes would be complete without a brief summary of water metabolism, sinced a major function of electrolytes is to assist in the maintenance of water balance within the body. Indeed, it is impossible to develop a functional discussion of one without the other.

Water

As part of the evolutionary process some 400 million years ago, when living organisms moved from an aquatic environment to dry land, they developed a layer of skin that provided a barrier between the fluid environment required by the cells and the surrounding air. At the same time, mechanisms evolved to regulate the composition of fluid inside the cell as well as that surrounding it (Maloiy, MacFarlane and Shkolnik, 1979). In essence, these organisms developed their own 'aquatic environment.'

Water is required to transport nutrients, gases, waste products and hormones about the body. It lubricates many moving parts and is essential in helping to maintain the delicate balance of acids and bases. It provides the medium in which the body's metabolic processes may occur and it assists in dissipating heat. Movement of digesta through the gastrointestinal tract could not occur without water, nor could the absorption of nutrients acorss the gut wall. Indeed, it is difficult to think of a single body function that does not require water, at least indirectly.

Not surprisingly, the chemical properties of water make it ideally suited to fulfil its central role in life. For example, water has a very high heat capacity, supporting its role in thermal homeostasis. It maximizes its electrolyte activities at 37°C which

Table 11.3 TYPICAL DISTRIBUTION OF BODY WATER

Compartment	As a percentage of total water	As a percentage of total body
Total body water	100	60
Extracellular	40	24
Interstitial	19	11.5
Plasma	7.5	4.5
Connective tissue	7.5	4.5
Bone	5.0	3.0
Intracellular	60	36

Adapted from Rose (1977)

is about body temperature and has very low viscosity. Furthermore, it has high surface tension, supporting capillary movement in the vascular system (Quinton, 1979).

Typically, total body water is divided into two major compartments: extracellular, including both interstitial fluid and blood plasma, and intracellular. A typical breakdown is illustrated in Table 11.3.

However, many factors can and do influence the proportion of the body that is water; for example, as the animal matures, body fat increases and body water content decreases. The reason for this is clear; adipose tissue contains only 7% water, while muscle is three-quarters water (Georgievskii, 1982a). Thus, as an animal matures, water content will vary in concert with changes in the proportions of fat and muscle.

Water is derived by the body from three sources, the primary one, of course, being drinking water. Common feed ingredients never contain 100% dry matter, so that eating food also provides some free water; most foodstuffs contain at least 8% water, and some much more. Finally, metabolic processes generate water as shown in equations 11.1, 11.2 and 11.3.

Carbohydrate (glucose)
$$C_6H_{12}O_6 + 6O_2 \rightarrow 6H_2O + 6CO_2 \tag{11.1}$$

Amino acid (methionine)
$$2C_5O_2NH_{11}S + 15O_2 \rightarrow 7H_2O + 9CO_2 + (NH_3)_2CO + 2H_2SO_4 \tag{11.2}$$

Fat (tripalmitin)
$$2C_{51}H_{98}O_6 + 145O_2 \rightarrow 98H_2O + 102CO_2 \tag{11.3}$$

It can be seen that oxidation of 100 g carbohydrate (equation 11.1) yields about 60 g water; a similar quantity of protein (equation 11.2) would yield about 44 g water, depending on the amino acid profile. In the case of methionine illustrated above, oxidation of 100 g would yield 42.3 g of water.

Interestingly, oxidation of fat (equation 11.3) yields more than its own weight in water, a fact exploited by the camel. It stores fat (not water) in its humps and then utilizes these deposits to generate water when required. The calculated water balance in a more typical farm animal, the growing pig, appears in Table 11.4.

Table 11.4 TYPICAL WATER BALANCE IN THE GROWING PIG

Intake	ml	Output	ml
Drinking water	4000	Urine	2930
Food water	200	Lungs, etc	1530
Water of oxidation	990	Faeces	250
		Growth	480
Total	5190		5190

Patience (unpublished data). This example pertains to a young pig, growing at the rate of about 700 g/day, eating 2000 g/day of a typical maize-soyabean meal diet in a thermoneutral environment

Obviously, urine is the primary vehicle for elimination of water, and is certainly the one that can be regulated most effectively according to physiological need. Fundamentally, urine volume is determined by the quantity of water requiring removal from the body. The composition of urine varies tremendously, being influenced by the quantity of solutes eliminated and by the quantity of water consumed relative to need. A complete consideration of water balance must also include losses due to sweat and respiration, which tend to be influenced in the first instance not by a need to maintain water balance, but rather to achieve some other homeostatic objective.

The major physiological regulators of water balance relate to intake (thirst) and excretion. Excretion is under the influence of hormones, including antidiuretic hormone (ADH), the renin-angiotensin-aldosterone complex and possibly prolactin. If plasma osmolality falls below 'osmotic threshold' ADH secretion is suppressed to very low levels and urine output rises; conversely if osmolality rises, ADH secretion increases. Thus, release of ADH results in the excretion of decreasing amounts of a urine that is increasing in concentration. The regulation of ADH is so sensitive that a change in osmolality of only 1% is sufficient to elicit a measurable change in ADH and subsequent adjustment in urinary excretion (Zerbe and Robertson, 1983). ADH release may also be influenced by haemodynamic factors, such as blood volume and pressure, which stimulate baroreceptors located in the heart, aorta and carotid artery (Robertson, 1977). The baroregulatory system is less sensitive to small changes than the osmotic system, but responds in a more dramatic fashion to large changes. Other factors, such as nausea and hypoglycaemia influence ADH release (Baylis, Zerbe and Robertson, 1981) but their practical application remains obscure.

Aldosterone is involved in water retention indirectly via its sodium-sparing action (Legros, 1979). Angiotensin and renin may act directly on the pituitary to stimulate ADH release. The role of prolactin in water balance is minimal, if it exists at all (Legros, 1979).

Thirst is an important primary regulator of water balance. Interestingly, thirst is stimulated by many of the same factors that increase the secretion of ADH, such as increased plasma osmolality (Thornton, Baldwin and Purdew, 1985). This is a sensitive response; only a 2–3% increase in osmolality will induce thirst (Fitzsimmons, 1972). Hypovolaemia will also initiate the sensation of thirst involving, in ways which remain unclear, the same stretch receptors involved in

ADH release (Thrasher, Keil and Ramsay, 1982). Thus, thirst and ADH appear to be closely linked.

However, it must also be noted that thirst can be stimulated by many factors unrelated to water requirement; important examples include hunger (Patience *et al.*, 1987d) and behaviour (Madec, Cariolet and Dantzer, 1986).

Electrolyte metabolism

SODIUM

Sodium plays a primary role in the regulation of water balance and, in particular, in extracellular fluid volume (Alcantara, Hanson and Smith, 1980). It also plays a critical role in cellular transport systems, due to its participation in the sodium-potassium ATPase. This system is responsible, in part, for the electrochemical gradient across cell membranes and thus is critical not only for both active and passive transport function but also for the creation of the electrical potential required for nervous system activity and muscle contraction. Clearly, the body's emphasis on sodium homeostasis is well founded.

The total amount of sodium in the body is divided into exchangeable and non-exchangeable compartments. About 43% is in bone, of which 68% is bound within the crystalline structure and is therefore unavailable for other metabolic use (Edelmon and Leibman, 1959). The remainder of bone sodium, plus that found in the various fluid compartments is considered to be exchangeable. In total, about 70% of total body sodium exists in the exchangeable pool (Rudd, Pailthorp and Nelp, 1972).

Sodium balance is monitored by a variety of systems. Receptors in various tissues detect changes in fluid colume, blood pressure or other phenomena which are related to sodium concentration (Gardenswartz and Schrier, 1982). Numerous hormones, including the renin-angiotensin system, aldosterone, some prostaglandins, the kallikrein-kinin system and natriuretic hormone are all involved to a lesser or greater extent. Clearly, the body appears to place a very high priority on the regulation of body sodium content.

Under normal circumstances, about 90% of the sodium in the diet will be absorbed; even when intake rises 4- or 5-fold, the proportion absorbed remains largely unchanged (Patience, Austic and Boyd, 1987c). Thus, the kidney remains the primary organ of sodium homeostasis, with the intestinal tract playing a relatively minor role.

Sodium metabolism can be influenced by a wide variety of nutritional circumstances, some of which might not be readily considered. For example, in ruminants, ingestion of a high fibre diet stimulates parotid secretion of sodium bicarbonate, impairing normal sodium and water circulation to the rumen. This generates a short-term acidaemia, hyperhydration, renal sodium retention and urinary acidification (Andersson *et al.*, 1986). Although such changes are transient, they underline the need to obtain blood samples for acid-base measurements with a full understanding of their relationship to such factors as time of feeding.

POTASSIUM

Potassium is the principal cation of the intracellular environment (Table 11.2) Almost 90% of total body potassium can be found in the intracellular spaces. The majority of extracellular potassium is present in bone (Tannen, 1986). Unlike sodium, potassium is largely an intracellular electrolyte, although exceptions do exist. Maintenance of constant extracellular potassium is absolutely critical for life. Hypokalaemia is defined by serum potassium <2.4 mEq/l and hyperkalaemia by serum levels >7.0 mEq/l. Hyperkalaemia appears to be particularly serious, due to the apparent vulnerability of cardiac tissue (Rose, 1977).

Potassium is readily absorbed from the gastrointestinal tract, although in the pig, the extent of absorption tends to be somewhat less than that for sodium or chloride (Patience *et al.*, 1987c). Potassium is involved in a large number of functions. Like sodium, it is part of the Na-K ATPase, mentioned above. Furthermore, it is a cofactor in a number of enzymes, participates in the maintenance of acid-base balance, gas transport by haemoglobin, cardiac function and protein synthesis (Fregly, 1981; Georgievskii, 1982a).

As in the case of sodium, potassium homeostasis is the responsibility of the renal tissue (Mason and Scott, 1972). However, the actual mechanisms in some cases differ substantially. Potassium is filtered by the glomerulii, but is largely reabsorbed in the proximal tubule. Thus, renal regulation of potassium is dependent on its secretion into the more distal portions of the tubules, largely under the influence of aldosterone (Brobst, 1986). Interestingly, glucocorticoids (Bastl, Binder and Hayslett, 1980) and insulin (Bia and DeFronzo, 1981) also play a role. Sodium concentration in the urine and acid-base status of the animal are also known to influence the extent of renal potassium secretion (Mason and Scott, 1972; Kem and Trachewsky, 1983). Because of the relationship between potassium excretion, sodium excretion and mineralocorticoid levels, urine and salivary sodium:potassium ratios are sometimes used as an indicator of mineralocorticoid effect (Kem and Trachewsky, 1983).

The role of insulin is an interesting one; the fact that such a critical component of the endocrine system responds to extracellular potassium concentration underlines the importance placed on its homeostasis. Increases in extracellular potassium stimulate insulin release resulting in enhanced potassium transport into the cell. The relationship between glucose and potassium, mediated via insulin must be understood by clinicians since any disturbance in carbohydrate metabolism may influence potassium, and vice versa. For example, infusion of glucose can cause hypokalemia (Rose, 1977) and similarly, potassium infusion could lead to a potentially lethal hypoglycaemia.

Acid-base disturbance has an interesting effect on potassium metabolism. Acidosis results in its translocation from within the cell to the extracellular fluid; alkalosis has the opposite effect (Adrogue and Madias, 1981). The basis of this effect is the buffering of excess hydrogen ions by the intracellular fluids; as the protons enter the cells, potassium leaves in order to maintain electrical neutrality. Burnell *et al.* (1956) have suggested that for each 0.1 unit decrease in extracellular pH, serum potassium would rise by 0.5 to 1.2 mEq/l. Thsi relationship is not universally valid, since organic (lactic) acidoses do not appear to apply (Lindeman and Pederson, 1983).

Another interesting phenomenon is the accumulation of basic amino acids in skeletal tissue of potassium deficient animals (Eckel, Norris and Pope, 1958). This response to potassium depletion is unique to the basic amino acids and the magnitude of change is substantial (Arnauld and Lachance, 1980a). A number of hypotheses have been proposed to explain this observation, including the action of basic amino acids as intracellular cations or the impairment of amino acid transport systems. However, experimental data refute both explanations.

CHLORIDE

Chloride tends to be overlooked in many discussions on electrolyte metabolism. This is quite surprising, since chloride represents approximately 65% of the total anions in the extracellular fluids; consequently, the regulation of total body chloride is also closely linked to extracellular water balance.

The close association between chloride and sodium has meant that it has traditionally been considered as part of sodium metabolism. More recently, unique aspects of chloride transport, for example, have received considerable attention (Alperin *et al.*, 1985).

Chloride participates in many critical metabolic functions. For example, as hydrochloric acid, it plays a major role in gastric digestion. In the blood, its transfer between the erythrocytes and surrounding plasma, in what is referred to as the 'chloride shift' helps to support more efficient transport of carbon dioxide from the cells to the lungs.

OTHER ELECTROLYTES

There are other important electrolytes in the body. However, from a nutritional perspective, those of greatest interest have been discussed above; sodium, potassium and chloride. The other inorganic electrolytes which are present in the body in substantial quantity include calcium, magnesium and phosphate.

Electrolyte disturbance

VOMITING

Vomiting can have a major effect on electrolyte balance. Vomitus generally contains more chloride than sodium, since it is rich in HCl, and a relatively large amount of potassium. Vomiting generally is associated with metabolic alkalosis, depletion of chloride reserves, potassium deficiency and dehydration. The most profound effects are associated with the hypochloraemia and hypokalaemia. Clinically, patients suffering from gastric fluid losses experience muscle weakness and cardiac arrhythmia, muscle tetany, respiratory depression and generalized weakness (Weinberg, 1986).

DIARRHOEA

Diarrhoea can have a profound influence on water and electrolyte metabolism. Whereas gastric contents are rich in acid, intestinal contents are highly alkaline, being rich in bicarbonate. Thus, whereas emesis results in alkalosis, diarrhoea is associated with metabolic acidosis. Diarrhoea also results in the abnormal loss of water, sodium, potassium and nitrogen (Krehl, 1966).

Diarrhoeas maybe categorized according to cause: osmotic, secretory or primary motility. Osmotic diarrhoea results from the ingestion of poorly absorbable solutes, such as magnesium sulphate, oxide or hydroxide, sodium phosphate, sulphate or citrate or mannitol. It may also be created by malabsorption or poor digestion of carbohydrates (fructose, lactose, etc.). Osmotic diarrhoea secondary to carbo-hydrate malabsorption can be differentiated from that due to poorly absorbed mineral salts by measuring faecal pH. The former is usually associated with low pH due to microbial fermentation, whereas the latter creates normal (7.0) or high pH.

Secretory diarrhoeas are due to depressed absorption of fluids secreted higher in the gastrointestinal tract, or due to chemicals that actually stimulate secretory activity in the gut. For example, ricinoleic acid, the active component in castor oil is an effective laxative because it stimulates secretory activity in the colon through a cAMP-dependent mechanism (Ammon, Thomas and Phillips, 1974). Previously, castor oil was thought to exert its effect by stimulating gut motility.

Oral rehydration therapy was a concept introduced into human medicine about two decades ago. The high cost and technical demands of conventional intravenous therapy limited its utility to cholera victims in remote areas. The development of oral treatment that was safe, simple and effective was a tremendous advancement. The key to success was the inclusion of glucose or similar carbohydrate that exploited the intestinal sodium:glucose cotransport system to increase sodium and thus water absorption (Cash, 1983). Nalin *et al.* (1970) clearly demonstrated the benefits of such oral therapy in reducing the requirement for intravenous fluids. Some typical rehydration formulae are shown in Table 11.5. Cash (1983) offered a

Table 11.5 ORAL ELECTROLYTE SOLUTIONS FOR CHOLERA PATIENTS

	Solution (mM)	
Component	*a*	*b*
Sodium	100	96
Potassium	10	25
Chloride	70	72
Bicarbonate	40	24
Citrate	–	25
Glucose	120	111

[a] Pierce *et al.* (1969)
[b] Palmer *et al.* (1977)

specific formula comprising (per litre of distilled water): 8 g glucose, 4 g sodium chloride, 6.5 g sodium acetate (or 5.4 g sodium lactate) and 1 g potassium chloride. Starch has been suggested as a replacement for glucose, since it will be gradually hydrolysed to glucose, and thus reduce the osmotic load on the intestine (Jelliffe *et al.*, 1987). Nalin *et al.* (1970) have proposed that combining a neutral amino acid with glucose will provide an even more effective rehydration solution.

Oral rehydration therapy represents an excellent example of applying fundamental knowledge on physiology to resolve a practical problem.

ENDOCRINE DISTURBANCE

Polyuria (excessive urine output) can result from endocrine disturbance, the two most common causes in humans being diabetes mellitus and diabetes insipidus. In the former, excessive glucose in the urine generates what is called a solute diuresis.

Thus, one can differentiate between the two forms of diabetes by measuring daily urine solute excretion. If it exceeds 1500 mOsm/kg, the polyuria is probably due to solute diuresis; an assay for glucose will determine if it is the problem substance. Conversely, if solute excretion is less than 1500 mOsm/kg, then the polyuria is probably due to water diuresis and is called diabetes insipidus.

Diabetes insipidus can be the result of one of three pathological situations: inadequate production of ADH, renal insensitivity to circulating ADH or reduced release of ADH due to high fluid intake (Vokes, 1987). In all cases, the final result is polyuria or excessive urine output (Vick, 1984).

STARVATION AND REFEEDING

Starvation has profound effects on water and electrolyte retention. During the early phases of fasting, renal excretion of sodium rises (Bloom and Mitchell, 1960). Extended fasting is associated with sodium conservation and subsequent refeeding is accompanied by sodium accumulation (Boulter, Hoffman and Arky, 1973). Previous salt restriction will reduce starvation-induced natriuresis, but will not eliminate it entirely (Schloeder and Stinebaugh, 1966).

The relationship with body weight gain or loss is interesting, and has been exploited in some human diets to effect rapid weight loss (unfortunately of water, not of actual body mass). During the early phases of fasting, increased sodium losses lead to reduced fluid osmolality with concomitant suppression of ADH; urine excretion increases and body weight loss is rapid. Refeeding increases body sodium reserves which in turn affects osmolality, stimulates ADH release and urine output falls with an accompanying rise in body water and thus weight gain (Vokes, 1987).

Starvation also results in kaliuresis; part of the loss is due to the loss of body mass and associated breakdown of muscle cells. However, there is also an impairment of renal conservation of potassium (Drenick *et al.*, 1966). Although other minerals, such as calcium, phosphorus and magnesium are also excreted in increasing amounts during fasting, serum levels rarely fall below normal and no adverse effects are reported. In all cases, including the sodium and potassium deficits noted above, oral supplements are most effective in reversing the problem (Weisner, 1971).

STRESS

The use of supplemental electrolytes during 'stressful' situations remains a somewhat confused topic, no doubt due in part to the difficulty in defining and quantifying 'stress'. There is some physiological basis for the practice, given that elevation of ACTH and/or glucocorticoids, a phenomenon associated with stress, has been shown to impair the utilization of many minerals including potassium (Kem and Trachewsky, 1983) and calcium (Kenny, 1981). For example, glucocorticoids have been shown to inhibit calcium absorption from the intestinal tract in the pig (Fox, Care and Marshall, 1978), chicken (Corradino, 1979) and humans (Klein *et al.*, 1977). Rude and Singer (1982) provided a thorough review of the role of adrenal corticoids in mineral metabolism.

Heroux (1981) discussed how cold stress increases renal excretion of magnesium and potassium, but not sodium. Apparently, the effects lessen over time, presumably as the animal acclimatizes to the environment. However, the challenge remains to determine the extent to which electrolyte requirements might be affected by environmental stress and whether normal diet composition meets the animal's needs under practical conditions.

Diagnosis of electrolyte disturbance

Diagnosis of mineral disturbance under clinical circumstances is quite difficult. Many overt symptoms of deficiency or imbalance are not specific to one mineral, so that differential diagnosis is particularly trying. Conversely, deficiencies of single minerals may occur, but often, inadequate dietary supplementation leads to multiple symptoms. Also, observable disturbance of dietary inadequacy is rarely acute, and chronic pathologies develop over extended periods of time. Clearly, more definitive means of evaluating electrolyte status than gross observation are required. The following is a list of some of the approaches that may be applied to evaluating the electrolyte status of farm animals. It is not intended to be inclusive, but rather provide a brief overview of some possible approaches.

HAEMATOCRIT

Haematocrit is a measure of the proportion of the blood represented by red blood cells. Thus, hypovolaemia (loss of water) or hypervolaemia (water accumulation) may be reflected not only in altered haematocrit, but in plasma protein values as well. In fact, plasma osmolality is a more sensitive indicator of water status than is haematocrit.

PLASMA SODIUM

This is generally reduced in primary sodium depletion. However, if sodium loss is accompanied by dehydration, changes in the concentration of sodium may not reflect the absolute loss of sodium from the body. Plasma sodium concentration may also be misleading when osmotically active solutes accumulate in the blood, drawing water from the cells into the plasma and diluting sodium concentration.

PLASMA POTASSIUM

This is not a good indicator of potassium status since most of the body potassium is found in the intracellular space. Depression of plasma potassium requires a severe depletion of potassium or a disturbance of water balance, in which case plasma potassium may reflect water, not potassium, disturbance.

PLASMA BICARBONATE AND CHLORIDE

These are useful indicators if evaluated in concert with plasma sodium and potassium. The anion gap can be calculated as shown in equation 11.4 (Cohen,

1984), identifying the nature of a disturbance or the success of treatment. By measuring known ion content, and assuming electroneutrality of gross body fluids, one can estimate the presence of other counterbalancing ions, such as lactate, sulphate, phosphate, ketoacids, etc.

$$\text{Anion gap} = Na^+ + K^+ - HCO_3^- - Cl^- \tag{11.4}$$

BLOOD pH, pCO$_2$ AND HCO$_3$

These define the acid-base status of the animal and therefore assist in determining if the disturbance includes an acidaemia or alkalaemia.

TISSUE CONTENT

Depending on the nature of the mineral and the suspected disturbance, tissue biopsy may prove useful. However, standard values are difficult to obtain and many non-pathological circumstances are known to alter analytical values.

New approaches

DIETARY UNDETERMINED ANION

The balance of macromineral cations and anions has recently become a topic of greater interest among animal nutritionists. The reports of Mongin (1981) in poultry, Leibholz *et al.* (1966), Yen, Pond and Prior (1981) in pigs, Thacker (1959) in rabbits, Chui, Austic and Rumsey (1984) in fish and Block (1984) in cattle demonstrate the importance of considering dietary electrolyte balance as an entity quite distinct from that involving the individual ions. All of the above reports revealed a close relationship between the balance of dietary cations and anions and the performance of the respective species. This broad application across so many species reflects the fundamental nature of the phenomenon.

Two estimates of electrolyte balance have been proposed (Austic and Patience, 1988); one, the more comprehensive is called dietary undetermined anion (dUA) and is calculated as:

$$dUA = (Na^+ + K^+ + Ca^{2+} + Mg^{2+}) - (Cl^- + H_2PO_4^- + HPO_4^{2-} + SO_4^{2-}) \tag{11.5}$$

Dietary electrolyte balance (dEB) considers only the monovalent ions and is calculated as:

$$dEB = Na^+ + K^+ - Cl^- \tag{11.6}$$

The relative advantages of each calculation have been discussed in detail elsewhere (Austic and Patience, 1988); suffice it to say that dUA is more comprehensive, considering all of the relevant fixed cations and anions, but is encumbered by the need for seven individual analyses in order to characterize a given diet or feedstuff. Dietary electrolyte balance is more convenient, requiring only three analyses but it ignores the potential contributions of the polyvalent ions.

The use of either must recognize the fact that other components of the diet, notably amino acid oxidation, can and do influence acid-base contributions (Brosnan and Brosnan, 1982; Chan, 1981). Also, adjustment for the relevant differences in digestibility is required (Lennon, Lemann and Litzow, 1966).

In essence, dUA and dEB are estimates of the dietary content of metabolizable anions and cations that consume or generate acid upon metabolism (Patience, Austic and Boyd, 1987b). Since feedstuffs are electrically neutral, any excess of positive or negative charges calculated from equations 11.5 or 11.6 above must be indicative of the presence of counterbalancing metabolizable anions or cations. In essence, dUA or dEB employ the assumption of electrical neutrality to quantify elements in the diet which would otherwise be difficult to analyse. These anions and cations generate base or acid, respectively, upon metabolism. Thus, a diet containing a relative excess of mineral cations will be more alkalinogenic than a diet containing a relative excess of mineral anions.

This phenomenon is independent of specific ion effects (Patience and Wolynetz, 1987). This was clearly demonstrated in an experiment in which pig performance was monitored in the presence of increasing dietary chloride (by replacing calcium carbonate with calcium chloride, so that calcium remained unchanged) at constant or varied dUA. To maintain constant dUA at elevated chloride, sodium and potassium, in a 2:1 molar ratio were added as their bicarbonates to the diet. Increasing dietary chloride at constant dUA had no effect on any of the variables measured; however, chloride-mediated changes in dUA depressed growth rate, feed intake and feed efficiency and also generated an apparent metabolic acidosis (Table 11.6).

Clearly, although most considerations of electrolyte nutrition deal with the function or metabolism of specific ions, this concept is more 'generic' and represents a combined effect. This occurs because the calculation of dUA and dEB is really a calculation of charges, and uses the macrominerals only as vehicles of this determination.

Alterations in dUA have been shown to influence the metabolism of a number of nutrients, including amino acids (Scott and Austic, 1978), vitamin D (Reddy *et al.*,

Table 11.6　EFFECT OF ELEVATED CHLORIDE AT CONSTANT OR VARIED DIETARY UNDETERMINED ANION

Diet content, assayed					
Sodium (g/kg)	2.1	2.0	6.5	2.4	7.3
Potassium (g/kg)	7.5	7.6	10.8	7.6	11.4
Chloride (g/kg)	2.7	10.5	10.5	12.7	12.7
dUA (mEq/kg)	388	172	412	98	431
Performance					
Growth rate (kg/day)	0.70	0.65	0.72	0.56	0.71
Feed intake (kg/day)	1.54	1.47	1.55	1.30	1.54
Feed conversion ratio (kg feed/kg liveweight gain)	0.46	0.44	0.47	0.43	0.46
Blood parameters					
pH	7.21	7.14	7.19	7.11	7.21
Bicarbonate (mmol/l)	29	26	29	23	28
Base excess (mmol/l)	0.6	−3.6	−0.2	−6.9	−0.5

Patience and Wolynetz (1987). Effect of dUA but not chloride significant ($P<0.05$) for all variables shown above

1982) and calcium (Block, 1984) as well as affecting growth in all species considered: poultry (Mongin, 1981), pigs (Yen *et al.,* 1981) and fish (Chui *et al.,* 1984). Furthermore, alteration in dUA has been exploited under practical conditions to alleviate the adverse effects of heat stress in poultry (Teeter *et al.,* 1985).

An example of dietary electrolyte balance influencing amino acid metabolism occurs in the lysine-arginine antagonism in the chick, which can be exacerbated or ameliorated by electrolyte additions to the diet (O'Dell and Savage, 1966; Savage, 1972). It has been suggested that alkaline salts stimulate lysine oxidation and thus reduce the effect of the antagonism (Scott and Austic, 1978).

Johnson and Farrell (1985) reported that electrolyte balance interacted with energy utilization in young growing broiler chicks. They concluded that the electrolyte balance affected the bird's maintenance energy requirements.

Specific ion effects are different from the more generalized dUA and dEB. For example, Patience, Austic and Boyd (1986) reported that alkaline salts of potassium but not sodium depressed the digestibility of energy in growing pigs (Table 11.7). In this experiment, pigs were fitted with simple T cannulae proximal

Table 11.7   EFFECT OF SODIUM AND POTASSIUM SUPPLEMENTATION ON APPARENT NUTRIENT DIGESTIBILITY IN PIGS

Diet content, assayed				
Sodium (g/kg)	2.4	7.5	10.2	2.2
Potassium (g/kg)	3.3	3.3	3.3	17.7
Chloride (g/kg)	3.5	3.1	3.4	3.5
Ileal digestibility (%)				
Nitrogen	82.3	83.1	82.2	80.1
Lysine	85.0	87.0	86.6	82.9
Tryptophan	66.9	65.0	65.5	62.2
Energy	76.7	77.1	76.7	74.0
Faecal digestibility (%)				
Nitrogen	84.9	83.1	82.2	80.1
Lysine	73.6	72.7	73.6	71.1
Tryptophan	78.7	77.0	77.8	73.9
Energy	85.2	84.8	85.1	82.7

Patience, Austic and Boyd (1986). The effect of potassium supplementation on apparent faecal energy digestibility significant ($P<0.05$)

to the ileo-caecal junction and fed diets based on maize and maize-gluten meal. Sodium or potassium were supplemented with their respective bicarbonate salts. There tended to be a generalized depression in nutrient absorption in the presence of excess potassium, but only the effect on energy was statistically significant ($P<0.05$). Since equivalent quantities of sodium and potassium had differing effects, it is clear that this was not related to electrolyte balance.

The relationship betwen dietary electrolytes and egg shell quality in the laying hen is well known (Hamilton and Thompson, 1980). Mongin (1981) has reviewed the subject thoroughly.

In dairy cattle, Block (1984) has confirmed the earlier observations of Dishington (1975) that electrolyte levels in the diet can have a profound effect on the incidence

Table 11.8   EFFECT OF ELECTROLYTE BALANCE
IN CATTLE DIET ON THE INCIDENCE OF MILK
FEVER

No. cows	Ration	Milk yield (kg)	Milk fever (%)
19	Anion	7142	0
19	Cation	6656	47

Source: Block (1984)

of milk fever. His results, summarized in Table 11.8, demonstrate that a diet rich in anions reduces the incidence of milk fever, while one rich in cations has the opposite effect.

Although electrolyte balance in the diet has been shown to influence growth, its application in evaluating the quality of 'novel' feed ingredients has been sketchy. Miller (1970) demonstrated that the balance of electrolytes in the diet could affect the perceived quality and therefore value of fish meal in poultry diets. Adjustment of mineral mixture composition used in chick bioassays of fish meal significantly altered growth performance.

It is well established that bone metabolism responds to chronic acid-base disturbance (Lemann, Litzow and Lennon, 1966) and further that dietary electrolyte balance can effect such changes in acid-base status (Patience, Austic and Boyd, 1987a). Consequently, researchers have investigated a direct link between dietary electrolyte levels and skeletal integrity. In poultry, it is now accepted that adjustment in dietary electrolyte balance will influence the incidence of tibial dischondroplasia (Sauveur, 1984). However, attempts to find such a relationship in pigs has met with limited success (Van der Wal *et al.*, 1986).

TAIL-BITING IN PIGS

Tail-biting and other cannibalistic activities of pigs has been a frustrating subject for many years. Although many environmental factors are known to encourage tail biting, a reasoned explanation has been lacking. Recently, however, Fraser (1987b) developed a working hypothesis that helps to focus much of the vague reasoning associated with tail-biting. By developing an experimental model for his research, involving pseudo-tails made from 1.3-cm thick sash cord, and studying the activity of pigs very intensively (one experiment alone employed over 211 000 observations), he has formulated an interesting description of the phenomenon.

He suggests in his model that tail-biting is a two-phase event. Confinement housing, devoid of straw and other chewable objects, causes the pig to direct his inate chewing behaviour to whatever is at hand, such as another pig's tail. Large pen size or overcrowding increases the opportunity for such behaviour. Other stressors may also contribute to the problem. Circumstances such as poor health, which creates a general apathy on the part of the victim, or poor environment, which causes general discomfort and associated hyperactivity, may lead to greater chewing activity.

Once the tail is wounded, and blood appears, phase two begins. Attraction to blood may be increased by what is often referred to as salt appetite, a phenomenon observed in many mammalian species (Mackay, 1979; Denton, 1982). An unbalanced diet, lack of salt or dietary monotony may contribute to the problem.

This model explains why many researchers have been unable to initiate tail-biting by feeding a low-salt diet, since phase one is required to create an outbreak, and salt deficiency only becomes a factor in phase two. Furthermore, it encompasses the many factors suggested to predispose pigs to tail-biting, such as overcrowding, poor ventilation, lack of feeder or waterer space, etc. (van Putten, 1969; Jericho and Church, 1972; Ewbank, 1973; Kelley, McGlone and Gaskins, 1980; Penny, Walters and Tredget, 1981; Lohr, 1983; Fraser, 1987a).

Fraser (1987b) proposes that stress may also contribute to the problem. For example, it has been shown that elevation of circulating ACTH, which can occur under the influence of 'stress', increases sodium appetite in other species such as sheep (Weisinger *et al.*, 1980) and rats (Weisinger *et al.*, 1978). Fraser hypothesizes that stress in pigs may also increase sodium appetite and thus exacerbate the attraction to blood as described above.

References

ADROGUE, H.J. and MADIAS, N.E. (1981). *American Journal of Medicine*, **71**, 456–467

ALCANTARA, P.F., HANSON, L.E. and SMITH, J.D. (1980). *Journal of Animal Science*, **50**, 1092–1101

ALPERIN, R.J., HOWLIN, J.J., PREISIG, P.A. and WONG, K.R. (1985). *Journal of Clinical Investigation*, **76**, 1360–1366

AMMON, H.V., THOMAS, P.J. and PHILLIPS, S.F. (1974). *Journal of Clinical Investigation*, **53**, 374–379

ANDERSSON, B., ANDERSSON, H., AUGUSTINSSON, O., FORSGREN, M., HOLST, H. and JONASSON, H. (1986). *Acta Physiologica Scandinavica*, **126**, 9–14

ARNAUD, J. and LACHANCE, P.A. (1980). *Journal of Nutrition*, **110**, 2480–2489

AUSTIC, R.E. and PATIENCE, J.F. (1988). *Critical Reviews in Poultry Biology*, **1**, 315–345

BASTL, C.P., BINDER, H.J. and HAYSLETT, J.P. (1980). *American Journal of Physiology*, **238**, F181–F186

BAYLIS, P.H., ZERBE, R.L. and ROBERTSON, G.L. (1981). *Journal of Clinical Endocrinology and Metabolism*, **53**, 935–940

BIA, M.J. and DeFRONZO, R.A. (1981). *American Journal of Physiology*, **240**, F257–F268

BLOCK, E. (1984). *Journal of Dairy Science*, **67**, 2939–2948

BLOOM, W.L. and MITCHELL, W. JR (1960). *Archives of Internal Medicine*, **106**, 321–326

BOULTER, P.R., HOFFMAN, R.S. and ARKY, R.A. (1973). *Metabolism*, **22**, 675–683

BROBST, D. (1986). *Journal of the American Veterinary Medical Association*, **188**, 1019–1025

BROSNAN, J.T. and BROSNAN, M.E. (1982). In *Advances in Nutrition Research*, pp. 77–105. Ed. Draper, H.H. Plenum Press, New York

BURNELL, J.M., VILLAMIL, M.F., UYENO, B.T. and SCRIBNER, B.H. (1956). *Journal of Clinical Investigation*, **35**, 935–939

CASH, R.A. (1983). In *Diarrhea and Malnutrition*, pp. 203–222. Ed. Chen, L.C. and Scrimshaw, N.S. Plenum Press, New York

CHAN, J.C.M. (1981). *Federation Proceedings,* **40**, 2423–2428

CHUI, Y.N., AUSTIC, R.E. and RUMSEY, G.L. (1984). *Comparative Biochemistry and Physiology,* **78B**, 777–783

COHEN, R.D. (1984). In *Clinical Physiology,* pp. 1–40. Ed. Campbell, E.J.M., Dickinson, C.J., Slater, J.D.H., Edwards, C.R.W. and Sikora, K. Blackwell Scientific, Oxford

CORRADINO, R.A. (1979). *Archives of Biochemistry and Biophysics,* **192**, 302–310

COX, M., STERNS, R.H. and SINGER, I. (1978). *New England Journal of Medicine,* **299**, 525–532

DENTON, D. (1982). *The Hunger for Salt.* Springer, Berlin

DISHINGTON, I.W. (1975). *Acta Veterinaria Scandinavica,* **16**, 503–512

DRENICK, E.J., BLAHD, W.H., SINGER, F.R. and LEDERER, M. (1966). *American Journal of Clinical Nutrition,* **18**, 278–285

ECKEL, R.E., NORRIS, J.E. and POPE, C.E. II (1958). *American Journal of Physiology,* **193**, 644–652

EDELMAN, I.S. and LEIBMAN, J. (1959). *American Journal of Medicine, 27,* 256–277

EWBANK, R. (1973). *British Veterinary Journal,* **129**, 366–369

FIELD, M.J., STANTON, B.A. and GIEBISCH, G.H. (1984). *Kidney International,* **25**, 502–511

FITZSIMMONS, J.T. (1972). *Physiological Reviews,* **52**, 468–561

FOX, J., CARE, A.D. and MARSHALL, D.H. (1978). *Journal of Endocrinology,* **78**, 187–194

FRASER, D. (1987a). *Applied Animal Behaviour Science,* **17**, 61–68

FRASER, D. (1987b). *Canadian Journal of Animal Science,* **67**, 909–918

FREGLY, M.J. (1981). *Annual Review of Nutrition,* **1**, 69–93

GARDENSWARTZ, M.H. and SCHRIER, R.W. (1982). In *Sodium: its Biological Significance,* pp. 19–71. Ed. Papper, S. CRC Press, Inc., Boca Raton

GEORGIEVSKII, V.I. (1982a). In *Mineral Nutrition of Animals,* pp. 91–170. Ed. Georgievskii, V.I., Annenkov, B.N. and Samokhin, V.T. Butterworths, London

GEORGIEVSKII, V.I. (1982b). In *Mineral Nutrition of Animals,* pp. 79–89. Ed. Georgievskii, V.I., Annenkov, B.N. and Samokhin, V.T. Butterworths, London

HAMILTON, R.M.G. and THOMPSON, B.K. (1980). *Poultry Science,* **59**, 1294–1303

HEROUX, O. (1981). In *Handbook of Nutritional Requirements in a Functional Context,* pp. 523–542. Ed. Rechcigl, M. CRC Press, Inc., Boca Raton

JELLIFFE, E.F.P., JELLIFFE, D.B., FELDIN, K. and NGOKWEY, N. (1987). *World Review of Nutrition and Dietetics,* **53**, 218–295

JERICHO, K.W.F. and CHURCH, T.L. (1972). *Canadian Veterinary Journal,* **13**, 156–159

JOHNSON, R.J. and FARRELL, D.J. (1985). In *Energy Metabolism of Farm Animals,* pp. 102–105. Ed. Moe, P.W., Tyrrell, H.F. and Reynolds, P.J. European Assoc. for Animal Production

KELLY, K. W., McGLONE, J.J. and GASKINS, C.T. (1980). *Journal of Animal Science,* **50**, 336–341

KEM, D.C. and TRACHEWSKY, D. (1983). In *Potassium: Its Biological Significance,* pp. 25–36. Ed. Whang, R. CRC Press, Inc., Boca Raton

KENNY, A.D. (1981). *Intestinal Calcium Absorption and its Regulation.* CRC Press, Inc., Boca Raton

KLEIN, R.G., ARNAUD, S.B., GALLAGHER, J.C., DeLUCA, H.F. and RIGGS, B.L. (1977). *Journal of Clinical Investigation,* **60**, 253–259

KREHL, W.A. (1966). *Nutrition Today,* **1**, 20–23

LEGROS, J.J. (1979). In *Mechanisms of Osmoregulation in Animals*, pp. 611–636. Ed. Gilles, R. John Wiley and Sons, Chichester

LEIBHOLZ, J.M., McCALL, J.T., HAYS, V.W. and SPEER, V.C. (1966). *Journal of Animal Science*, **25**, 37–43

LEMANN, J. JR, LITZOW, J.R. and LENNON, E.J. (1966). *Journal of Clinical Investigation*, **45**, 1608–1614

LENNON, E.J., LEMANN, J. JR, and LITZOW, J.R. (1966). *Journal of Clinical Investigation*, **45**, 1601–1607

LINDEMAN, R.D. and PEDERSON, J.A. (1983). In *Potassium: Its Biological Significance*, pp. 45–75. Ed. Whang, R. CRC Press, Inc., Boca Raton

LOHR, J.E. (1983). *New Zealand Veterinary Journal*, **31**, 205

MacKAY, W.C. (1979). *Comparative Animal Nutrition*, **3**, 80–99

MADEC, F., CARIOLET, R. and DANTZER, R. (1986). *Annales de Recherches Veterinaires*, **17**, 177–184

MALOIY, G.M.O., MacFARLANE, W.V. and SHKOLNIK, A. (1979). In *Comparative Physiology of Osmoregulation in Animals*, pp. 185–209. Ed. Maloiy, G.M.O. New York, Academic Press

MASON, G.D. and SCOTT, D. (1972). *Quarterly Journal of Experimental Physiology*, **57**, 393–403

MILLER, D. (1970). *Poultry Science*, **49**, 1535–1540

MONGIN, P. (1981). *Proceedings of the Nutrition Society*, **40**, 285–294

NALIN, D.R., CASH, R.A., RAHMAN, M. and YUNUS, M.D. (1970). *Gut*, **11**, 768–772

O'DELL, B.L. and SAVAGE, J.E. (1966). *Journal of Nutrition*, **90**, 364–370

PALMER, D.L., KOSTER, F.T., ISLAM, A.F.M.R., RAHMAN, A.S.M.M. and SACK, R.B. (1977). *New England Journal of Medicine*, **297**, 1107–1110

PATIENCE, J.F. and WOLYNETZ, M.S. (1987). *Journal of Animal Science*, **65** (Supp. 1): 303–304

PATIENCE, J.F., AUSTIC, R.E. and BOYD, R.D. (1986). *Nutrition Research*, **6**, 263–273

PATIENCE, J.F., AUSTIC, R.E. and BOYD, R.D. (1987a). *Feedstuffs*, **59**(27), 13 and 15–18

PATIENCE, J.F., AUSTIC, R.E. and BOYD, R.D. (1987b). *Journal of Animal Science*, **64**, 457–466

PATIENCE, J.F., AUSTIC, R.E. and BOYD, R.D. (1987c). *Journal of Animal Science*, **64**, 1079–1085

PATIENCE, J.F., WOLYNETZ, M.S., FRIEND, D.W. and HARTIN, K.E. (1987d). *Canadian Journal of Animal Science*, **67**, 859–863

PENNY, R.H.C., WALTERS, J.R. and TREDGET, S.J. (1981). *The Veterinary Record*, **108**, 35

PIERCE, N.F., SACK, S.B., MITRA, R.C., BANWELL, J.C., BRIGHAM, K.L., FEDSON, D.S. and MONDAL, A. (1969). *Annals of Internal Medicine*, **70**, 1173–1181

POND, W.G. and HOUPT, K.A. (1978). *The Biology of the Pig*. Cornell University Press, Ithaca

QUINTON, P.M. (1979). *Comparative Animal Nutrition*, **3**, 100

REDDY, G.S., JONES, G., KOOH, S.W. and FRASER, D. (1982). *American Journal of Physiology*, **243**, E265–E271

ROBERTSON, G.L. (1977). *Recent Progress in Hormone Research*, **33**, 333–374

ROBERTSON, G.L. and BERL, T. (1985). In *The Kidney*, pp. 385–432. Ed. Brenner, B.M. and Rector, F.C. W. Saunders Co., Philadelphia

ROSE, D.B. (1977). *Clinical Physiology of Acid-Base and Electrolyte Disorders*. McGraw-Hill Book Co., New York

RUDD, T.G., PAILTHORP, K.G. and NELP, W.B. (1972). *Journal of Laboratory and Clinical Medicine*, **80**, 442–448

RUDE, R.K. and SINGER, F.R. (1982). In *Disorders of Mineral Metabolism*, pp. 482–556. Ed. Bonner, F. and Coburn, J.W. Academic Press, New York

SAUVEUR, B. (1984). *World's Poultry Science*, **40**, 195–206

SAVAGE, J.E. (1972). *Poultry Science*, **51**, 35–43

SCHLOEDER, F.X. and STINEBAUGH, B.J. (1966). *Metabolism*, **15**, 838–846

SCOTT, R.L. and AUSTIC, R.E. (1978). *Journal of Nutrition*, **108**, 137–144

SIENKO, M.J. and PLANE, R.A. (1979). *Chemistry Principles and Applications*. McGraw-Hill Book Co., New York

STERN, P.H. (1981). *Calcified Tissue International*, **33**, 1–4

TANNEN, R.L. (1986). In *Fluids and Electrolytes*, pp. 150–228. Ed. Kokko, J.P. and Tannen, R.L. W.B. Saunders Co., Philadelphia

TEETER, R.G., SMITH, M.O., OWENS, F.N., ARP, S.C., SANGIAH, S. and BREAZILE, J.E. (1985). *Poultry Science*, **64**, 1060–1064

THACKER, E.J. (1959). *Journal of Nutrition*, **69**, 28–32

THORNTON, S.N., BLADWIN, B.A. and PURDEW, T. (1985). *Quarterly Journal of Experimental Physiology*, **70**, 549

THRASHER, T.N., KEIL, L.C. and RAMSAY, D.J. (1982). *American Journal of Physiology*, **243**, R354–R362

VAN DER WAL, P.G., HEMMINGA, H., GOEDEGEBUURE, S.A. and VAN DER VALK, P.C. (1986). *Veterinary Quarterly*, **8**, 136–144

VAN PUTTEN, G. (1969). *British Veterinary Journal*, **125**, 511–517

VICK, R.L. (1984). *Contemporary Medical Physiology*. Addison-Wesley Publ. Co., Menlo Park

VOKES, T. (1987). *Annual Review of Nutrition*, **7**, 383–406

WEINBERG, J.M. (1986). In *Fluids and Electrolytes*, pp. 742–759. Ed. Kokko, J.P. and Tannen, R.L. W.B. Saunders Co., Philadelphia

WEISNER, R.L. (1971). *American Journal of Medicine*, **50**, 233–240

WEISINGER, R.S., DENTON, D.A., McKINLEY, M.J. and NELSON, J.F. (1978). *Pharmacology, Biochemistry and Behaviour*, **8**, 339–342

WEISINGER, R.S., COGHLAN, J.P., DENTON, D.A., FAN, J.S.K., HATZIKOSTAS, S., McKINLEY, M.J., NELSON, J.F. and SCOGGINS, B.A. (1980). *American Journal of Physiology*, **239**, E45–E50

YEN, J.T., POND, W.G. and PRIOR, R.L. (1981). *Journal of Animal Science*, **52**, 778–782

ZERBE, R.L. and ROBERTSON, G.L. (1983). *American Journal of Physiology*, **224**, E607–E614

12

PREDICTING NUTRIENT RESPONSES OF THE LACTATING SOW

B. P. MULLAN, W. H. CLOSE
AFRC Institute for Grassland and Animal Production, Pig Department, Church Lane, Shinfield, Reading, RG2 9AQ, UK

and

D. J. A. COLE
University of Nottingham School of Agriculture, Sutton Bonington, Loughborough, Leics, LE12 5RD, UK

Introduction

During the past two decades there have been numerous extensive reviews concerned with either establishing the nutritional requirements of the pregnant and lactating sow or assessing the consequences of nutrition on various aspects of sow performance (Lodge, 1962; Elsley, 1971; Elsley and MacPherson, 1972; Lodge, 1972; O'Grady, 1980; Agricultural Research Council, 1981; Cole, 1982; National Research Council, 1988). In general, these have shown that it is possible to make reasonable estimates of the short-term nutritional requirements of the sow during either a single pregnancy or lactation period, but only a few have considered the consequences of applying these short-term feeding recommendations to long-term nutritional or productive needs. It has been an explicit assumption that having established these requirements, it is possible to develop appropriate feeding strategies and that there are no restrictions to limit their application in practice. Whereas this may be true for the pregnant sow, which is generally fed on a restricted basis, it is not the case for the lactating animal which may not have the capacity to consume sufficient feed to meet its calculated daily needs. This, therefore, raises a number of questions. What are the consequences for the lactating sow if nutrient demand exceeds nutrient supply; how does the animal respond in both the short and long term? How may recent knowledge of nutritional, metabolic and physiological responses be interpreted and applied to provide a better understanding of the long-term nutritional needs and responses of the lactating sow?

The need to establish the response of the modern lactating sow to specified nutritional inputs is generally recognized and indeed, such reviews as ARC (1981) used information which was actually obtained in the 1960s and 1970s when the type of animal, its reproductive capacity, nutritional management and the husbandry and environmental conditions under which it was kept were markedly different from today. Modern sows are mated earlier and at a lighter body weight, they now begin their reproductive life with less body reserves and are expected to sustain a higher level of productivity for a longer period of time than hitherto. They are weaned earlier, kept individually in controlled environmental conditions and often creep feed is not provided with the consequence that growth and development of the suckling piglets are entirely dependent upon milk production and hence appetite of the sow. As a result, the nutritional needs and responses of the lactating sow must be considered more carefully and the consequences for not meeting these needs must be known. However, in assessing requirements it is recognized that responses during lactation cannot be treated in isolation from other events within the same or other parities. For example, both the level and composition of the diet fed during pregnancy influence appetite during lactation and this may have indirect consequences for subsequent reproductive performance, such as reduced ovulation rate and an extended weaning-to-oestrus interval. To take account of these assumptions Cole (1982) has suggested a strategy of maximum conservation which involves a low feeding level to control and limit weight gain during pregnancy followed by a high feeding level to minimize body weight loss and, as far as possible, maintain body condition during lactation.

The factorial estimation of nutrient requirements

The objective of the sow during lactation is to produce sufficient milk to wean an adequate number of piglets of an acceptable body weight, with minimum variation, yet without utilizing excessive body reserves to prejudice subsequent reproductive performance. However, examination of the relationships between maternal feed intake and such aspects of productivity as piglet growth, change in body weight or backfat thickness of the sow during lactation, show the great variability that exists in the ability of the sow to nurture her young or maintain body condition (Figure 12.1). This suggests that many factors, which are not always included as experimental variables, have to be taken into account when assessing responses to nutrients and indicates the limited use that can be made of many empirical relationships to describe nutritional requirements and responses. The alternative approach described in this chapter is therefore based on the factorial procedure which allows the prediction of requirements and the determination and interpolation of changes in productivity according to metabolic and physiological principles. This complements a similar exercise carried out for the pregnant sow which has been reported at a previous University of Nottingham Feed Manufacturers Conference (Williams, Close and Cole, 1985) and which forms the basis of the new AFRC review of the nutrient requirements of sows (Agricultural and Food Research Council, unpublished). The overall objective is to provide a greater scientific basis to interpret changes in performance and to establish guidelines and indices which may then be used to develop feeding strategies to sustain optimum sow productivity.

The primary components necessary to establish the factorial assessment of the nutritional requirements of the lactating sow are:

(1) determination of the nutrient requirements of the suckling piglets at various growth rates, litter sizes and stages of lactation,
(2) calculation of the energy and nitrogen requirements for milk production necessary to sustain different rates of piglet gain,

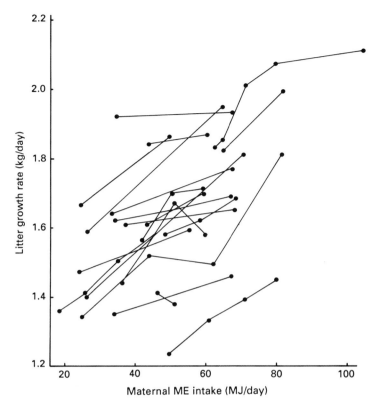

Figure 12.1 The influence of maternal energy intake on growth rate of the litter, taken from various sources (O'Grady *et al.*, 1973; Reese *et al.*, 1982; Danielsen and Nielsen, 1984; King and Williams, 1984a and 1984b; King, Williams and Barker, 1984; Nelssen *et al.*, 1985; Harker, 1986; Johnston *et al.*, 1986; King and Dunkin, 1986a and 1986b; Noblet and Etienne, 1987; Eastham *et al.*, 1988; Yang *et al.*, 1988; Mullan and Close, 1989; Mullan and Williams, 1989)

(3) assessment of the maintenance energy and nitrogen needs of the lactating sow and hence the total nutrient requirements for both maintenance and milk production,

(4) estimation of the energy and nitrogen intake of the sow under *ad libitum* feeding conditions and those factors which influence it,

(5) calculation of the change in lean and fat and hence body weight of the sow together with corresponding changes in backfat thickness.

The factorial procedure assumes that it is possible to partition nutrient needs according to the requirements for maintenance, for tissue deposition within the body and for products formed but subsequently lost from the body. It assumes that sufficient experimentation exists to derive the relationship between nutrient intake and nutrient output according to the scheme presented in Figure 12.2. The maintenance requirement is therefore equivalent to that intake which maintains an animal in a state of equilibrium neither gaining nor losing nutrients. It is normally

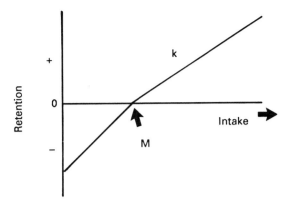

Figure 12.2 Diagrammatic representation of the relationship between nutrient intake (I) and retention (R) in animals. M = maintenance requirement, where $R = 0$; k = net efficiency of nutrient utilization, that is $\Delta R/\Delta I$; total requirement (T) = $M + R/k$

expressed on a metabolic body size basis to take account of variations in body weight, age and maturity. Above maintenance intake the requirement for tissue deposition or product formation is a function of the rate of nutrient retention (or loss) and the net efficiency of nutrient utilization. This varies with the level of animal performance and increases as feeding level increases. Thus, in the diagrammatic representation in Figure 12.2, the total nutrient requirement (T) at any level of performance can be calculated as

$$T = M + R/k \tag{12.1}$$

where M is the requirement for maintenance, R is the nutrient content of the product formed and k is the net efficiency of nutrient utilization. This procedure has been used to calculate the energy and nitrogen needs and hence requires information on various aspects of the energy and nitrogen metabolism of both the suckling piglet and the lactating sow.

THE NUTRITIONAL REQUIREMENTS OF THE SUCKLING PIGLET

A major prerequisite for the factorial estimation of the nutrient requirements of the suckling piglet is knowledge of the pattern and the rate of tissue accretion during lactation and those factors which influence it. If changes in growth rate are to be accommodated, then variation in the rate of protein and fat deposition must be known and these have been calculated from the experiments of Berge and Imdrebø (1954), Brooks *et al.* (1964) and Elliot and Lodge (1977). These show that the composition of the piglet tends to stabilize after the first week of life, but that the rate of fat accretion is much greater than that of protein (Figure 12.3). Thus the mean protein content was calculated to increase from 12% at birth to 14 and 15% at 1 and 2 weeks of age, respectively, and thereafter remained constant; the corresponding increases in mean fat content were from 1.3% at birth to 9, 13, 15 and 16% at 1, 2, 3 and 4 weeks, respectively. These values have therefore been

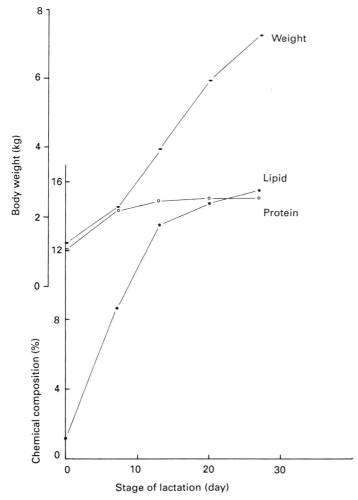

Figure 12.3 The mean change in body weight and body composition of the suckling piglet, compiled from several sources (see text)

used to calculate the rates of protein and fat deposition and hence energy and nitrogen contents of suckling piglets at various stages of lactation and rates of growth.

From the above values of protein and fat content the daily rates of tissue accretion can be calculated and the energy requirements of the suckling pig determined from the knowledge that 1 g protein contains 23.8 kJ, 1 g fat contains 39.8 kJ, assuming that the net efficiency of energy utilization is 0.78 and that the maintenance energy requirement is 498 kJ/kg body weight$^{0.75}$ per day (Close, unpublished). Similarly, the nitrogen requirements have been calculated on the assumptions that 1 g protein contains 0.16 g nitrogen (N), the nitrogen requirement for maintenance is 0.34 g digestible N/kg body weight$^{0.75}$ per day and the efficiency of nitrogen accretion is 0.90 (Burlacu, Iliescu and Cárámida, 1986; Beyer, 1986).

Table 12.1 THE ENERGY AND NITROGEN REQUIREMENTS OF SUCKLING PIGLETS (LITTER SIZE = 10; NO CREEP FEED PROVIDED)

	Stage of lactation			
	Week 1	*Week 2*	*Week 3*	*Week 4*
Litter weight gain (kg/day)	1.50	2.00	2.50	2.40
Nitrogen deposition in gain (g/day)	40.0	53.6	60.0	58.4
Fat deposition in gain (g/day)	280	390	480	470
Maintenance				
Energy (MJ ME/day)	7.7	11.7	15.7	19.9
Nitrogen (g N/day)	5.3	8.0	10.7	13.6
Tissue gain				
Energy content (MJ ME/day)	17.1	23.5	28.0	27.4
Nitrogen content (g N/day)	40.0	53.6	60.0	58.4
Requirement for tissue gain				
Energy (MJ ME/day)	21.9	30.1	35.9	35.1
Nitrogen (g digestible N/day)	44.4	59.6	66.7	64.9
Total requirement from milk				
Energy (MJ ME/day)	29.6	41.8	51.6	55.0
Nitrogen (g digestible N/day)	49.7	67.6	77.4	78.5

These values relate to the energy and nitrogen requirement for tissue gain in the piglets and can be converted into that required from milk on the basis that the digestibility of nutrients in milk is 0.95 (ARC, 1981) (Table 12.1).

THE REQUIREMENTS FOR MAINTENANCE AND MILK PRODUCTION

If the maintenance requirement of the sow is known then it is possible to determine the total requirements for both milk production and maintenance. Similarly, if dietary nutrient intake is known the rate of gain or loss of energy and nitrogen during lactation can be determined by difference. From this the rate and composition of the change in body weight of the sow can be determined.

The best estimate of the maintenance energy requirement of the lactating sow has been calculated from several recent experiments to be 471 kJ ME/kg body weight$^{0.75}$ per day (Beyer, 1986; Burlacu, Iliescu and Cáramida, 1983; Kirchgessner, 1987; Noblet and Etienne, 1987; Verstegen *et al.*, 1985). The corresponding values for the net efficiency of energy utilization were 0.72 when dietary energy was in excess, or 0.87 when dietary energy intake was insufficient to meet nutrient needs and body tissue was mobilized to provide the deficit. For estimation of the dietary nitrogen requirements, the mean nitrogen for maintenance was calculated to be 0.38 g digestible N/kg body weight$^{0.75}$ per day, with a corresponding efficiency of utilization of 0.70 (Beyer, 1986; Burlacu, Iliescu and Cáramida, 1983, 1986). It has been further assumed that the digestibility of dietary nitrogen for the lactating sow is 0.88 (Noblet and Etienne, 1987). Details of these calculations are presented in Tables 12.1 and 12.2 and show the changes in the requirements of the sow during lactation at rates of piglet gain likely to be

Table 12.2 NUTRIENT REQUIREMENTS AND BODY TISSUE
CHANGE DURING LACTATION (LITTER SIZE = 10; NO CREEP FEED
PROVIDED; 160 kg SOW POSTPARTUM)

	Stage of lactation			
	Week 1	*Week 2*	*Week 3*	*Week 4*
Maintenance				
Energy (MJ ME/day)	21.2	20.9	20.6	20.2
Nitrogen (g digestible N/day)	17.1	16.9	16.6	16.3
Milk production				
Energy (MJ ME/day)	41.6	58.9	72.4	77.2
Nitrogen (g digestible N/day)	74.7	101.7	116.4	118.0
Total requirements				
Energy (MJ ME/day)	62.8	79.8	93.0	97.4
Nitrogen (g digestible N/day)	91.8	118.6	133.0	134.3
Intake				
Energy (MJ ME/day)	51	67	74	78
Nitrogen (g digestible N/day)	87.0	113.2	125.8	132.7
Deficit (from body reserves)				
Energy (MJ ME/day)	−12	−13	−19	−19
Nitrogen (g digestible N/day)	−4.8	−5.4	−7.2	−1.6
Loss of body tissue				
Lean (kg/day)	0.16	0.18	0.23	0.06
Fat (kg/day)	0.24	0.26	0.36	0.41
Body weight (kg/day)	0.42	0.46	0.62	0.49

achieved in practice. However the question which arises is whether the sow has the capacity to consume sufficient nutrients in the food to meet total requirements and, if not, to what extent must body tissue, that is both lean and fat, be mobilized to meet the deficit. This requires knowledge of factors influencing the voluntary feed intake of the lactating sow.

Voluntary feed intake

There are few estimates of the true voluntary feed intake of the lactating sow since by definition it refers to *ad libitum* feeding where the animal has continuous access to a supply of fresh food and water (Cole, 1984). However the National Research Council (1987) recently summarized the available data and reported that the average feed intake of gilts was 15% less than that for sows (two or more parities), 4.36 compared with 5.17 kg/day respectively, but commented on the great variability between estimates which reflects the multitude of factors that can influence feed intake. From limited experimentation they suggested that the change in energy intake (DE, MJ/day) with time after farrowing (t, days) was best estimated as

$$DE = 56.066 + 2.494t - 0.072t^2 \tag{12.2}$$

with the relationship being valid for lactation periods of 28 days or less. However, this equation has been derived from experiments which did not necessarily practise

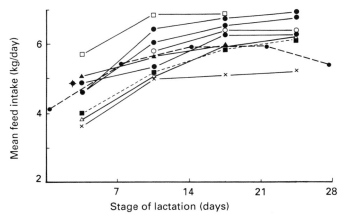

Figure 12.4 The effect of stage of lactation on the voluntary feed intake of the sow taken from several sources (O—O, Mahan and Mangan (1975); x—x, O'Grady and Lynch (1978); ▲—▲, Stahly, Cromwell and Simpson (1979); △—△, Boyd *et al.* (1985); ◆ Moser *et al.* (1987); ●---●, NRC (1987); □—□, Mullan and Close (unpublished); ●—●, Zhu and Cole (unpublished); ■---■, predicted

true *ad libitum* feeding and does not take account of differences in body weight, level of performance or suckling intensity. On the basis of the data summarized in Figure 12.4, the approach adopted in this chapter has been to assume that a sow fed *ad libitum* will attain its maximum feed intake during the fourth week of lactation, and that mean feed intake in the preceding weeks is then a percentage of that maximum value, corresponding to 65, 85 and 95% for weeks 1, 2 and 3, respectively (Pettigrew *et al.*, 1986). In order to take account of the increase in maintenance energy requirement and hence increase in feed intake with increase in body weight, maximum intakes of 5.5, 6.1 and 6.7 kg/day were set for sows weighing 140, 160 and 180 kg postpartum, respectively, corresponding to sows in their first, second and subsequent parities and fed conventional cereal-based diets.

Although these values relate to sows suckling 10 piglets throughout lactation, it is recognized that milk yield varies with litter size (ARC, 1981). O'Grady, Lynch and Kearney (1985) indicated that the animals' voluntary feed intake increased by 0.2 kg per piglet per day during lactation, somewhat less than that of 0.5–0.6 kg per additional piglet per day calculated by Verstegen *et al.* (1985) from energy balance studies. The increase is, however, likely to be greater the lower the litter size so that at larger litter sizes the increased feed intake does not meet the additional nutrient demand and hence the growth rate of individual piglets will be reduced and/or the rate of body tissue mobilized from the sow increased.

There are a number of factors which may influence the voluntary feed intake of the lactating sow, the primary one being the environmental conditions to which the animals are exposed (O'Grady and Lynch, 1978). The environmental temperature is particularly important since both Lynch (1977) and Stansbury, McGlone and Tribble (1987) have shown that each 1°C increase in temperature above 21 and 18°C reduced feed intake by approximately 0.1 and 0.2 kg/day, respectively, with concomitant effects upon both the liveweight loss of the sow and the growth rate of the piglets. The reduced performance of the piglets resulted from the reduced milk yield of the sow and these effects illustrate the complexity of assessing and arranging the environment within the farrowing house to meet the needs of both the sow and her piglets.

There are also important associations both within and between parities which can affect feed intake of the lactating sow. It is well established that the more feed an animal consumes during pregnancy the lower its voluntary feed intake during lactation (Salmon-Legagneur and Rerat, 1962; Baker *et al.*, 1969; Harker, 1986; Mullan and Williams, 1989). Similarly, Mahan and Mangan (1975) demonstrated a close interaction between protein level fed during both pregnancy and lactation; the lower the crude protein content of the diet fed during both pregnancy and lactation the lower the voluntary feed intake during lactation. Maximum intake during lactation was achieved when a diet containing 120 g crude protein/kg was fed during pregnancy followed by one containing 180 g/kg during lactation. However, it may not necessarily be the food intake *per se* during pregnancy which influences subsequent voluntary feed intake but the body weight, rate of gain or level of fatness of the sow at farrowing (O'Grady, Lynch and Kearney, 1985; Mullan and Williams, 1989). If this is the case then feeding a diet to enhance the deposition of fat rather than body lean during pregnancy may have an adverse effect on feed intake during lactation.

Voluntary feed intake of sows may fluctuate considerably during lactation (McGrath, 1981) and there is considerable variation between animals of similar parity (Lynch and O'Grady, 1988). Data from the Moorepark herd indicate that 40% of sows during their first lactation failed to consume more than 4 kg/day whereas 45% consumed between 4.1 and 5.0 kg/day. Although the large range in individual intakes and concomitant changes in metabolism may provide an explanation of some of the variation in sow productivity in both the short and long term, there is need for an improved understanding of the control of feed intake in the lactating sow.

Predicted changes in body weight, body composition and backfat thickness

Because nutrient intake, even under *ad libitum* feeding conditions, is often inadequate to meet metabolic needs, body tissue is mobilized to compensate for the nutrient deficit. For the practical levels of piglet performance indicated in Table 12.2, both energy and nitrogen intake were inadequate to meet requirements and hence both lean and fat reserves of the sow were mobilized throughout the 4-week lactation. The extent to which this reflects a reduction in body weight was calculated on the basis that each g of nitrogen is equivalent to 6.25 g protein and that protein gain represents 22% of the lean (Shields and Mahan, 1983). Since the energy deficit represents the loss of both protein and fat, the energy contained as fat can be calculated since each g of protein contains 23.8 kJ. The weight of fat mobilized is then calculated assuming an energetic efficiency of 0.85 and an energy content for fat of 39.7 kJ/g. Total body-weight change was calculated on the basis that lean and fat represent 95% of the total (Mullan, 1987). A more detailed account of the loss of body tissue which occurs during lactation is presented in Table 12.2. In this example, a sow weighing 160 kg postpartum and suckling 10 piglets, each gaining 200 g/day, would lose 13.8 kg body weight during a 4-week lactation, comprising a 4.4 kg reduction in the lean and an 8.9 kg reduction in the fat content of the body.

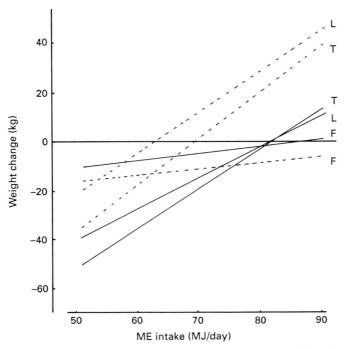

Figure 12.5 Predicted changes in body weight and body composition of sows over a 28-day lactation when fed an isoenergetic diet containing either 140 (——) or 180 (– – –) g CP/kg at different levels of intake. T, L and F are the respective changes in body weight, body lean and body fat

Such procedures as those presented can be used to calculate the change in both body weight and body composition during lactation for a range of different nutritional circumstances and Figure 12.5 has been compiled to illustrate the effects of feeding diets of similar energy but varying protein content. Increasing the protein content of the diet leads to a higher rate of body gain at any given level of intake. At similar levels of bodyweight loss there were differences in the type and extent of body tissue mobilized so that more lean and less fat was catabolized at the lower dietary protein content. Such procedures therefore allow changes in both body weight, and body composition, to be predicted and interpolated for any given dietary circumstance, suckling intensity and body weight, and hence provide not only a greater appreciation of maternal responses but also of how changes in body composition and the dynamics of nutrient utilization during lactation may influence subsequent sow productivity.

While the model can be useful in assessing the likely consequences of a change in feeding practice, its application may be extended if the procedures can be used to predict the changes in body composition, and especially fat reserves, for additional production guidelines such as change in backfat thickness. This would provide a practical framework for testing the suggestion that it is the loss of body fat during lactation which may influence subsequent reproductive performance (Aherne and Kirkwood, 1985). However such procedures require knowledge of the relationship between body weight, backfat thickness and body fat content.

Quantitative data on the body composition of the lactating sow were not presented by ARC (1981). Indeed, in that review maternal weight loss during lactation was considered to be exclusively fat, although subsequent experimentation (King and Williams, 1984b; Etienne, Noblet and Desmoulin, 1985; Mullan and Williams, 1988), as well as the current model predictions, have shown that considerable quantities of maternal lean tissue are also mobilized during lactation. In addition, there have been a number of subsequent experiments which have examined the relationship between body lipid content and backfat thickness. Harker (1986) slaughtered gilts at mating, at days 90 and 110 of pregnancy and at weaning and found that generally there was a consistent relationship between total body lipid and backfat thickness at the different stages of the reproductive cycle. Consequently, it has been assumed in the present chapter that it is valid to pool data from sows slaughtered at different stages of reproduction (Harker, 1986; King, Spiers and Eckerman, 1986; King and Dove, unpublished; Mullan, 1987). The relationship between total chemical body lipid (L, kg), body weight (W, kg) and depth of backfat measured at the P_2 position (P_2, mm) derived from these data was:

$$L = W (0.128 + 0.0088 P_2); RSD = 3.46, R^2 = 0.76, n = 131 \qquad (12.3)$$

Equation 12.3 has been derived exclusively from first-parity gilts. Whittemore, Franklin and Pearce (1980) and Lee (1989) slaughtered sows following their second and third lactations, respectively, and at the same level of backfat the proportion of lipid in the body was less than that for the young sow. This may be due to a change in the chemical composition of the backfat with increased hydration of the fat depots during periods of fat mobilization (Lee, Close and Wood, 1989) or because backfat thickness and hence body reserves are lower in older sows.

From the above equation it may be calculated that a sow weighing 160 kg body weight postpartum and having a P_2 measurement of 20 mm would have a body fat content of 46.7 kg. On the basis of the results presented in Table 12.2 this would be reduced to 37.8 kg after a 4-week lactation, equivalent to a 5 mm decrease in P_2. This reduction in P_2 is consistent with the results of Eastham *et al.* (1988) who measured a 6.1 mm change in P_2 of sows which lost 15 kg body weight over a 28-day lactation, those of King and Dunkin (1986a) where a 19.6 kg loss in body weight was associated with a 5.7 mm change in P_2 and is also comparable with that reported by King (1987). It is important to note, however, that the above relationship has been derived from first-parity sows fed conventional diets and hence any predictions with older sows should be interpreted cautiously.

Lactation and subsequent reproduction

Reproductive failure is the major reason for sows being culled prematurely from the breeding herd (Dagorn and Aumaitre, 1979). A major component of this can be attributed to a delayed resumption of oestrus activity after weaning particularly with young sows (King, 1987) and to the low levels of body reserves in sows at the start of lactation (King and Dunkin, 1986a; Mullan and Williams, 1989). The loss of body fat (Aherne and Kirkwood, 1985) and body protein (King, 1987) during lactation has been implicated as being responsible for the cessation of reproductive function, but it is not clear what signals (hormonal or metabolic) mediate the response to the nutritional state of the sow (Britt, Armstrong and Cox, 1988).

The metabolic state of the sow during lactation may also influence subsequent litter size. The failure of young sows to revert from a catabolic phase during lactation to an anabolic phase immediately after weaning has been suggested by Brooks (1982) to be the primary reason for some sows producing smaller litters in their second parity compared with the first. The beneficial effect on litter size of a delay in the age at which a gilt is first mated (Mercer and Francis, 1988) and a delay in the time of remating after the first lactation may indicate an association between the nutritional state of the animal and ovulation rate. Similarly the practice of split-weaning, that is weaning of the heaviest piglets some days before their smaller litter mates, has been shown to be beneficial in reducing the weaning to oestrus interval (Britt, 1988). This again demonstrates the possible association between the nutritional, metabolic and physiological state of the sow and its subsequent reproductive performance.

Conclusions

The purpose of this chapter has been to describe a factorial procedure to predict the response of the lactating sow to nutritional manipulation. This objective has been achieved but there is little empirical information to validate the predictions and this must await further experimentation. However, while such exercises are useful it is

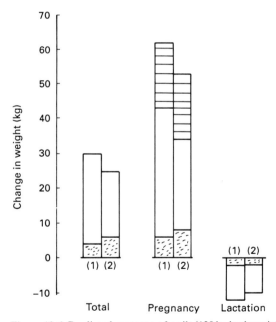

Figure 12.6 Predicted response of a gilt (120 kg body weight at mating) during its first parity, and fed 2.0 kg/day throughout pregnancy and *ad libitum* during lactation: (1) a diet containing 160 g CP/kg during both pregnancy and lactation or (2) a diet containing 130 g CP/kg during pregnancy and 160 g CP/kg during lactation. All diets contain 13 MJ DE/kg. Total is the combined net response during both pregnancy and lactation. □, body lean; ▣, body fat; ▤, products of conception

recognized that the response of the sow during lactation cannot be considered in isolation from events within the same or other parities. It is therefore important to consider the reproductive cycle as a whole, that is pregnancy together with lactation, since O'Grady (1980) and Cole (1982) have suggested that it is the balance of nutrients throughout the whole reproductive period that is important. This has been attempted by extending the predictive model developed by Williams, Close and Cole (1985) for the pregnant sow to include that developed in this chapter for the lactating animal. A typical example of the response obtained is presented in Figure 12.6 and shows the extent to which body weight, and lean and fat reserves, change during both pregnancy and lactation under different dietary situations. In this particular example the changes which result when a single diet is fed during the whole reproductive cycle (160 g CP/kg) are compared with those which occur when a two-diet system in which separate diets are fed in both pregnancy (130 g CP/kg) and lactation (160 g CP/kg). Feeding the single diet system resulted in a higher gain in body lean and a lower gain in body fat, but a greater total bodyweight gain than the two-diet system. However when extended over four parities the predicted increases in body weights were similar, 111 and 107 kg, although lean gain was approximately 10 kg higher and fat gain 10 kg lower with the single compared with the two-diet system. While the consequences of such changes in body composition are not yet known, relating them to changes in reproductive performance should allow additional practical indices and guidelines to be developed which facilitate a better understanding of the associations between nutrition, metabolism and reproductive performance of sows.

References

AGRICULTURAL RESEARCH COUNCIL (1981). *The Nutrient Requirements of Pigs.* Commonwealth Agricultural Bureaux, Slough

AHERNE, F.X. and KIRKWOOD, R.N. (1985). *Journal of Reproduction and Fertility,* (Supplement) **33**, 169–183

BAKER, D.H., BECKER, D.E., NORTON, H.W., SASSE, C.E., JENSEN, A.H. and HARMON, B.G. (1969). *Journal of Nutrition,* **97**, 489–495

BERGE, S. and INDREBØ, T. (1954). *Meldinger fra Norges Landbrukshøgskole,* **34**, 481–500

BEYER, M. (1986). *Untersuchungen zum Energie- und Stoffumsatz von graviden und laktierenden Sauen sowie Saugferkeln – ein Beitrag zur Prazisierung des Energie- und Proteinbedarfes.* Promotionsarbeit, aus dem Forschungszentrum fur Tierproduktion, Dummerstorf-Rostock

BOYD, R.D., HARKINS, M., BAUMAN, D.E. and BUTLER, W.R. (1985). *Proceedings of the 1985 Cornell Nutrition Conference,* pp. 10–19. Cornell University, Ithaca, New York

BROOKS, C.C., FONTENOT, J.P., VIPPERMAN, P.E., THOMAS, H.R. and GRAHAM, P.P. (1964). *Journal of Animal Science,* **23**, 1022–1026

BROOKS, P.H. (1982). In *Control of Pig Reproduction,* pp. 211–224. Ed. Cole, D.J.A. and Foxcroft, G.R. Butterworths, London

BRITT, J.H. (1988). *Journal of Animal Science,* **63**, 1288–1296

BRITT, J.H., ARMSTRONG, J.D. and COX, N.M. (1988). In *11th International Congress on Animal Reproduction and Artificial Insemination,* pp. 117–125. Dublin

BURLACU, G., ILIESCU, M. and CÁRÁMIDA, P. (1983). *Archiv für Tierernahrung*, **33**, 23–45

BURLACU, G., ILESCU, M. and CÁRÁMIDA, P. (1986). *Archiv for Animal Nutrition*, **36**, 803–825

COLE, D.J.A. (1982). In *Control of Pig Reproduction*, pp. 603–619. Ed. Cole, D.J.A. and Foxcroft, G.R. Butterworths, London

COLE, D.J.A. (1984). In *Fats in Animal Nutrition*, pp. 301–312. Ed. Wiseman, J. Butterworths, London

DAGORN, J. and AUMAITRE, A. (1979). *Livestock Production Science*, **6**, 167–177

DANIELSEN, V. and NIELSEN, H.E. (1984). *Paper presented at the 35th Annual Meeting of the European Association for Animal Production*. The Hague

EASTHAM, P.R., SMITH, W.C., WHITTEMORE, C.T. and PHILLIPS, P. (1988). *Animal Production*, **46**, 71–77

ELLIOT, J.I. and LODGE, G.A. (1977). *Canadian Journal of Animal Science*, **57**, 141–150

ELSLEY, F.W.H. (1971). In *Lactation*, pp. 393–411. Ed. Falconer, I.R. Butterworths, London

ELSLEY, F.W.H. and MacPHERSON, R.M. (1972). In *Pig Production*, pp. 417–434. Ed. Cole, D.J.A. Butterworths, London

ETIENNE, M., NOBLET, J. and DESMOULIN, B. (1985). *Reproduction Nutrition Development*, **25**, 341–344

HARKER, A.J. (1986). *Nutrition of the Sow*, PhD Thesis, University of Nottingham

JOHNSTON, L.J., ORR, D.E., TRIBBLE, L.F. and CLARK, J.R. (1986). *Journal of Animal Science*, **63**, 804–814

KING, R.H. (1987). *Pig News and Information*, **8**, 15–22

KING, R.H. and DUNKIN, A.C. (1986a). *Animal Production*, **42**, 119–125

KING, R.H. and DUNKIN, A.C. (1986b). *Animal Production*, **43**, 319–325

KING, R.H. and WILLIAMS, I.H. (1984a). *Animal Production*, **38**, 241–247

KING, R.H. and WILLIAMS, I.H. (1984b). *Animal Production*, **38**, 249–256

KING, R.H., SPIERS, E. and ECKERMAN, P. (1986). *Animal Production*, **43**, 167–170

KING, R.H., WILLIAMS, I.H. and BARKER, I. (1984). *Proceedings of the Australian Society of Animal Production*, **15**, 412–415

KIRCHGESSNER, M. (1987). In *European Association for Animal Production – Tenth Symposium on Energy Metabolism*, pp. 362–367. Ed. Moe, P.W., Tyrrell, H.F. and Reynolds, P.J. Rowman and Littlefield, Virginia

LEE, P.A. (1989). *Animal Production* (in press)

LEE, P.A., CLOSE, W.H. and WOOD, J.D. (1989). *Animal Production* (in press)

LODGE, G.A. (1962). In *Nutrition of Pigs and Poultry*, pp. 224–237. Ed. Morgan, J.T. and Lewis, D. Butterworths, London

LODGE, G.A. (1972). In *Pig Production*, pp. 399–416. Ed. Cole, D.J.A. Butterworths, London

LYNCH, P.B. (1977). *Irish Journal of Agricultural Research*, **16**, 123–130

LYNCH, P.B. and O'GRADY, J.F. (1988). *Proceedings of the International Conference – Improvement of Reproductive Efficiency in Pigs*, Italy (in press)

McGRATH, F. (1981). *Level of Feeding of Lactating Sows*, Thesis, University of Nottingham

MAHAN, D.C. and MANGAN, L.T. (1975). *Journal of Nutrition*, **105**, 1291–1298

MERCER, J.T. and FRANCIS, M.J.H. (1988). *Animal Production*, **46**, 493

MOSER, R.L., CORNELIUS, S.G., PETTIGREW, J.E., HANKE, H.E., HEEG, T.R. and MILLER, K.P. (1987). *Livestock Production Science*, **16**, 91–99

MULLAN, B.P. (1987). *The Effect of Body Reserves on the Reproductive Performance of First-Litter Sows*, PhD Thesis, University of Western Australia

MULLAN, B.P. and CLOSE, W.H. (1989). *Animal Production* (in press)

MULLAN, B.P. and WILLIAMS, I.H. (1988). *Animal Production*, **46**, 495

MULLAN, B.P. and WILLIAMS, I.H. (1989). *Animal Production*, **48**, 449–457

NATIONAL RESEARCH COUNCIL (1987). *Predicting Feed Intake of Food-Producing Animals*. National Academy Press, Washington, DC

NATIONAL RESEARCH COUNCIL (1988). *Nutrient Requirements of Swine – Ninth Revised Edition*. National Academy Press, Washington, DC

NELSSEN, J.L., LEWIS, A.J., PEO, E.R. and CRENSHAW, J.D. (1985). *Journal of Animal Science*, **61**, 1164–1171

NOBLET, J. and ETIENNE, M. (1987). *Journal of Animal Science*, **64**, 774–781

O'GRADY, J.F. (1980). In *Recent Advances in Animal Nutrition – 1980*, pp. 121–131. Ed. Haresign, W. Butterworths, London

O'GRADY, J.F., ELSLEY, F.W.H., MacPHERSON, R.M. and McDONALD, I. (1973). *Animal Production*, **17**, 65–74

O'GRADY, J.F. and LYNCH, P.B. (1978). *Irish Journal of Agricultural Research*, **17**, 1–5

O'GRADY, J.F., LYNCH, P.B. and KEARNEY, P.A. (1985). *Livestock Production Science*, **12**, 355–365

PETTIGREW, J.E., CORNELIUS, S.G., EIDMAN, V.R. and MOSER, R.L. (1986). *Journal of Animal Science*, **63**, 1314–1321

REESE, D.E., MOSER, B.D., PEO, E.R., LEWIS, A.J., ZIMMERMAN, D.R., KINDER, J.E. and STROUP, W.W. (1982). *Journal of Animal Science*, **55**, 590–598

REESE, D.E., PEO, E.R. and LEWIS, A.J. (1984). *Journal of Animal Science*, **58**, 1236–1244

SALMON-LEGAGNEUR, E. and RERAT, A. (1962). In *Nutrition of Pigs and Poultry*, pp. 207–223. Ed. Morgan, J.T. and Lewis, D. Butterworths, London

SHIELDS, R.G. and MAHAN, D.C. (1983). *Journal of Animal Science*, **57**, 594–603

STAHLY, T.S., CROMWELL, G.L. and SIMPSON, W.S. (1979). *Journal of Animal Science*, **49**, 50–54

STANSBURY, W.F., McGLONE, J.J. and TRIBBLE, L.F. (1987). *Journal of Animal Science*, **65**, 1507–1513

VERSTEGEN, M.W.A., MESU, J., VAN KEMPEN, G.J.M. and GEERSE, C. (1985). *Journal of Animal Science*, **60**, 731–740

WHITTEMORE, C.T., FRANKLIN, M.F. and PEARCE, B.S. (1980). *Animal Production*, **31**, 25–31

WILLIAMS, I.H., CLOSE, W.H. and COLE, D.J.A. (1985). In *Recent Advances in Animal Nutrition – 1985*, pp. 133–147. Ed. Haresign, W. and Cole, D.J.A. Butterworths, London

YANG, H., PHILLIPS, P., WHITTEMORE, C.T. and EASTHAM, P.R. (1988). *Animal Production*, **46**, 494

13

AMINO ACID NUTRITION OF PIGS AND POULTRY

D. H. BAKER
Department of Animal Sciences, University of Illinois, Urbana, Illinois, USA

Introduction

Over the past decade a transition has been made by animal nutritionists to the formulation of diets on an amino acid rather than on a protein basis. Formulation for amino acids is nonetheless anything but simple or straightforward. This review attempts to discuss many of the factors complicating studies of amino acid requirements and their utilization.

In the simplest terms, amino acid needs of the pig (or any other species), can be depicted in a simple flow diagram shown in Figure 13.1.

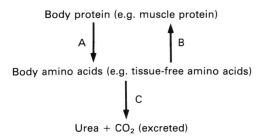

Figure 13.1 Flow diagram of amino acid requirements where A = protein degradation, B = protein synthesis and C = amino acid oxidation

All of the body processes (A, B and C) in Figure 13.1 go on continuously. It is important to note that anywhere from 60% (young animal) to 80% (adult animal) of the amino acid needs for body protein synthesis come from endogenous protein degradation. Hence, the remaining 20–40% must be supplied in the diet. Because individual amino acids have different turnover rates (lysine slow, methionine high) and are therefore depleted via amino acid oxidation from tissue pools at different rates, one cannot assume that the amino acid composition of muscle tissue is predictive of the dietary amino acid requirement pattern. It is for this reason that requirements for individual amino acids must be determined by experiment.

Table 13.1   RESPONSE OF YOUNG CHICKS TO DIETARY LYSINE[a]

Lysine[b] level (%)	Weight gain (g)	Feed intake (g)	Lysine intake (mg)	Feed conversion ratio[d]	Lysine conversion ratio[e]
0	−2.43	39.8	0	—	—
0.1	0.13	45.7	46	352	354
0.2	4.56	50.0	100	10.96	21.9
0.3	16.50	69.6	209	4.22	12.7
0.4	26.60	82.4	330	3.10	12.4
0.5	47.20	111.1	543	2.35	11.5
0.6	68.60	132.6	797	1.93	11.6
0.7	97.60	164.6	1147	1.69	11.7
0.8	118.00	182.0	1456	1.54	12.3
0.9	118.90	174.8	1574	1.47	13.2
1.0	116.50	169.5	1702	1.45	14.6
Pooled SEM	4.92	7.2	NA[c]	NA[c]	NA[c]

[a] Data (Baker, 1978) represent means of triplicate groups of five male chicks per treatment during the period 8–16 days post-hatching; birds averaged 82 g at day 8 post-hatching
[b] Lysine provided as L-lysine.HCl (80% lysine); the basal diet contained 3.8 Mcal ME (15.9 MJ) kg and 15% crude protein
[c] NA = not available due to heterogeneous variance
[d] feed/g liveweight gain
[e] mg lysine intake/g liveweight gain

Before addressing problems of assessing amino acid requirements, some comments should be made about response curves that result from incremental amino acid addition to a diet designed to be specifically deficient in a single amino acid.

Response curves

Few studies have been carried out with pigs in which the entire growth curve has been covered in an amino acid supplementation study. Thus, the points to be made will be based upon an actual lysine experiment conducted with young growing chicks, and it is likely that pigs would respond similarly. Data in Table 13.1 show a lysine response curve. The purified crystalline amino acid diet (Baker, Robbins and Buck, 1979) to which lysine was added contained 3.8 Mcal ME/kg (15.9 MJ/kg) and 15% crude protein (N × 6.25).

A plot of the gain and feed intake data shows that a sigmoidal response occurs, incremental responses per unit of lysine intake being less in both the lower and higher portions of the curve than in the linear response area (between 0.5 and 0.7% lysine). The chicks used in this experiment were very homogeneous. In practice, animals for which requirements are sought are far more heterogeneous. The slope of the line below 0.5% lysine is lower than that between 0.5 and 0.7% lysine because in this region of the growth curve the lysine required for maintenance (zero gain requires about 45 mg lysine) is exerting a substantial effect on the overall efficiency of lysine utilization. At and above 0.7% lysine, some of the animals in the population are beginning to have their lysine requirement satisfied. The closer the lysine level gets to 0.8%, the greater will be the portion of the population that has been satiated insofar as lysine intake is concerned. Because of these phenomena, slope (gain per unit of lysine) should (and does) decrease at an increasing rate until the animal in the population requiring the most lysine has been satiated, at which point no further gain response will occur. This type of response is seen with all amino acids, as has been illustrated and discussed with histidine (Baker, 1986).

While growth of chicks is maximized at 0.80% dietary lysine, higher levels result in the same weight gain but at lower levels of dietary intake under *ad libitum* feeding conditions. Thus, a higher lysine level (0.90%) is required to maximize feed efficiency than is required to maximize weight gain. The same phenomenon occurs with pigs (Brown, Harmon and Jensen, 1973; Lin and Jensen, 1985), optimum feed conversion ratio or maximum carcass leanness requiring higher levels of a limiting amino acid (or crude protein) than that required for maximum weight gain. Because gilts and boars are leaner than barrows, the former have higher amino acid and crude protein requirements than the latter (Baker, 1986; Baker *et al.*, 1967). Also, exogenous protein anabolic substances such as Ractopamine (oral) or porcine growth hormone (injected) have a marked effect in increasing protein and amino acid requirements of pigs (Anderson *et al.*, 1987; Easter 1987).

Factors complicating accurate assessment of amino acid requirements

Because amino acid requirements are not easy to determine with accuracy and precision and because a multitude of factors may influence the 'requirements', it has become popular to calculate amino acid requirements using mathematical models instead of actually determining the requirements by experiment. Factors underlying the modelling of amino acid requirements have been reviewed in a recent paper by Black *et al.* (1986). Certainly, calculated amino acid requirements are interesting, but the projected requirement estimates can only be as good, or as accurate, as the validity of the assumed factors going into construction of the model. Many of the assumptions themselves are in need of verification and further study. Some of the factors known to affect amino acid requirements of pigs and poultry are discussed below.

CRITERIA OF RESPONSE

Requirements for amino acids are generally best defined in growing animals by growth data in *ad libitum* feeding studies. While pair or equal feeding circumvents several interpretive problems, it creates other problems (Baker, 1984). Thus, part of the growth response to an essential amino acid is due to its favourable effect on metabolism; but another part of its efficacy resides in its effect on food intake.

STATISTICAL CONSIDERATIONS

With growth data, both weight gain over a defined period of time and efficiency of diet or nutrient utilization are the usual criteria of response. Requirement and bioavailability studies necessitate feeding graded levels of the nutrient in question. Statistical calculation is easier in these studies if increments are equally spaced. As a general rule there is only minimum utility in applying any kind of a range or paired-comparison test to these data (Peterson, 1977; Little, 1978). Indeed, such tests often lead to misinterpretation of the results. With bioavailability assays, the linear response area (i.e. constant utilization of the nutrient in question) is defined by a minimum of three levels of the independent variable (i.e. the nutrient)*. The significance of a difference between any two adjacent points is essentially meaningless. Instead the linearity of the response should be established together with the 'fit' (r^2 value) such that the determined slope of the response line has veracity. With requirement studies, a minimum of four levels is required (preferably six or more) such that the data can be fitted to a descriptive response curve (e.g. sigmoidal, asymptotic or broken line), thus facilitating objective assessment of the requirement (Robbins, Norton and Baker, 1979). It is improper procedure to use paired-comparison tests of adjacent points in requirement studies. Thus, some investigators select a requirement on the basis of a maximum level that no longer is statistically different from the level immediately below it. This is not a defensible procedure from a statistical standpoint. Moreover, all levels yielding growth performance data 'appearing' to reside on the plateau area of the growth curve must be evaluated carefully in order to arrive properly at what should be considered the true maximum response.

Aberrant results often occur in the upper curvilinear area of a growth curve where animal-to-animal variability is generally greatest. It is thus not uncommon to find a rat gaining 4 g/day at (ascending) level 3, 5 g/day at level 4, 6 g/day at level 5 and 5 g/day at level 6. Even though the gain at level 5 may be statistically better than the gain at level 4, this should not be taken as the requirement level of that nutrient. Because gain at level 6 was less than that at level 5, one must question whether level 4 might be a more proper selection as the minimum effective dose for maximum response. While curve fitting can be of great assistance in taking the subjectivity out of decisions such as these, a repeat of the experiment with additional, and possibly different, levels would seem called for.

Expressions of efficiency present more complicated problems because they involve both growth and consumption data. Unlike rodent studies where the animals are generally fed individually, studies with other species usually involve group feeding. While individual animal data are available for weight gain calculations, only group data are available for diet or nutrient intake calculations. This, of course, presents special problems when an animal dies or must be removed from a given pen for other reasons, since consumption by the animal removed cannot be determined accurately. Proper procedure for group feeding experiments dictates that both weight gain and diet efficiency data be based upon pen means. Hence, all statistical calculations should be made on a pen-means basis, with the total degrees-of-freedom equal to one less than the total number of pens.

EXPRESSIONS OF NUTRIENT REQUIREMENTS

Requirements are generally defined for animals of a given age and for a specific physiological function (i.e. maintenance, growth, reproduction or production).

---

*Requirement studies can legitimately use growth as the dependent variable and concentration of the nutrient as the independent variable, but slope-ratio bioavailability studies should use growth rate regressed on absolute intake of the nutrient. The latter should be done in the constant slope region (i.e. generally between 30 and 70% of the requirement) of the growth curve.

Because amino acid requirements are most useful if defined for groups or populations, requirement expressions are generally based upon dietary concentration rather than absolute intake. Hence, a pen of 100 newly hatched broiler chicks may vary by 10–20% in body weight (e.g. from 100 to 120 g). The heavier chicks obviously require greater quantities of each essential amino acid than the lighter chicks, but the available evidence suggests that requirements expressed as a percentage of diet, percent of metabolizable energy (ME) or mg/kg do not vary greatly among heavy and light animals of the same age. The heavier animal merely meets its requirements by consuming more diet.

One opinion is that 'concentration' requirements are best expressed as g or mg/kJ ME. This is based upon the long-standing nutritional principle that animals will eat to meet their energy needs. While this principle is generally true in practice, there are many cases, some practical, where another factor overrides an animal's tendency to eat to a given energy need. To use just one example of practical relevance, (Parsons, Edmonds and Baker, 1984; Edmonds, Parsons and Baker, 1985), broiler chicks increase rather than decrease their voluntary feed intake when dietary protein level in a corn-soyabean meal diet is reduced in decrements from 24% (normal level) to 16% by replacing soyabean meal (10.2 MJ ME/kg) with corn (14.1 MJ ME/kg). Thus, despite a considerable increase in energy density, the low protein diet is actually consumed in greater quantity than the high-protein diet. In effect, therefore, the birds appear to be trying to eat to meet their protein-amino acid needs rather than their energy needs. In so doing, they overeat energy in relation to 'effective' protein and deposit more body fat (Edmonds, Parsons and Baker, 1985). Low-protein or minimum soyabean meal diets have become a practical reality in the USA, but also in other countries whose growing seasons and climatic conditions are not amenable to production of high-lysine oilseeds such as soyabeans. The availability of economically priced methionine, lysine, threonine and tryptophan have served to encourage the use of amino acid fortified low-protein diets (Baker, 1985; Baker and Parsons, 1985). There are, therefore, instances in which expressing dietary requirements per unit of ME can result in over- or underformulation of a given diet.

BODY COMPOSITIONAL FACTORS

Requirements for amino acids in growing animals, expressed in terms of dietary concentration, decrease as age and weight of the animal increase (NRC, 1979; Boomgaardt and Baker, 1973a). This occurs because body composition changes (e.g. more fat and less protein in the weight gain) as a growing animal matures. Whether protein percentage in the weight gain decreases in a straight line from weaning to market weight is questionable, although calculations of Whittemore (1987) assume a linear decrease does occur. For amino acids, weight gain generally correlates well with nitrogen retention in young, rapidly growing animals. After secondary sex characteristics have developed, however, body composition factors come into play such that a higher requirement is often predicted for maximal nitrogen retention than for maximum weight gain. Also, by the same reasoning, use of criteria such as maximum carcass leanness or optimum feed efficiency results in a higher requirement than that predicted for maximum weight gain (Baker, 1977; Baker *et al.,* 1967; Smith, Clawson and Barrick, 1967).

In species like the pig, the sex of the animal is a very important consideration. Thus, after secondary sex characteristics have developed, gilts, and boars, because they deposit more lean in relation to fat, exhibit higher protein and amino acid requirements than is the case for castrated males. By the same token, pigs bred for leanness require higher concentrations of amino acids in their diets than those not similarly possessing a high lean:fat ratio. Hence, should the use of materials such as recombinant porcine growth hormone become a practical reality in pig production (Boyd *et al.*, 1985; Etherton and Kensinger, 1984; Easter, 1987) then it is likely dietary requirements for amino acids will increase.

FOOD COMPOSITION AND FOOD INTAKE FACTORS

Dietary protein level affects requirements for essential amino acids (Baker, 1977; Baker, Katz and Easter, 1975; Boomgaardt and Baker, 1973b). Thus, as protein level decreases, as would be the case with a minimum-soyabean meal, amino acid-fortified diet, food intake generally increases as the animal attempts to meet its protein or amino acid needs. Just as important, however, is that with low-protein diets fortified with a limiting amino acid in pure form, a greater portion of the total concentration of the limiting amino acid is bioavailable, since pure sources of amino acids are more bioavailable than intact-protein sources of the same amino acid.

The contribution of nutrients from protein- or energy-yielding ingredients relative to those from pure sources presents subtle and perplexing problems in diet formulation and nutrient requirement assessment. Most nutrients present in feed ingredients are bound in one form or another (e.g. amino acids in polymeric peptide linkage). Ideally, amino acids should be characterized in basal diets as to their bioavailability before a requirement study is undertaken.

FREQUENCY OF FEEDING

The conditions of feeding studies can create bioavailability problems that contribute to confusion in amino acid requirement assessment. Dog, cat and adult pig feeding studies can be used as an example here. These animals are generally considered 'meal eating', as opposed to 'nibbling', species. As such, many requirement studies with them are conducted using a one-meal-per-day feeding regimen. Can it be assumed, however, that pure sources of the nutrient in question will be absorbed at the same time and rate as bound sources of the same nutrient? It probably cannot and pure lysine, for example, is absorbed much faster than intact-protein sources of lysine that require time in the gut for proteolysis to occur (Batterham, 1984; Baker, 1984; Baker and Izquierdo, 1985). As a result, pure lysine, generally assumed to be 100% bioavailable, may become only 50% bioavailable in a once-per-day feeding regimen (Batterham, 1984). Hence, much of the crystalline lysine may be wasted if it is absorbed from the gut 1–2 h before the rest of the intact-protein bound amino acids are similarly made available at the proper sites of tissue protein synthesis. Moreover, it cannot be assumed that the same magnitude of unavailability existing with lysine in a once-per-day feeding regimen will also occur with other essential amino acids. Thus, lysine is conserved in tissue pools during a deficit and its turnover rate is slow compared with that of, say, threonine and methionine. As a result, the magnitude of methionine or threonine wastage may exceed that of lysine wastage when pure sources of each are used to supplement intact protein diets in a once-per-day feeding regimen.

Selection of a proper feeding regimen for amino acid requirement studies is important. The researcher is presented with a dilemma, however, in that to be of greatest value, requirement studies should be carried out using feeding regimens typifying those used in practice. Yet, using these feeding regimens (e.g. one meal per day) may present problems in interpreting the results because, generally speaking, amino acid requirement assays, of necessity, require supplementing a deficient semi-purified diet (containing the deficient amino acid in bound form) with a free form of the amino acid in question. Yet, in practice, the nutrient in question may be provided in a grain-soyabean meal diet totally in the bound form. If such is likely to be the case, a perspicacious research scientist will probably decide that a two- or three-times-per-day feeding regimen is preferable to a once-per-day feeding for the requirement bioassay.

PRECURSOR MATERIALS

With several amino acids, precursor materials may be present in the diet used to assess a requirement. Classic examples are (precursor/product) methionine/cysteine, phenylalanine/tyrosine, cysteine/taurine, glutathione/cysteine, and carnosine/histidine (Czarnecki, Halpin and Baker, 1983).

Errors are frequently made in assessing the maximum portion of the total sulphur amino acid requirement that can be furnished by cysteine. Knowledge of this exact value is important in least-cost diet formulation for several species, but particularly for poultry. A series of three experiments is necessary to quantify accurately the percentage of the sulphur amino acid requirement met by cysteine (Table 13.2). Unfortunately, many investigators have merely done assays 1 and 2 without doing assay 3. They have concluded, therefore, that 55% (w/w) of the sulphur amino acid requirement can be furnished by cysteine instead of the actual value of 50%. Moreover, some have even suggested, based upon a two-assay approach, that the sulphur amino acid requirement is 0.60% of the diet (methionine alone) rather than the correct requirement of 0.54% (methionine + cysteine, 1:1). Thus, because the molecular weight of cysteine is only about 80% that of methionine, the total sulphur amino acid requirement (i.e. methionine + cysteine) is lower with what would be considered a proper combination of methionine + cysteine than with

Table 13.2 EXPERIMENTS REQUIRED TO QUANTIFY THE MAXIMUM PERCENTAGE OF THE CHICK'S SULPHUR AMINO ACID REQUIREMENT MET BY CYSTEINE AND THE EFFICIENCY OF METHIONINE AS A CYSTEINE PRECURSOR – AN EXAMPLE

Assay number and order	Requirement study	Dietary condition	Resulting requirement (% of diet)
1	Methionine	0 cysteine	0.60
2	Methionine	Excess cysteine	0.27
3	Cysteine	0.27% methionine	0.27

methionine alone furnishing the sulphur amino acids (Graber and Baker, 1971; Halpin and Baker, 1984). Failure to consider carefully precursor:product interrelationships and their implications can also result in improper assessment of the bioefficacy of analogue materials that may be metabolized into one or more useful products (e.g. methionine hydroxy analogue serving as a precursor of methionine, cysteine, or both (Baker, 1976; Boebel and Baker, 1982).

MATHEMATICAL AND JUDGEMENTAL CONSIDERATIONS

Requirement bioassays should be designed in such a way as to allow objective estimation of desired maxima, e.g. maximum weight gain (Robbins *et al.*, 1977; 1979) or minima, e.g. minimum plasma urea concentration (Lewis *et al.*, 1977). Some investigators have assumed that an objective requirement can be obtained by calculating a linear regression line through the selected points or levels appearing to reside in the linear response surface, then determining the intercept of this line with a horizontal (zero slope) line calculated from points appearing to reside in the plateau region of the response curve. This procedure is unacceptably marred by subjectivity in that it involves some form of selection of the points or levels to use in the linear and plateau regression lines. A continuous broken line calculated by least squares, is the preferred method because it objectively selects the breakpoint in the response line (Robbins, Baker and Norton, 1977; Robbins, 1986).

Broken lines, even if properly calculated, may not adequately describe the response. In previous work from our laboratory with histidine, for example, involving 10 increments of L-histidine ranging from 0 to 0.57% of the diet, the response was best described by a sigmoidal fit (Robbins, Norton and Baker, 1979). The problem with fitted curvilinear response lines (e.g. sigmoid or asymptotes) is that breakpoints (i.e. an abrupt change in slope, representing a requirement) are not a component. Only if an arbitrary point is selected, such as 95% of the upper asymptotic value, can one allow the computer rather than the investigator to select an objective requirement value.

Bias can and should be taken out of the requirement-selection process. It is thus not defensible to construct two independent straight lines by selecting specific points for the ascending linear line and the zero-slope horizontal line. The subjective selection process can and generally does have a material effect on where the two lines intersect (i.e. the assumed requirement) which can lead to unintentional, or intentional, bias.

Extrapolating literature amino acid requirements to practice

Assuming a proper objective method is used to assess levels of an amino acid that result in optimum measures of animal performance (e.g. growth, feed efficiency, carcass yield), and assuming the animals used in the experiments are representative of the defined population as a whole (e.g. UK meat-type pigs between 20 and 50 kg body weight), how can one translate literature requirement estimates to the real world (i.e. on-farm conditions)? The author has found that including the following as factors often leads to an accurate translation of a requirement determined experimentally to one that would apply to on-farm conditions: (1) dietary metabolizable energy level, (2) dietary protein level and (3) bioavailability.

It is known that both dietary protein level and energy density of the diet exert effects on amino acid requirements expressed in terms of dietary concentration, i.e. % of diet (Mitchell *et al.*, 1965; Boomgaardt and Baker, 1973b, c; Baker, Katz and Easter, 1975). Research in this laboratory indicates, however, that the relationships are not directly proportional, i.e. doubling energy level does not double the requirement, nor does a doubling of the crude protein level double the requirement for an essential amino acid. Thus, while the directional nature is apparent, the quantitative relationships have not been clearly delineated nor have associative effects, either positive or negative, between energy level (and source) and protein level (and source) been defined. Sometimes animals appear to eat to meet energy needs, but sometimes they appear to eat to meet amino acid needs.

Some have suggested that amino acid requirements are affected by environmental temperature extremes. Although there is some logic in this reasoning, there also is a basic fallacy in the relationship. Pigs or chickens do eat less feed in a warm environment (e.g. 35°C) than in a cold environment (e.g. −10°C). In so doing they eat less of the limiting amino acid. Does adding more of the first-limiting amino acid (e.g. methionine for poultry, lysine for pigs) result in increased response? The author's experience would suggest that an increase in gain will occur only if addition of the first-limiting amino acid stimulates voluntary feed intake. Thus, given the almost perfect relationship between feed intake and weight gain (Allen *et al.*, 1972), how can more of the limiting amino acid bring about more weight gain unless more of the remaining growth-requiring raw materials (especially energy) are not present as well? In reality, a 20% reduction in voluntary feed intake due to heat stress may require only a 10% increase in the dietary concentration of the limiting amino acid in order to obtain an approximate 5% increase in weight gain of the heat-stressed animals. Certainly, the 20% reduction in feed intake will not necessitate a 20% increase in the need for the limiting amino acid. The best that can be hoped for is that increasing the limiting amino acid by 10% may increase feed intake by 10% and weight gain by 5%.

Most purified diets used to establish amino acid requirements are lower in crude protein but higher in metabolizable energy than conventional practical diets. Thus, an adjustment upwards needs to be made to translate the lower crude protein purified diet work to the higher crude protein practical diet basis. An adjustment downwards needs to be made, on the other hand, to correct for the higher metabolizable energy level present in the purified diet. In the author's experience, whatever errors are present in the linear relationship between crude protein level and the amino acid requirement, and between metabolizable energy level and the same requirement, these tend to cancel out when both factors are included in the adjustment equation together.

An additional factor must be considered before a proposed correction equation can be established. Purified diets frequently contain amino acids that are 100% available while practical diets do not. Thus, an average amino acid bioavailability factor of 85% will be used for the correction equation that follows:

$$R_p = \frac{R_e}{BV_p} \times \frac{CP_p}{CP_e} \times \frac{ME_p}{ME_e} \tag{13.1}$$

where: R_p = Amino acid requirement in practical diet (%).
R_e = Requirement determined with experimental diet (%).
BV_p = Bioavailability in practical diet relative to that in purified experimental diet (%).

CP_p = Crude protein level in practical diet (%).
CP_e = Crude protein level in experimental diet (%).
ME_p = Metabolizable energy level in practical diet (MJ/kg).
ME_e = Metabolizable energy level in experimental diet (MJ/kg).

This equation almost exactly translates broiler-chick requirements established for sulphur amino acids and lysine using the Illinois amino acid diet (15% CP, 15.9 MJ ME/kg) to the practical setting (i.e. NRC = 23% CP, 13.4 MJ ME/kg). Thus, for sulphur amino acids where the purified diet requirement is 0.60%:

$$\frac{0.60}{0.85} \times \frac{23}{15} \times \frac{13.4}{15.9} = 0.91\% = R_p \tag{13.2}$$

The estimate of 0.91% sulphur amino acids is very close to the NRC estimate of 0.93% sulphur amino acids for broiler chicks fed a 23% crude protein corn-soyabean meal diet containing 13.4 MJ ME/kg.

For lyseine, the purified diet requirement estimate is 0.80% (cf. Table 13.1). Using the same translation formula, a practical diet requirement of 1.21% lysine is predicted, very close to the NRC estimate of 1.20%. Work in our laboratory with arginine suggests that this amino acid, too, fits well into the practical-diet prediction equation.

For young chicks and poults, sulphur amino acids, lysine and arginine are the first-, second- and third-limiting amino acids in lower protein corn–soyabean meal diets. Thus, the aforementioned prediction equation may provide a useful means of interpreting purified diet requirements for these amino acids in terms of relating them to the practical setting.

In lieu of resorting to some form of prediction equation, one could attempt to convince all pig nutritionists doing requirement work that amino acid requirement experiments should consist of the following features (Izquierdo, Wedekind and Baker, 1988).

(1) A defined age and weight range of pigs should be covered in the experiments.
(2) The pigs should be identified as to sex and meat characteristics.
(3) The experimental diets employed should contain both protein and metabolizable energy levels consistent with those used in on-farm commercial pig production.
(4) The experimental diet should be frankly and singly deficient in the amino acid under investigation so that a substantial growth response can be demonstrated.
(5) The diet when adequately fortified with the amino acid under study should promote near optimum (maximum?) growth in *ad libitum*-fed pigs.
(6) Both the linear and plateau portions of the growth curve should be covered in the amino acid levels employed in the growth trials.
(7) Both *total* and *bioavailable* levels of the limiting amino acids in the basal experimental diet must be known.

It would indeed be helpful if pig experiments involved many amino acid levels (i.e. six or more) such that an objective curve-fitting procedure could be used to identify the dose that yields optimum (most profitable?) response. Because the purified and semipurified diets necessarily required for pig amino acid requirement studies are often very costly, most investigators compromise on the 'ideal' insofar as dosage levels are concerned.

AN EXAMPLE

An attempt has recently been made at Illinois to establish a meaningful histidine requirement for pigs in the weight range of 10–20 kg (Izquierdo, Wedekind and Baker, 1988). A somewhat bizarre and far-from-practical basal diet (based upon corn, dried whey and feather meal) had to be used in order to achieve a diet that was deficient in histidine. After proving experimentally that 10-kg pigs would grow well on this diet when it contained adequate histidine, analytical and bioassay studies were carried out to establish both the total (0.22% of diet) and bioavailable (0.19% of diet) histidine concentration in the basal experimental diet. The next step involved experiments where graded levels of histidine (covering both deficiency and adequacy) were fed. Because the basal diet was costly (it contained crystalline methionine, tryptophan and lysine in addition to corn starch and the aforementioned protein-supplying ingredients), only four carefully chosen doses were employed in the first experiment (Table 13.3). Moreover, based upon the results of experiment 1, it was reasoned that only three doses would be sufficient in experiment 2. Thus, experiment 2 was really needed to clear up confusion associated with experiment 1. A brief summary of experiments 1 and 2 is presented in Table 13.3.

If experiment 1 had been the only requirement study conducted, we would have been forced to conclude that the histidine requirement was greater than 0.25% but not greater than 0.31%. Thus, since the magnitude of the gain and gain:feed response between 0.19% and 0.25% bioavailable histidine was substantially greater than that between 0.25% and 0.31, one could suggest that maximal growth might

Table 13.3 GROWTH PERFORMANCE OF PIGS FED GRADED LEVELS OF HISTIDINE[a]

Diet[b]	Bioavailable histidine (%)	Daily weight gain (g)	Food conversion ratio[e]
Exp. 1			
1. Basal diet	0.19	188	2.46
2. As 1 + 0.06% L-histidine	0.25	383	1.80
3. As 1 + 0.12% L-histidine	0.31	467	1.59
4. As 1 + 0.18% L-histidine	0.37	438	1.68
Pooled SEM		22[c]	0.08[c]
Exp. 2			
2. Basal + 0.06% L-histidine	0.25	254	2.21
2. As 1 + 0.03% L-histidine	0.28	322	2.02
3. As 1 + 0.06% L-histidine	0.31	458	1.69
Pooled SEM		29[d]	0.08[d]

[a] Data (Izquierdo, Wedekind and Baker, 1988) represent means of six individually fed pigs during the period 5–8 weeks of age; average initial weight was 10 kg
[b] The basal histidine-deficient diet contained 20% feather meal, 20% corn and 10% dried whey; it was also fortified with methionine, lysine and tryptophan. Supplemental L-histidine was provided as L-histidine·HCl·H$_2$O (74% histidine)
[c] Histidine linear and quadratic effect ($P<0.01$)
[d] Histidine linear effect ($P<0.01$)
[e] kg feed/kg liveweight gain

have occurred somewhere between 0.25% and 0.31%. Experiment 2 was conducted, therefore, to establish whether this was the case. In this trial involving narrower increments, the gain and feed efficiency responses between 0.28% and 0.31% bioavailable histidine were as great or greater than those occurring between 0.25% and 0.28%. Hence, when the two histidine requirement trials are viewed as a whole, 0.31% bioavailable histidine emerges as a defendable requirement estimate.

Because pig amino acid requirements are based upon those levels needed for optimum rate and efficiency of weight gain of pigs consuming grain-soyabean meal type diets, extrapolation of the bioavailable histidine requirement to a grain-soyabean meal basis is necessary. If ileal histidine digestibility data are available for dietary ingredients used in the grain-soyabean meal diet, these should be used to adjust the requirement estimate. If unavailable, an educated bioavailability estimate should be made, e.g. 85%. Using 85%, one then can correct the bioavailable histidine requirement estimate to a more meaningful value of 0.36% (0.31% ÷ 0.85 = 0.36%).

Factors complicating assessment of amino acid limitations

There are pitfalls in chemical scoring as well as biological evaluative methods of determining limiting amino acids in feedstuffs or diets. Quite obviously, chemical scoring methods such as 'chemical score' (whole egg standard) or amino acid indexes do not take account of either palatability or bioavailability. As such, methods like these often predict wrongly the order of limiting amino acids in a feedstuff or diet (Harper, 1959; Boomgaardt and Baker, 1972). Less obvious are the pitfalls inherent in using animal tests to assess amino acid limitations. Because of the increase in commercial availability of synthetic amino acids, determination of limiting indispensable amino acids in low-protein grain-soyabean meal diets is becoming very important to the poultry and pig industries. The two most common methods used to assess limiting amino acids in diets are amino acid addition and amino acid deletion assays. Amino acid addition studies involve adding amino acids individually and in combination to a low-protein or amino acid-deficient diet. In amino acid deletion studies, individual amino acids are deleted sequentially from a diet containing a full complement of supplemented amino acids.

Amino acid addition trials, though less efficient in experimental design, are probably more relevant to practical pig and poultry production (Edmonds, Parsons and Baker, 1985). Indeed, sequential supplementation studies have led to the conclusion that the three most limiting amino acids in reduced protein (i.e. lower soyabean meal) corn-soya diets, in order of limitation, are (1) lysine, (2) tryptophan and (3) threonine for growing pigs (Russell, Cromwell and Stahly, 1983; Russell *et al.*, 1986) but are (1) methionine, and (2) lysine and arginine (equally second-limiting) for young chicks (Edmonds, Parsons and Baker, 1985). Using barley/wheat, or sorghum, in place of most of the corn in grain-soya diets for growing pigs results in lysine being first-limiting, threonine being second-limiting and tryptophan or methionine being third-limiting.

Amino acid deletion experiments conducted with chicks have yielded conclusions different from amino acid addition experiments as to the order of limiting aamino acids in low-protein corn-soya diets (Edmonds, Parsons and Baker, 1985). These authors added eight essential amino acids plus glutamic acid to a 16% protein corn-soya diet fed to 1-week-old chicks (23% CP corn-soya diets are normally fed to chicks of this age). Deletion of methionine from the amino acid supplement produced the same growth depression as that obtained from deletion of lysine. Thus, that these two amino acid deletions produced the greatest growth depressions (which were equal) led to the conclusion that these two amino acids were equally first-limiting. Three other amino acids (arginine, valine and threonine), upon individual deletion from the complete amino acid supplement, also caused growth depressions that were approximately equal but of lesser magnitude than those resulting from either methionine or lysine deletion. Hence, the conclusion was that arginine, valine and threonine were equally third-limiting – after methionine and lysine.

It is clear when examining the order of amino acid limitation resulting from 'addition' vs 'deletion' experiments in the work of Edmonds *et al.* (1985) that different conclusions resulted from the two methods. Deletion experiments have a far more efficient experimental design. The key question asks if this approach accurately predicts the situation as it would occur in commercial production? The answer is, that it probably does not. Deletion experiments potentially impose amino acid imbalances on amino acid deficiencies. While the amino acid excesses over and above each deficiency (i.e. each individual amino acid deletion) are not likely to cause metabolic inefficiency, they may, particularly with certain amino acid deficiencies, reduce voluntary food intake and therefore growth rate. Thus, addition experiments, because they more nearly simulate the practical setting, should be considered the more appropriate method of determining the order of limiting amino acids in a feedstuff or diet.

Amino acid excesses

The scientific literature contains many articles on effects of excesses of individual amino acids, or amino acid mixtures, on performance of rats and chicks. Most of the studies, however, have used semi-purified or purified diets. Only recently has information been generated using pigs and poultry where individual amino acid excesses have been evaluated in animals fed practical grain-soya diets (Southern and Baker, 1982; Hagemeier, Libal and Wahlstrom, 1983; Anderson *et al.*, 1984a,b; Rosell and Zimmerman, 1984a,b; Edmonds and Baker, 1987a,b,c; Edmonds, Gonyou and Baker, 1987).

A brief summary of findings in experiments where excess amino acids were added to practical diets for pigs and poultry is:

(1) A 1% excess addition of any single essential amino acid depresses neither growth nor feed efficiency in pigs or chicks. Methionine is the only essential amino acid that depresses weight gain in pigs at a supplemental level of 2%.
(2) At a supplemental level of 4%, leucine, isoleucine and valine are not growth depressing.
(3) At supplemental levels of 4%, methionine is the most growth depressing essential amino acid. In choice studies, however, diets containing 4% excess tryptophan are rejected to a greater extent than diets containing 4% excess methionine.

(4) A 4% supplementary level of lysine is tolerated better by pigs than by chicks; the reverse is true with arginine, which is far more growth depressing in pigs than in poultry.

(5) A 4% supplementary level of threonine is growth depressing in chicks but not in pigs.

(6) Excess essential amino acids [i.e. arginine (pigs), leucine, phenylalanine-tyrosine] found in practical grain-soya diets for pigs and poultry are not deleterious to animal performance.

(7) Excess amino acids were not found to increase heat production in pigs by Fuller *et al.* (1987), although Kerr (1988) did observe an increase in heat production in pigs fed excess amino acids during summer months but not during winter months.

References

ALLEN, N.K., BAKER, D.H., SCOTT, H.M. and NORTON, H.W. (1972). *Journal of Nutrition,* **102**, 171–180

ANDERSON, D.B., VEENHUIZEN, E.L., WAITT, W.P., PAXTON, R.E. and YOUNG, S.S. (1987). *Federation Proceedings,* **46**, 1021 (abstract)

ANDERSON, L.C., LEWIS, A.J., PEO, E.R. and CRENSHAW, J.D. (1984a). *Journal of Animal Science,* **58**, 362–368

ANDERSON, L.C., LEWIS, A.J., PEO, E.R. and CRENSHAW, J.D. (1984b). *Journal of Animal Science,* **58**, 369–377

ANONYMOUS (1985). *Nutrition Reviews,* **43**, 88–90

BAKER, D.H. (1976). *Journal of Nutrition,* **106**, 1376–1377

BAKER, D.H. (1977). In *Advances in Nutrition Research*, Vol. I, pp. 229–335. Ed. Draper, H.H. Plenum Publishing Company, New York, NY

BAKER, D.H. (1978). *Proceedings of the Georgia Nutrition Conference*, pp. 1–12

BAKER, D.H. (1984). *Nutrition Reviews,* **42**, 269–273

BAKER, D.H. (1986). *Journal of Nutrition,* **116**, 2339–2349

BAKER, D.H. and IZQUIERDO, O.A. (1985). *Nutrition Research,* **5**, 1103–1112

BAKER, D.H. and PARSONS, C.M. (1985). *Recent Advances in Amino Acid Nutrition*, 48 pp. Ajinomoto Publishing Company, Tokyo, Japan

BAKER, D.H., JORDAN, C.E., WAITT, W.P. and GOUWENS, D.W. (1967). *Journal of Animal Science,* **26**, 1059–1066

BAKER, D.H., KATZ, R.S. and EASTER, R.A. (1975). *Journal of Animal Science,* **40**, 851–856

BAKER, D.H., ROBBINS, K.R. and BUCK, J.S. (1979). *Poultry Science,* **58**, 749–750

BATTERHAM, E.S. (1984). *Pig News and Information,* **5**, 85–88

BLACK, J.L., CAMPBELL, R.G., WILLIAMS, I.H., JAMES, K.J. and DAVIES, G.T. (1986). *Research and Development in Agriculture,* **3**, 121–145

BOEBEL, K.P. and BAKER, D.H. (1982). *Poultry Science,* **61**, 1167–1175

BOOMGAARDT, J. and BAKER, D.H. (1972). *Poultry Science,* **51**, 1650–1655

BOOMGAARDT, J. and BAKER, D.H. (1973a). *Poultry Science,* **52**, 592–599

BOOMGAARDT, J. and BAKER, D.H. (1973b). *Poultry Science,* **52**, 586–592

BOOMGAARDT, J. and BAKER, D.H. (1973c). *Journal of Animal Science,* **36**, 307–311

BOYD, R.D., HARKENS, M., BAUMAN, D.E. and BUTLER, W.R. (1985). *Proceedings of the Cornell Nutrition Conference*, pp. 10–19

BROWN, H.D., HARMON, B.G. and JENSEN, A.H. (1973). *Journal of Animal Science,* **37**, 708–712

CZARNECKI, G.L., HALPIN, K.M. and BAKER, D.H. (1983). *Poultry Science,* **62**, 371–375

EASTER, R.A. (1987). In *Repartitioning Resolution: Impact of Somatotropin and Beta Adrenergic Agonists on Future Pork Production*, pp. 193–199. University of Illinois Pork Industry Conference

EDMONDS, M.S. and BAKER, D.H. (1987a). *Journal of Animal Science,* **64**, 1664–1671

EDMONDS, M.S. and BAKER, D.H. (1987b). *Journal of Animal Science,* **65**, 699–705

EDMONDS, M.S. and BAKER, D.H. (1987c). *Journal of Nutrition,* **117**, 1396–1401

EDMONDS, M.S., GONYOU, H.W. and BAKER, D.H. (1987). *Journal of Animal Science,* **65**, 179–185

EDMONDS, M.S., PARSONS, C.M. and BAKER, D.H. (1985). *Poultry Science,* **64**, 1519–1526

ETHERTON, T.D. and KENSINGER, R.S. (1984). *Journal of Animal Science,* **59**, 511–528

FULLER, M.F., CADENHEAD, A., MOLLISON, G. and SEVE, B. (1987). *British Journal of Nutrition,* **58**, 277–285

GRABER, G. and BAKER, D.H. (1971). *Journal of Animal Science,* **33**, 1005–1011

HAGEMEIER, D.L., LIBAL, G.W. and WAHLSTROM, R.C. (1983). *Journal of Animal Science,* **57**, 99–105

HALPIN, K.M. and BAKER, D.H. (1984). *Journal of Nutrition,* **114**, 606–612

HARPER, A.E. (1959). *Journal of Nutrition,* **67**, 109–122

IZQUIERDO, O.A., WEDEKIND, K.J. and BAKER, D.H. (1988). *Journal of Animal Science,* **66**, 2886–2892

KERR, B.J. (1988). PhD Thesis. University of Illinois, Urbana, Illinois, USA

LEWIS, A.J., PEO, E.R., CUNNINGHAM, P.J. and MOSER, B.D. (1977). *Journal of Nutrition,* **107**, 1369–1376

LIN, C.C. and JENSEN, A.H. (1985). *Journal of Animal Science,* **61** (Supplement 1), 298–299 (abstract)

LITTLE, T.M. (1978). *Hortscience,* **13**, 504–506

MITCHELL, J.R., BECKER, D.E., JENSEN, A.H., NORTON, H.W. and HARMON, B.G. (1965). *Journal of Animal Science,* **24**, 977–980

NRC (1979). *Nutrient Requirements of Swine.* National Academy of Science, Washington DC

PARSONS, C.M., EDMONDS, M.S. and BAKER, D.H. (1984). *Poultry Science,* **63**, 2438–2443

PETERSON, R.G. (1977). *Agronomy Journal,* **69**, 205–208

ROBBINS, K.R. (1986). *University of Tennessee Agriculture Experiment Station Research Report, 86-09,* pp. 1–8

ROBBINS, K.R., BAKER, D.H. and NORTON, H.W. (1977). *Journal of Nutrition,* **107**, 2055–2061

ROBBINS, K.R., NORTON, H.W. and BAKER, D.H. (1979). *Journal of Nutrition,* **109**, 1710–1714

ROSELL, V.L. and ZIMMERMAN, D.R. (1984a). *Journal of Animal Science,* **59**, 135–140

ROSELL, V.L. and ZIMMERMAN, D.R. (1984b). *Nutrition Reports International,* **39**, 1345–1352

RUSSELL, L.E., CROMSELL, G.L. and STAHLY, T.S. (1983). *Journal of Animal Science,* **56**, 1115–1123

RUSSELL, L.E., EASTER, R.A., GOMEZ-ROJAS, V., CROMWELL, G.L. and STAHLY, T.S. (1986). *Animal Production,* **42**, 291–295

SMITH, J., CLAWSON, A.J. and BARRICK, E.R. (1967). *Journal of Animal Science,* **26**, 752–758

SOUTHERN L.L. and BAKER, D.H. (1982). *Journal of Animal Science,* **55**, 857–866

WHITTEMORE, C.T. (1987). *Elements of Pig Science*, p. 43. Longman Group UK Limited, Essex, England

14

AETIOLOGY OF DIARRHOEA IN PIGS AND PRE-RUMINANT CALVES

J. W. SISSONS
AFRC Institute for Grassland and Animal Production, Shinfield, Reading, Berkshire, UK

Introduction

Post-weaning diarrhoea amongst piglets and calves continues to be a serious and frustrating problem. The pathogenesis of the disease is complex and its aetiology is multi-factorial. Although severe and persistent diarrhoea undoubtedly involves intestinal microbial pathogens, it is clear that diet and the act of weaning itself are widely regarded as factors which predispose young animals to the condition. It is not known if the problem is increasing in prevalence, but in recent times certain limitations in feeding and husbandry practice may have made the disorder more difficult to control. These include an acute shortage of skim-milk, (the source of choice protein for weaner diets), the use of alternative proteins which may have antinutritional properties, production pressures to wean at younger ages and growing public concern over the use of oral antibiotics. It is therefore appropriate to review current understanding of diarrhoea from a dietary point of view, to examine the problems associated with using other feedstuffs as replacements for milk protein, to consider the effects of digestive disorders leading to diarrhoea on animal performance and to appraise possible remedies for overcoming or preventing nutritional scours amongst young farm animals.

The pathogenesis of diarrhoea

As a symptom of a disorder in gastrointestinal function, diarrhoea is easily recognized as an increase in stool water excretion. Quantitatively it may be regarded as a decrease in faecal dry matter to below 20%. The problem is to identify the source and cause of this additional water. Physiologically it can arise through failings in fluid handling in either the small or large bowel or both. Fluxes of water and solute move through the intestinal mucosa in opposite directions. Normally these fluxes are large; in calves it has been estimated that intestinal bi-directional fluxes of fluid can be about 80 l/day. But in healthy animals, the net difference between them is very small. For example, as calves develop diarrhoea the net difference increases from about 50 ml/day to 1 l/day, and if this level reaches 2.5 l/day, dehydration and death will follow (Bywater, 1973; Bywater and Logan, 1974).

The process of fluid transport is governed by both active and passive driving forces. The former is a metabolic process which is governed by mucosal cyclic adenosine monophosphate (cAMP) and can result in electrolyte passage from the blood into the lumen. In contrast passive movements are driven by electrical and chemical potentials and hydrostatic pressure gradients across the mucosa (see review by Argenzio and Whipp, 1980). Clearly the aetiology of diarrhoea should be considered from the standpoint of processes which alter the bi-directional balance of fluid transport in favour of its accumulation in the lumen of the intestine. Mechanisms likely to bring this about include changes in the permeability of the gut wall, deranged contractile activity of the smooth muscle coat, altered osmolality of digesta, and mucosal inflammation. Physiological and immunological studies have shown that these dysfunctions can occur as a result of weaning calves and piglets onto whole milk substitutes or dry starter meals which contain large amounts of vegetable protein (Sissons, 1982; Newby *et al.*, 1985).

WITHDRAWAL OF WHOLE MILK

Protective effects of milk against pathogenic *E. coli* have been demonstrated in piglets weaned onto a diet supplemented with sows' milk (Deprez *et al.*, 1986). The mechanism is not well characterized. Antibody and binding proteins of whole milk may prevent the attachment of microorganisms to enterocyte receptors and thereby minimize the stimulation of active secretion by bacterial exotoxins. Both *E. coli* and *Salmonella* have the ability to activate the enzyme adenylate cyclase located in the enterocyte membrane (Ooms and Degryse, 1986). The abundance of lactobacilli in milk may add further protection against coliforms by competitive growth or secretion of acid and bacteriocides (Fuller, 1986). These protective effects may account, in part, for the lower incidence of diarrhoea when skim-milk provides the major source of protein in the weaner diet. But these beneficial properties could be destroyed by overheating during processing of milk protein (see Roy, 1984).

Piglets which are abruptly weaned onto solid food usually starve for several hours before learning to eat dry meal or pellets. A subsequent tendency to over-eat so as to satisfy their extreme hunger may be the root cause of digestive upsets. So far there have been no satisfactory studies of digestive processes in piglets to test this notion. Electromyography in older pigs has shown that during fasting there is an increase in the numbers of migrating waves of pulsatile contractions passing along the small intestine (Laplace, 1984). If this happens in the early weaned and starving animal, then when food is consumed a combination of intestinal hypermotility and excessive feed intake could lead to an abnormal load of partially digested food filling the lumen of the small bowel. This digesta may provide substrates for an undesirable microflora whilst soluble components would increase osmotic pressure within the lumen. Thus, there may be value in developing feed additives which reduce motor activity of the distal stomach during the first few days after introducing solid food to piglets.

Post-weaning starvation is less of a problem for animals reared on liquid diets. But Roy (1984) has emphasized the importance of coagulation of casein in the true stomach of calves as a means of preventing the premature passage of undigested protein and fat to the duodenum. This property is impaired by severe heat treatment of milk protein. However, as a result or extensive studies by Roy and co-workers (see Roy, 1984) the need to control temperature during the processing of spray-dried milk powder for calf feeding in order to avoid diarrhoea is widely accepted.

MATURATION OF DIGESTIVE FUNCTION

Secretion of digestive enzymes does not reach a maximum until piglets and calves reach about 5 weeks of age (Corring, Aumaitre and Durand, 1978; Kidder and Manners, 1980; Toullec, Guilloteau and Villette, 1984). It is also important to note that variation between animals in the quantities of specific enzymes secreted may be considerable. For example, Henschel (1973) showed that in some calves rennin, and in others pepsin, predominates during the first few weeks of life. For some animals, a slow maturation of enzyme synthesis and secretion probably limits their ability to digest early weaner diets especially when the composition deviates from that of whole milk. This is exemplified by data given in Table 14.1 showing that digestibility of nitrogen in milk substitutes containing non-milk protein increases with age and decreases with level of milk protein replacement. Part of the problem concerning level of substitution could be due to increased intake of antinutritional factors associated with the alternative protein source, or simply a lack of enzymes having the correct specificity for breaking down unusual structures. However, adaptation to diet is a phenomenon recognized in the digestion of some carbohydrates such as lactose (Huber, Rifkin and Keith, 1964). There is no evidence that this applies to protein.

Another weakness of early weaning is that development of the digestive processes may be halted or even depressed by radically changing the diet. Several workers have reported reductions in pancreatic enzyme activity in piglets weaned onto dry feeds at ages of less than 1 month (Hartmann *et al.*, 1961; Efird, Armstrong and Herman, 1982; Lindemann *et al.*, 1986). The extent of this depression may depend on the nature of the dietary protein since partial replacement of dried skim-milk with soyabean flour or fish protein concentrate in a milk substitute for calves led to reduced activity of pancreatic enzymes (Ternouth *et al.*, 1975). Likewise, soya or single cell protein was shown to reduce chymosin and pepsin activity in the calf abomasum, but this did not happen when fish protein was used in the calf diet (Ternouth *et al.*, 1975; Sedgman, 1980; Williams, Roy and Gillies, 1976). The exact nutritional implications arising from inhibited digestive enzyme activity by diet is not known, but these limitations would be expected to favour conditions leading to diarrhoea.

ANTINUTRITIONAL FACTORS

Dried skim-milk, provided it has not been over-heated, has for long been regarded as a highly digestible protein source for both calves and piglets (Toullec, Mathieu and Pion, 1974; Roy, 1984; Armstrong and Clawson, 1980). But, because of restricted production of cows' milk within the European Economic Community, it has become both expensive and scarce. Considerable interest is now being given to cheaper alternatives such as protein from oil seeds (e.g. soyabeans and rapeseed) and arable legumes (e.g. peas and beans). Unfortunately the growth of piglets and calves reared on diets containing large amounts of vegetable protein is generally

Table 14.1a   APPARENT DIGESTIBILITY OF NITROGEN BY CALVES GIVEN WEANER
DIETS CONTAINING DIFFERENT PROTEIN SOURCES

Protein source	Proportion of milk protein replaced	Age (days)	Digestibility of total nitrogen of the diet	Reference
Cows' milk		28–35	0.93	Roy *et al.*, 1964
Skim milk	1.00	28–35	0.95	Roy *et al.*, 1970
Whey	0.05	7–14	0.88	Stobo and Roy, 1977
		28–35	0.90	
Soyabean				
full fat, unheated	0.43	15–19	0.53	Nishimatsu and Kumeno, 1966
		35–39	0.76	
defatted, heated	0.35	21–26	0.76	McGillard *et al.*, 1970
		49–54	0.85	
	0.50	21–26	0.68	McGilliard *et al.*, 1970
		59–54	0.74	
Concentrate, alcohol extracted				
unheated	0.82	28	0.76	Nitsan *et al.*, 1971
heated	0.70	21–28	0.82	Gorill and Nicholson, 1969
		42–49	0.82	
	0.82	28	0.90	Nitsan *et al.*, 1971
Pea				
flour	0.52	7–12	0.53	Bell *et al.*, 1974
		19–24	0.74	
concentrate	0.50	7–12	0.50	Bell *et al.*, 1974
		19–24	0.79	

much lower than animals fed milk based diets, in part because of much lower
digestibility (Meade, 1967; Combs *et al.*, 1963; McGilliard *et al.*, 1970; Bell, Royan
and Young, 1974). Another reason is that these protein sources contain several
biologically active substances which have antinutritional properties. Descriptions of
these components and their effects on animal performance were given in a recent
review by Wiseman and Cole (1988). Some of these substances, such as protease
inhibitors, lectins and antigens, have been implicated in the onset of diarrhoea.

Enzyme inhibitors

Protease inhibitors are widely distributed amongst plants used as protein sources
for animal feed (Liener and Kakade, 1969). Their blockage of protein digestion
may indirectly contribute to the bulkiness of digesta by increasing the load of
undigested constituents in the gut lumen. They also possess the ability to stimulate
hypersecretion of the pancreas in rats and chicks (Lepkovsky, Bingham and
Pencharz, 1959; De Muelenaere, 1964), although this physiological effect has not
been confirmed in calves (Gorrill *et al.*, 1967). Indeed, studies of relations between
protease activity or trypsin inhibitor and growth are conflicting. For example,
whilst Ducharme (1982) showed a positive correlation between activities of trypsin
and chymotrypsin and growth in soya fed calves, others were unable to depress

Table 14.1b APPARENT DIGESTIBILITY OF NITROGEN BY PIGLETS GIVEN WEANER
DIETS CONTAINING DIFFERENT PROTEIN SOURCES

Protein source	Proportion of milk protein replaced	Age (days)	Digestibility of total nitrogen of the diet	Reference
Skim-milk	1.00	7–11	0.98	Pettigrew *et al.*, 1977
		11–15	0.97	
	1.00	14–25	0.96	Seve *et al.*, 1983
		26–32	0.95	
	0.76	28	0.80	Christison and Parra de Solona, 1982
		41	0.86	
Whey	0.24	7–11	0.99	Pettigrew *et al.*, 1977
		11–15	0.97	
	0.25	14–25	0.95	Seve *et al.*, 1983
		26–32	0.96	
Soyabean				
defatted	1.00	21–28	0.74	Combs *et al.*, 1963
		29–35	0.76	
		36–42	0.85	
defatted	0.77	35–36	0.87	Neport and Keal, 1983
	1.00	35–56	0.86	Neport and Keal, 1983
defatted	0.80	28	0.72	Christison and Parra de Solano, 1982
		41	0.78	
Pea				
ground, heated	0.31	15–22	0.91	Seve *et al.*, 1985
		23–29	0.94	
concentrate	0.81	28	0.65	Christison and Parra de Solano, 1982
		41	0.75	

performance by adding trypsin inhibitor to calf milk substitutes (Kakade *et al.*, 1974).

Steam treatment of soyabean meal is claimed to substantially reduce trypsin and chymotrypsin activity (Circle and Smith, 1972). Even so, attempts to rear pre-ruminant calves on liquid diets containing more than 30–40% of the protein as heated soyabean flour have led to diarrhoea, poor growth and sometimes death (Gorrill and Thomas, 1967; Nitsan *et al.*, 1972; Kwiatkowska, 1973). This suggests that either the steam process does not destroy all forms of the inhibitor, or that heat stable factors are involved in the adverse response. Protease inhibitors exist in several forms and some of these are only unstable to heat in the presence of alkali (Obara and Watanabe, 1971). Alkali treatment has been reported to improve the utilization of heated soyabean flour when given in liquid feeds to young pigs (Lennon *et al.*, 1971). But this was not reproduced in all experiments in which fatal diarrhoea occurred. Some workers (De Groot and Slump, 1969; Gorrill, 1970) have warned that alkali treatment of soya protein may destroy certain amino acids and could yield lysinoalanine which has toxic effects in the kidneys of rats (Gorrill, 1970; Woodward, 1972). However, since there is much variation between soyabean varieties and content of protease inhibitor (Kakade *et al.*, 1972) in the longer term it may be possible to reduce enzyme inhibitors in legume seeds through plant breeding.

Lectins

Lectins are characterized for their property of agglutinating red blood cells. There is doubt, however, as to whether this accounts for their fatal effects when injected into rodents, since non-toxic fractions exhibiting relatively high haemagglutinating activity have been isolated from different varieties of beans (Jaffe, 1969). Nevertheless, studies in rats have indicated that these glycoproteins strongly bind with membranes of enterocytes and could disrupt processes of intestinal absorption (Jaffe, 1969). Because the haemagglutinating property is possessed by many bacteria that show specific adherence to mucosal cells (Firon, Ofek and Sharon, 1984), it is feasible that plant lectins may compete with gut microorganisms for binding sites at the brush border. Thus, lectins could prevent beneficial organisms, such as lactobacilli, from colonizing the gastric and gut epithelium. This may lead to an imbalance in gut flora and allow the establishment of pathogenic strains of bacteria. However, the effects of lectins on intestinal function of young pigs and calves awaits investigation.

Studies *in vitro* have shown that soyabean lectins are readily inactivated by pepsin digestion and hydrochloric acid (Liener, 1955; Birk and Gertler, 1961). It is feasible, therefore, that lectins are rendered harmless by digestion. Many reports have also shown that haemagglutinating and toxic activities of lectins are also very susceptible to treatment with moist steam (Circle and Smith, 1972). Thus, at least for heated sources of protein, lectins may be of little antinutritional importance.

Dietary antigens

Despite processing with heat to destroy enzyme inhibitors and lectins, diarrhoea still occurs amongst piglets and calves fed non-milk protein. Recently much interest has focused on the idea that some animals may be intolerant of immunologically active globulins in legume protein and suffer from diarrhoea caused by inflammatory reactions in mucosal tissue.

Calves   Early work in calves first pointed to the possibility that immune responses to dietary soyabean protein might be linked with disorders in gastrointestinal function (Smith and Sissons, 1975; Sissons and Smith, 1976). Severe diarrhoea linked with disturbances in digesta movement and nutrient absorption were observed in calves given a succession of feeds containing defatted and steamed soyabean flour. These disorders were not seen on the first occasion that heated soya flour was given but developed with successive feeds. Replacement of soya by feeding casein eliminated the digestive upsets and demonstrated that the problem of diarrhoea was temporarily associated with the ingestion of soyabean flour. However, microbiological examination of ileal digesta revealed no abnormalities in gut microflora after consumption of soya. On the other hand, this protocol of oral sensitization to soyabean flour did induce the development of high titres of circulatory anti-soya antibodies. These findings led to the hypothesis that calves fed soyabean flour suffered from diarrhoea because they developed a gastrointestinal allergy to some factor present in soyabeans.

From studies in humans and rodents it is known that intestinal hypersensitivity to food involves stimulation of the immune system. This requires penetration of the mucosal barrier by antigens of dietary origin and their interaction with lymphoid tissue. Subsequent systemic and/or mucosal immune reactions leads to inflammation of the intestinal epithelium and disruption of processes controlling fluid movement through the gut wall (Walker, 1987). Similar events are postulated to mediate intolerance of calves to soya protein. Morphological and histological studies of mucosal tissue of calves orally sensitized to soyabean flour have revealed loss of normal villous architecture (see Figure 14.1), oedema, increased numbers of intra-epithelial lymphocytes and crypt hyperplasia (Barratt, Strachan and Porter, 1978; Sissons, Pedersen and Wells, 1984; Pedersen *et al.*, 1984). Such damage to the villi would, by inhibiting the maturation of enterocytes, impair the digestive and absorptive functions of the gut. Indeed, unusual gut permeability to milk protein and depressed xylose absorption has been demonstrated in soya fed calves (Kilshaw and Slade, 1982; Seegraber and Morrill, 1979).

Although mucosal tissue damage and fluid accumulation in the small bowel was seen to occur concurrently with a rise in systemic anti-soya antibodies, the immune mechanism mediating the inflammatory reaction is not fully understood. Most calves fed soya flour showed high titres of IgG antibodies which were specifically against native glycinin and β-conglycinin, the major storage proteins of soyabeans (Smith and Sissons, 1975; Kilshaw and Sissons, 1979a). Barratt, Strachan and Porter (1978) attributed the mucosal injury to activation of the complement system. This involves complex formation between IgG antibodies and soya antigens and is known as a type-3 complex-mediated hypersensitivity (Roitt, 1974). However, IgE antibodies have also been detected in the sera of some soya fed calves (Kilshaw and Sissons, 1979a). These antibodies are known to passively sensitize mucosal mast cells. Antigen reacting with the sensitized cells can then trigger the release of intracellular granules containing inflammation producing substances (Coombs, 1984). Reactions involving IgE antibody occur shortly after an antigen challenge and are known as immediate hypersensitivity (Roitt, 1974).

It is possible that more than one immune reactions brings about the gut hypersensitivity to soya in calves. Besides those attributed to IgG and IgE against soya protein there is indirect evidence of immune mechanisms not involving antibodies. From observations of the time course of changes in villous and crypt architecture after feeding calves on soya protein, Kilshaw and Slade (1982) concluded that hypersensitivity resulted from a delayed mechanism which implied that a cell-mediated response to dietary antigen could have been involved. This finding contrasts with evidence reported by Pedersen *et al.* (1984) of a more rapid occurrence of mucosal injury seen within 1–2 h of oral challenge with antigenic soya flour. Tubular fibrin casts have been collected from the ileum of some calves within 1 h of giving a feed of soyabean flour which suggests that a dramatic change in vascular permeability must have been invoked shortly after the feed had entered the gut lumen. In another study immediate hypersensitivity, but not systemic cell-mediated immunity, was observed in cutaneous reactions to intradermal injections of a soluble extract of soya flour in calves previously fed the same antigenic proteins (Heppell *et al.*, 1987). More recently Dawson *et al.* (1988) reported that native soyabean proteins did not stimulate mitogenic activity of circulatory lymphocytes obtained from calves fed milk substitutes containing soyabean products. This result was also taken to indicate an absence of a cell-mediated immune response.

Although there is uncertainty as to whether the gut inflammation and disturbed digesta passage linked with feeding soya involves more than one immune mechanism, there is strong evidence that the adverse reaction is stimulated by

268

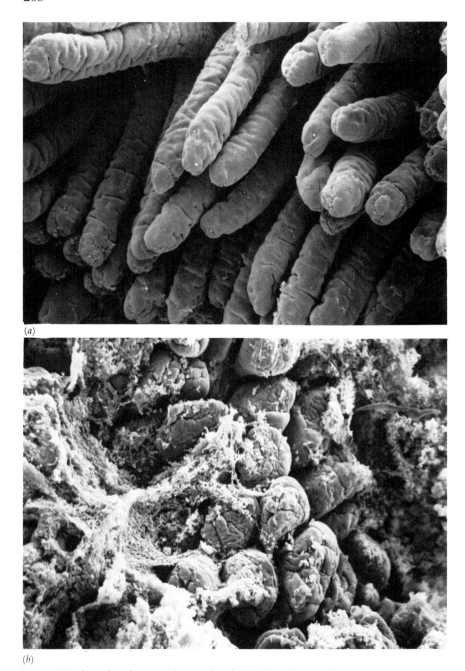

(a)

(b)

Figure 14.1 Scanning electron micrographs of villi in the jejunum of calves which had been orally sensitized to heated soyabean flour and then 2 days later given a challenge feed containing milk protein (a, control) or soya protein (b). Note after challenge with soya, the severe shortening of villi and the web of fibrin coating their surface

antigenic protein in the diet. Therefore, if large amounts of soya are to be used in milk substitutes for calves some form of processing is required to eliminate its immunological activity. This view may also apply to other sources of legume protein such as peas and beans which contain globulins having similar structures and biochemical properties to those in soya (Gueguen and Azanza, 1984).

Piglets Intestinal hypersensitivity to dietary antigens has also been postulated to predispose diarrhoea which sometimes occurs when piglets are weaned onto soya diets (Miller *et al.*, 1983). However, in contrast to the persistence of diarrhoea in soya fed calves the intolerance to soyabean meal by piglets is transient. Diarrhoea occurs about 1 week after the introduction of the weaner diet and, in the absence of a secondary pathogen, faecal stools usually become firm again by 2–3 weeks post-weaning. During this transient period of hypersensitivity the gut mucosa of the piglet undergoes similar tissue inflammatory reactions to those seen in the calf. At the same time absorption is impaired and there is a reduction in the activity of brush-border disacccharides (Miller *et al.*, 1984; Miller *et al.*, 1986; Ratcliffe *et al.*, 1987). Studies of cutaneous reactions to intradermal injections of soya protein extract in piglets weaned onto a diet based on soyabean meal showed evidence of a delayed hypersensitivity response. The skin reactions were noted to coincide with observations of reduced capacity by the gut to absorb xylose (Newby *et al.*, 1985). The observed cutaneous response to soya implies that intestinal inflammation in weaned piglets could involve a cell mediated response to dietary antigens. Delayed hypersensitivity reactions begin with the primary exposure of a T-lymphocyte to a specific antigen resulting in T-cell activation. Upon secondary exposure to the same antigen the T-lymphocyte is induced to release lymphokines which have been linked with inflammation of mucosal tissue (Mowat and Ferguson, 1981).

According to Newby *et al.* (1985) the post-weaning immune response may be modified by primary exposure of the piglet to the same antigenic material during 'suckling', for example, through ingestion of creep feed. Some support for this view was obtained from experiments with cows' milk protein showing that the incidence and severity of the diarrhoea in piglets weaned onto native casein was increased if they ingested a small 'priming' dose of the cows' milk protein during suckling. Extending the pre-weaning exposure of the supposedly antigenic protein from 3 to 7 days reduced or prevented post-weaning scours (Miller *et al.*, 1984). Thus it was proposed that ingestion of a creep feed could either 'tolerize' or 'sensitize' the piglet, depending on the amount of antigen consumed prior to re-exposure of the same antigen at weaning.

This interesting hypothesis of pre-weaning sensitization to dietary antigens has serious implications for the value of creep feeding. But it must be noted that, so far, the idea has not been validated with protein sources other than that of cows' milk. When the oral tolerance experiments were repeated using soya, rather than casein, as the antigen source, Miller and co-workers were not able to sensitize the piglets by feeding a small amount of soya protein during the suckling phase (Miller, private communication). Although, feeding the animals relatively large amounts of soya antigen did prevent both xylose malabsorption and diarrhoea (Miller *et al.*, 1985). Since piglets are commonly born to sows which receive soyabean meal in their diet, small amounts of soya antigen passing to the suckling animal in colostrum and milk may have sensitized them in advance of a so-called 'priming' dose ingested as creep feed. But further doubt has been cast on the idea of sensitization during suckling in studies which showed that the architecture and function of mucosal tissue of piglets which had been forcibly 'primed', or offered creep feed, were not different from those in abruptly weaned animals (Hampson and Kidder, 1986; Hampson and Smith, 1986; Hampson, Fu and Smith, 1988). Unfortunately there is a lack of information about the precise immune status of the animals in these studies. There is a need, for example, to understand whether factors other than diet could, through mechanisms of immune suppression, concurrently modulate the response of piglets to antigens in their feed.

OSMOTIC COMPONENTS OF DIET

Raised osmotic pressure through the build up of non-absorbable solutes in the gut lumen can result in a considerable driving force for the movement of water. For instance, feeding young calves an excess of glucose or lactose (i.e. a daily intake of more than 9 g/kg live weight of hexose equivalent) will induce diarrhoea (Roy, 1969). A problem could also arise in circumstances where the activities of pancreatic and brush border enzymes are impaired or lack specificity. About 30% of soyabean meal comprises sucrose, stachyose, raffinose and complex polysaccharides (Rackis *et al.,* 1970). It is unlikely (though not proven) that any of these carbohydrates are digested by the calf since it does not possess suitable enzymes (Sissons, 1981). Except perhaps for the digestion of sucrose, the pig probably has the same digestive limitation (Kidder and Manners, 1978). However, studies of intestinal digesta flow in calves fed soya molasses (Sissons and Smith, 1976) have indicated that the induction of abnormal water movement by these carbohydrates is relatively small compared with the disturbed handling of fluid attributed to inflammation of the gut mucosa (see Table 14.2). Recent evidence (unpublished observations) supporting this view has been obtained from recordings of intestinal myoelectric activity. This work showed that the pattern of gut motility in calves suffering from diarrhoea which had been induced by feeding heated soyabean flour was distinctly different from motor activity in the same animals fed sufficient

Table 14.2   EFFECT OF MILK OR SOYA PROTEIN ON THE MOVEMENT OF DIGESTA THROUGH, AND APPARENT NITROGEN ABSORPTION FROM, THE SMALL INTESTINE OF CALVES ORALLY SENSITIZED TO HEATED SOYA FLOUR

Order of giving feeds:	Small gut transit time (h)	Flow rate of digesta (g/h)	Net nitrogren absorption (%)
HSF 1st feed (unsensitized)	3.1	77	57
HSF challenge feed (sensitized)	1.4	165	25
Casein	3.5	48	85
ETHSC	3.1	69	74
ETHSC + 30 g soya molasses	2.2	82	77
ETHSC + 30 g sucrose	2.3	61	83

HSF = heated soyabean flour
ETHSC = ethanol treated and heated soya concentrate
After Sissons and Smith (1976)

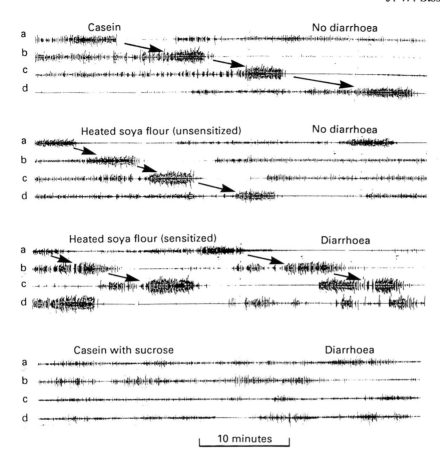

Figure 14.2 Typical recordings of myoelectric activity made from the small intestine of a calf given feeds containing milk protein with and without a sucrose load (200 g) or heated soyabean flour on a first (unsensitized) or fifth (sensitized) occasion. Recording sites a–d were spaced about 2 m apart on the jejunum. Arrows denoted a propagating complex of contractions corresponding to peristalsis

sucrose to cause osmotic diarrhoea (see Figure 14.2). These recordings show that normal intestinal motility on the small intestine is characterized by cyclical phases of weak myoelectric activity (associated with motions which mix digesta) followed by stronger activity (associated with peristaltic contractions). Inducing a hyperosmotic condition with sucrose led to the replacement of the cyclic motor pattern with a mixing type of activity, whilst feeding antigenic soya protein had the effect of stimulating considerable peristalsis. Under these circumstances enhanced peristalsis might have a protective action by facilitating the rapid removal of offending antigens. Even though soluble molasses may not normally have a large effect on the osmotic pressure of digesta, they may be important in a state of malabsorption especially where there could be added contributions from soluble fatty acids and electrolytes.

Effect of diarrhoea on performance

Despite concern to prevent diarrhoea there is little information on the effects of scouring on animal performance. Ball and Aherne (1982) showed that diarrhoea caused a significant slowing of growth of early weaned pigs between 21 and 35 days compared with animals not suffering from loose faeces. They also found that piglets with diarrhoea required 65–72% more feed per unit of gain than healthy animals. However, the degree of post-weaning diarrhoea was reported not to have significantly affected the growth rate to 90 kg live weight. In studies of pre-ruminant calves fed heated soyabean flour, flow rates of digesta at the distal ileum were negatively correlated with net nitrogen absorption (Sissons and Smith, 1976). On occasions when digesta flow was very high (about 4 litres/day) the amounts of nitrogen appearing at the ileum exceeded the amount given in a liquid feed. Much of this nitrogen loss probably arose from mucosal tissue and vascular exudation of fibrin as a result of inflammation.

Dietary approaches to the control of diarrhoea

Methods of controlling diarrhoea should take account of the pathogenesis of the disorder, the species and whether the animal is being weaned onto a liquid or solid diet. Many studies indicate that provided animals are not severely infected with enteric microbial pathogens, dietary manipulation can be used to avoid or reduce diarrhoea. Some success has been achieved by altering intake or composition of feeds and through processing of dietary ingredients.

RESTRICTION OF FEED INTAKE

Early weaned piglets show increased susceptibility to diarrhoea when they overeat. This is more likely to occur when solid food is given *ad libitum* than when intake is restricted during the first few weeks post-weaning (Palmer and Hulland, 1965; Smith and Halls, 1968; Hampson and Smith, 1986; Ball and Aherne, 1982). The precise physiological benefits of restricted feeding have, however, not been established. Reducing feed consumption may ensure that nutrient intake is consistent with a limited digestive capacity. At the same time, small meals are less likely to cause derangements in gastric and intestinal motor activity (Ruckebusch and Bueno, 1976) which might otherwise lead to the accumulation of digesta in the small intestine. Excessive food intake may also upset mechanisms which regulate gastric secretion. Failure of the calf abomasum to secrete sufficient acid and protease could be a reason why the drinking of large volumes of milk predisposes some calves to diarrhoea (Roy, 1984). Physiological studies in milk fed calves have shown that the enterogastric reflex control of abomasal emptying is impaired when liquid intake exceeds the normal capacity of the abomasum (Sissons and Smith, 1979; Sissons, 1983). This results in an abnormally high rate of abomasal digesta flow to the duodenum.

ADDING FIBRE TO THE WEANER DIET

Several reports indicated that adding dietary fibre to a weaner diet for piglets had a beneficial effect by reducing the incidence and severity of diarrhoea (Palmer and

Hulland, 1965; Smith and Halls, 1968; Armstrong and Cline, 1976). The ameliorating effects of fibre on digestive function are unclear. But, dietary fibre stimulates the secretion of saliva, gastric juice, bile and pancreatic juice (Low, 1985). This, together with its effect of reducing nutrient density and increasing satiety may help to maintain a satisfactory balance between feed intake and the digestive capacity of the young pig. Another possible effect is that the yield of volatile fatty acids from fermentation of fibre in the colon alters the absorption of water (Crump, Argenzio and Whipp, 1980). Because fibre has a high water holding capacity it may increase the firmness of faecal stools. However, if there is a serious alteration in fluid handling by the small and, or large bowel, water retention by fibre could disguise critical losses of water and nutrients.

Studies in humans have investigated the relation between colonic motor processes and the passage of digesta during disease of the digestive tract (Connell, 1962). Curiously a diarrhoeal state has been linked with hypomotor activity of the colon whilst constipation is associated with hypermotility. It is well established that consumption of dietary fibre is an effective way of shortening transit time, and that fibre acts on colonic motility. Electromyographic studies in growing pigs showed that adding bran to a milk substitute diet stimulated increased numbers of propulsive contractions (Fioramonti and Bueno, 1980). It seems unlikely, however, that the beneficial effects of fibre in reducing diarrhoea are related to its effects on colonic motility, but rather its high water holding capacity.

FAT LEVELS IN THE WEANER DIET

Milk substitutes with high levels of glucose or lactose have been shown to invoke diarrhoea in calves if the diet is low in fat (Blaxter and Wood, 1953; Mathieu and De Tugny, 1965). Fat is known to reduce the rate of abomasal emptying of digesta (Smith and Sissons, 1975) and this effect may slow the release of soluble carbohydrates so as to prevent a hyperosmotic condition developing in the lumen of the small intestine. Interactions between fat and carbohydrate on the digestive physiology of the young pig do not appear to have been investigated.

PROCESSING TO REDUCE ANTINUTRITIONAL FACTORS

Recognition of a number of antinutritional factors in feedstuffs implicated in the genesis of diarrhoea and slowing of growth has stimulated a search for treatments for inactivation or removing offending constituents.

Heat treatment

As discussed earlier the need to treat protein sources containing lectins and protease inhibitors with heat is widely accepted. Heating brings about aggregate formation and loss of water solubility. Studies of soya protein indicate that molecules with few disulphide bridges are most susceptible to heat treatment. Thus, the Kunitz trypsin inhibitor (molecular weight 21 500) with two disulphide bridges in its polypeptide chain is less stable than the Bowman-Birk inhibitor (molecular weight almost 8000) which has seven disulphide linkages (Steiner, 1965; Bidlingmeyer, Leary and Laskowski, 1972). Even so, several studies have shown that steam treatment alone is not sufficient to prevent diarrhoea or overcome negative effects of soya flour on digestibility and growth of calves (Gorill and Thomas, 1967; Nitsan *et al.,* 1971; Sudweeks and Ramsey, 1972). Similarly, additional treatments appear to be necessary to improve the nutritive value of soyabean products for very young piglets. Newport (1980) reported severe scouring and high mortality amongst piglets weaned at 2 days of age onto a liquid diet containing 74% of the protein in the form of a soyabean isolate. This soya product was claimed to have undetectable levels of urease and trypsin inhibitor activity. However, older pigs may benefit from legume protein which has been steam treated. Recently Van der Poel and Huisman (1988) studied growing pigs given meals based on steamed beans (*Phaseolus vulgaris*) and showed that apparent digestibility of nitrogen, dry matter and fat measured at the ileum improved with duration of treatment.

Steam or toasting at 100°C has little effect on the antigenic activity of the major globulins of soya beans (Kilshaw and Sissons, 1979b). Inactivation of these immunological structures requires temperatures of at least 145°C (Srihara, 1984). Unfortunately, processing soyabean meal at such high temperatures may not yield a product suitable for feeding to calves (Srihara, 1984) or piglets, since the resulting polymerization of the protein molecules could impair digestibility (Hagar, 1984).

Alcohol treatment

Some improvement in the utilization of soya protein by calves has been achieved when defatted soyabean flour was extracted with aqueous ethanol before toasting the product (Gorrill and Thomas, 1967; Gorrill and Nicholson, 1969; Nitsan *et al.,* 1971; Stobo, Ganderton and Connors, 1983). Studies of digestive processes in the small intestine suggest that the beneficial effects of processing soya flour with ethanol may be linked with alterations in protein structure rather than the removal of sucrose and oligosaccharides during extraction (Sissons and Smith, 1976; Sissons, Smith and Hewitt, 1979). Aqueous ethanol treatment appears to destroy the antigenic integrity of glycinin and β-conglycinin, the major globulins implicated in the gastrointestinal disorders of soya fed calves (Kilshaw and Sissons, 1979a). But the conditions of alcohol concentration and temperature required to prepare products with minimal levels of native antigenic activity are critical (see Figure 14.3); effective treatment was achieved using between 65–70% ethanol at 78°C (Sissons *et al.,* 1982a; Sissons *et al.,* 1982b). Aqueous organic mixtures are necessary to disrupt globular proteins as their structures are stabilized by hydrophobic side-chain residues located towards the centre of the molecule and hydrophilic groups positioned at the surface (Fukushima, 1969).

Although hot aqueous ethanol extraction of soya flour reduces antigenicity of native globulins and substantially improves its feeding value for calves, the product remains somewhat inferior to milk protein. For example, Stobo, Ganderton and Connors (1983) reported liveweight gains of 0.9, 0.7 and 0.2 kg/day during a 14-day period after weaning 2-day-old calves onto liquid diets in which all of the protein was derived from cows' milk or with 65% replaced by ethanol extracted soya concentrate or soyabean flour respectively. It is possible that new protein structures having immunological activity could be exposed through partial digestion. However, gut inflammatory responses to ethanol extracted soyabean concentrate were not seen in calves previously observed to react adversely to antigenic soya flour (Pedersen, 1986).

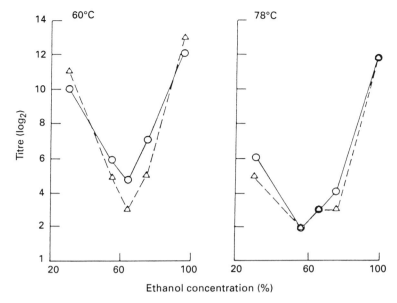

Figure 14.3 Antigenically active glycinin (o) and β-conglycinin (△) contents of defatted soyabean flour extracted with different concentrations of ethanol at 60°C and 78°C (after Sissons *et al.*, 1982)

An attempt to wean 2-day-old piglets onto a liquid milk substitute containing large amounts of ethanol extracted soya concentrate led to severe scours and high mortality. The cause of death was not established. Decreasing the proportion of protein supplied by soya concentrate from 70 to 35% reduced deaths and scouring. Nevertheless, results for liveweight gain and feed conversion efficiency were much lower than performances observed in piglets fed a control diet based entirely on cows' milk protein (Newport and Keal, 1982). Measurements of digestive enzymes in these animals suggested that proteolysis of milk and soyabean proteins were equally efficient when the piglets reached 28 days of age.

Predigestion of protein

Survival of antinutritional factors in the gastrointestinal tract of young animals could be influenced by maturity of the digestive function. It is also possible that endogenous enzymes may be ineffective in digesting some components of milk substitutes. For instance, proteases of the calf abomasum and small intestine have been reported to be less effective in their hydrolysis of soya protein compared with actions on milk protein (Jenkins, Mahadevan and Emmons, 1980). Ideally enzymes used in predigestion should be complementary to the actions of endogenous enzymes. Thus, there is a need to appreciate the character of structures of residual protein and carbohydrates of weaner feeds which resist digestion. *In vitro* investigations of the stability of soya antigens to actions by porcine and bovine proteases indicated that, under optimal conditions of pH for protease activity,

Table 14.3 RELATIVE QUANTITIES OF IMMUNOLOGICALLY ACTIVE GLYCININ AND β-CONGLYCININ IN HEATED SOYA FLOUR AND SOYA ISOLATE PROTEIN

Enzyme treatment	Haemagglutination inhibition titre (log 2)			
	Anti-glycinin		Anti-β-conglycinin	
	Soya flour	Soya isolate	Soya flour	Soya isolate
No enzyme, pH 1.8	9	10	12	12
Pepsin, pH 1.8	>1	>1	12	12
No enzyme, pH 7.8	12	11	12	12
Trypsin, pH 7.8	12	7	7	9
Pepsin then trypsin	>1	>1	7	5

After Sissons and Thurston (1984)

β-conglycinin but not glycinin was unaffected by pepsin, whilst both proteins retained their antigenic activity after treatment with trypsin (see Table 14.3).

The possibility that digestive difficulties might be overcome by predigestion of ingredients used in a weaner diet was demonstrated in experiments which showed that intestinal damage and diarrhoea amongst early weaned piglets fed casein did not occur when animals were given hydrolysed cows' milk protein (Miller *et al.,* 1983). The improvement was attributed to the absence of antigenic material in the hydrolysed product. In another study piglets weaned onto diets containing pre-digested milk or soya isolate protein absorbed more nitrogen compared with animals fed native protein (Leibholz, 1981). In contrast, supplementary enzyme treatments of soya protein did not result in improved growth performance of pre-ruminant calves (Fries, Lassiter and Huffman, 1958; Colvin and Ramsay, 1968). This disappointing result may have been related to the poor specification of the enzymes used in the pre-digestion treatment. As a note of caution, products arising from predigestion of non-milk sources should be assessed to ensure that the process does not increase thet activity of antinutritional components.

PROBIOTICS

The normal microflora of the alimentary tract is thought to complement the digestive functions of the host and provide protection against invading pathogens. At times of stress, such as weaning, the 'balance' of gut bacteria may become disturbed in favour of enterotoxic bacteria (Barrow, Fuller and Newport, 1977). Recently there has been much interest in the possibility of using bacterial feed additives to stabilize the indigenous flora and prevent diarrhoea in young animals. Many claims for the efficacy of these preparations have been made by commercial companies. Whilst some published studies support these claims (Bechman, Chambers and Cunningham, 1977; Muralidhara *et al.,* 1977) others have failed to demonstrate a beneficial effect of probiotic organisms (Hatch, Thomas and Thayne, 1973; Elinger, Muller and Ganzt, 1978; Morrill, Dayton and Mickelsen, 1977). In a recent review of probiotic efficacy, Fuller (1986) proposed that organisms may fail to achieve a beneficial response because of non-adherence to gastric and gut epithelial tissue, inability to grow in the gut environment and a lack of specificity for the host. Clearly further work is needed on modes of action and desirable features of probiotic cultures before bacterial additives can replace antimicrobial agents as a means of controlling diarrhoea.

PHARMACOLOGICAL TREATMENTS

Several pharmacological treatments have proved effective in preventing the adverse immune responses involved in certain human food allergies. The drug disodium cromoglycate has been used to prevent histamine release possibility by acting to stabilize mast cell membranes (Dahl and Zetterstrom, 1978). Other reports suggest that indomethacin can alleviate symptoms of food intolerance in humans (Buisserat *et al.*, 1978) and protect calves against cardiovascular anaphylactic shock (Burka and Eyre, 1974). However, neither drug showed any beneficial effects on intestinal reactions of calves which had been orally sensitized with heated soya flour (Kilshaw and Sissons, 1979a; Kilshaw and Slade, 1980). The absence of a response may have been due to a lack of specific action on mucosal mast cells by these treatments. But if suitable agents could be found for controlling intestinal allergic disorders in young farm animals they could be more appropriate for piglets where intolerance to dietary protein seems to be transient and therefore administration of a drug need not be continuous.

Conclusions

Diarrhoea of nutritional origin amongst calves and piglets is fundamentally either a problem of mucosal inflammation and/or increased luminal osmolality, whilst deranged motility may well aggravate the primary cause. Tissue damage linked with adverse immune responses to dietary antigens appears to be the most serious problem and this can lead to considerable loss of nutrients. Much recent research has focused attention on immaturity and limitations of digestive and immune function.

Further studies are needed to clarify the effect of creep feed in predisposing piglets to dietary intolerance at weaning, together with maternal factors which may influence mechanisms regulating the balance between sensitivity and tolerance to immunologically active components. Also more knowledge is required of molecular aspects of digestion, especially of antinutritional structures which may be revealed by partial hydrolysis. Besides heat processing for eliminating lectins and protease inhibitors in legume protein, further treatment is required to destroy antigenic activity of globulins. This can be achieved with hot aqueous ethanol, but the exact conditions are critical. This treatment has so far only proved successful for calves.

At the present time, strategies for preventing diarrhoea in young farm animals continue to be based on formulating weaner diets with highly digestible ingredients. But the risk of digestive disorders will increase as nutrient sources of weaner diets deviate from those in whole milk. Possible control measures include restrictions in intake during the first week of weaning, lowering nutrient density and processing to eliminate antinutritional factors. Predigestion with novel enzymes and a putative protective and digestive role for probiotics need evaluation and therefore remain as future options.

References

ARGENZIO, R.A. and WHIPP, S.C. (1980). In *Veterinary Gastroenterology – 1980*, pp. 220–232. Ed. Anderson, N.V. Lea and Febiger, Philadelphia

ARMSTRONG, W.D. and CLAWSON, A.J. (1980). *Journal of Animal Science,* **50**, 377–384

ARMSTRONG, W.D. and CLINE, T.R. (1976). *Journal of Animal Science,* **42**, 592–598

BALL, R.O. and AHERNE, F.X. (1982). *Canadian Journal of Animal Science,* **62**, 907–913

BARRATT, M.E.J., STRACHAN, P.J. and PORTER, P. (1978). *Clinical Experimental Immunology,* **31**, 305–312

BARROW, P.A., FULLER, R. and NEWPORT, M.J. (1977). *Infection and Immunity,* **18**, 586–595

BECHMAN, T.J., CHAMBERS, J.V. and CUNNINGHAM, M.D. (1977). *Journal of Dairy Science,* **60**, 74

BELL, J.M., ROYAN, G.F. and YOUNG, C.E. (1974). *Canadian Journal of Animal Science,* **54**, 355–362

BIDLINGMEYER, U.D., LEARY, T.R. and LASKOWSKI, M. (1972). *Biochemistry,* **11**, 3303–3310

BIRK, Y. and GERTLER, A. (1961). *Journal of Nutrition,* **75**, 379–387

BLAXTER, K.L. and WOOD, W.A. (1953). *Veterinary Record,* **65**, 889–892

BUISSERAT, P.D., HEINZELMANN, D.I., YOULTEN, L.J.F. and LESSOF, M.H. (1978). *Lancet,* **i**, 906–908

BURKA, J.F. and EYRE, P. (1974). *Canadian Journal of Physiology and Pharmacology,* **52**, 942–951

BYWATER, R.J. (1973). *Research in Veterinary Science,* **14**, 35–41

BYWATER, R.J. and LOGAN, E.F. (1974). *Journal of Comparative Pathology,* **84**, 599–610

CHRISTISON, G.I. and PARRA DE SOLANO, N.M. (1982). *Canadian Journal of Animal Science,* **62**, 899–905

CIRCLE, S.J. and SMITH, A.K. (1972). In *Soybeans: Chemistry and Technology*, Vol. 1, *Proteins*, pp. 294–338. Ed. Smith, A.K. and Circle, S.J. The Avi Publishing Company, Connecticut, USA

COLVIN, B.M. and RAMSAY, H.A. (1968). *Journal of Dairy Science,* **51**, 898–904

COMBS, G.E., OSEGUEDA, F.L., WALLACE, H.D. and AMMERMAN, C.B. (1963). *Journal of Animal Science,* **22**, 396–398

CONNELL, A.M. (1962). *Gut,* **3**, 342–348

COOMBS, R.R.A. (1984). *Chemistry and Industry,* **3**, 87–90

CORRING, T., AUMAITRE, A. and DURAND, G. (1978). *Nutrition and Metabolism,* **22**, 231–243

CRUMP, M.H., ARGENZIO, R.A. and WHIPP, S.C. (1980). *American Journal of Veterinary Research,* **41**, 1565–1568

DAHL, R. and ZETTERSTROM, O. (1978). *Clinical Allergy,* **8**, 419–422

DAWSON, D.P., MORRILL, J.L., REDDY, P.G., MINOCHA, H.C. and RAMSEY, H.A. (1988). *Journal of Dairy Science,* **71**, 1301–1309

DE GROOT, A.P. and SLUMP, P. (1969). *Journal of Nutrition,* **98**, 45–56

DE MEULENAERE, H.L.H. (1964). *Journal of Nutrition,* **82**, 197–205

DEPREZ, P., HENDE, C., VAN D., MUYLLE, E. and OYAERT, W. (1986). *Veterinary Research Communications,* **10**, 469–478

DUCHARME, G.A. (1982). In *Effect of Soybean Trypsin Inhibitor on Growth, Protein Digestibility, and Pancreatic Enzyme Activity in the Young Calf*, PhD Thesis, North Carolina State University

EFIRD, R.C., ARMSTRONG, W.D. and HERMAN, D.L. (1982). *Journal of Animal Science*, **55**, 1370–1379

ELLINGER, D.K., MULLER, L.D. and GANTZ, P.J. (1978). *Journal of Dairy Science*, **61**, 126

FIORAMONTI, J. and BUENO, L. (1980). *British Journal of Nutrition*, **43**, 155–162

FIRON, N., OFEK, I. and SHARON, N. (1984). *Infection and Immunity*, **43**, 1088–1090

FRIES, G.F., LASSITER, C.A. and HUFFMAN, C.F. (1958). *Journal of Dairy Science*, **41**, 1081–1087

FUKUSHIMA, D. (1969). *Cereal Chemistry*, **46**, 156–163

FULLER, R. (1986). *Journal of Applied Bacteriology*, 1S–7S

GORRILL, A.D.L. (1970). *Canadian Journal of Animal Science*, **50**, 745–747

GORRILL, A.D.L. and NICHOLSON, J.W.G. (1969). *Canadian Journal of Animal Science*, **49**, 315–321

GORRILL, A.D.L. and THOMAS, J.W. (1967). *Journal of Nutrition*, **92**, 215–223

GORRILL, A.D.L., THOMAS, J.W., STEWART, W.E. and MORRILL, J.L. (1967). *Journal of Nutrition*, **92**, 86–92

GUEGEN, J. and AZANZA, J.L. (1984). In *Les Proteines Vegetales – 1984*, pp. 135–159. Ed. Godon, B. Lavoisier, Apria

HAGAR, D.F. (1984). *Journal of Agriculture and Food Chemistry*, **32**, 293–296

HAMPSON, D.J. and KIDDER, D.E. (1986). *Research in Veterinary Science*, **40**, 24–31

HAMPSON, D.J. and SMITH, W.C. (1986). *Research in Veterinary Science*, **41**, 63–69

HAMPSON, D.J., FU, Z.F. and SMITH, W.C. (1988). *Research in Veterinary Science*, **44**, 309–314

HARTMANN, P.A., HAYS, V.W., BAKER, R.O., NEAGLE, L.H. and CATRON, D.V. (1961). *Journal of Animal Science*, **20**, 114–123

HATCH, R.C., THOMAS, R.O. and THAYNE, W.V. (1973). *Journal of Dairy Science*, **56**, 682

HENSCHEL, M.J. (1973). *British Journal of Nutrition*, **30**, 285–295

HEPPELL, L.M.J., SISSONS, J.W., STOBO, I.J.F., THURSTON, S.M. and DUVAUX, C. (1987). In *Food Allergy – 1987*, pp. 109–115. Ed. Chandra, R.K. Nutrition Research Education Foundation, St John's, Newfoundland

HUBER, J.T., RIFKIN, R.J. and KEITH, J.M. (1964). *Journal of Dairy Science*, **47**, 789–792

JAFFE, W.G. (1969). In *Toxic Constituents of Plant Foodstuffs – 1969*. Ed. Liener, I.E. Academic Press, New York and London

JENKINS, K.J., MAHADEVAN, S. and EMMONS, D.B. (1980). *Canadian Journal of Animal Science*, **60**, 907–914

KAKADE, M.L., SIMONS, N.R., LIENER, I.E. and LAMBERT, J.W. (1972). *Journal of Agriculture and Food Chemistry*, **20**, 87–90

KAKADE, M.L., THOMPSON, R.M., ENGELSTAD, W.E., BEHRENS, G.C. and YODER, R.D. (1974). *Journal of Dairy Science*, **57**, 650

KIDDER, D.E. and MANNERS, M.J. (1978). In *Digestion in the Pig – 1978*. Ed. Kidder, D.E. and Manners, M.J. Scientechnica, Bristol

KIDDER, D.E. and MANNERS, M.J. (1980). *British Journal of Nutrition*, **43**, 141–153

KILSHAW, P.K. and SISSONS, J.W. (1979a). *Research in Veterinary Science*, **27**, 361–365

KILSHAW, P.J. and SISSONS, J.W. (1979b). *Research in Veterinary Science*, **27**, 366–371

KILSHAW, P.K. and SLADE, H. (1980). *Clinical and Experimental Immunology*, **41**, 575–582

KILSHAW, P.K. and SLADE, H. (1982). *Research in Veterinary Science*, **33**, 305–308

KWIATKOWSKA, A. (1973). *Prace i Materialy Zootech.*, **3**, 63–75

LAPLACE, J.P. (1984). In *Function and Dysfunction of the Small Intestine – 1984*, pp. 1–20. Ed. Batt, R.M. and Lawrence, T.L.J. Liverpool University Press, UK

LENNON, A.M., RAMSEY, H.A., ALSMEYER, W.L., CLAWSON, A.J. and BARRICK, E.R. (1971). *Journal of Animal Science*, **33**, 514–519

LEIBHOLZ, J. (1981). *British Journal of Nutrition*, **46**, 59–69

LEPKOVSKY, S., BINGHAM, E. and PENCHARZ, R. (1959). *Poultry Science*, **38**, 1289–1295

LIENER, I.E. (1955). *Archives of Biochemistry and Biophysics*, **54**, 223–231

LIENER, I.E. and KAKADE, M.L. (1969). In *Toxic Constituents of Plant Foodstuffs – 1969*, pp. 7–68. Ed. Liener, I.E. Academic Press, New York and London

LINDEMANN, M.D., CORNELIUS, S.G., EL KANDELGY, S.M., MOSER, R.L. and PETTIGREW, J.E. (1986). *Journal of Animal Science*, **62**, 1298–1307

LOW, A.G. (1985). In *Proceedings of the 3rd International Seminar on Digestive Physiology in the Pig – 1985*, pp. 157–179. Ed. Just, A., Jorgensen, H. and Fernandez, J.A. National Institute of Animal Science, Copenhagen, Denmark

McGILLIARD, A.D., BRYANT, J.M., BRYANT, A.B., JACOBSON, N.L. and FOREMAN, C.F. (1970). *Iowa State Journal of Science*, **45**, 185–195

MATHIEU, C.M. and DE TUGNEY, H. (1965). *Annales de Biologie Animale Biochimie Biophysique*, **5**, 21–39

MEADE, R.J. (1967). *Feedstuffs*, **39**, 18–21

MILLER, B.G., JAMES, P.S., SMITH, M.V. and BOURNE, F.J. (1986). *Journal of Agricultural Science*, **107**, 579–589

MILLER, B.G., NEWBY, T.J., STOKES, C.R., HAMPSON, D. and BOURNE, F.J. (1983). *Annales de Recherches Veterinarie*, **14**, 487–492

MILLER, B.G., NEWBY, T.J., STOKES, C.R., HAMPSON, D., BROWN, P.J. and BOURNE, F.J. (1984). *American Journal of Veterinary Research*, **45**, 1730–1733

MILLER, B.G., PHILLIPS, A., NEWBY, T.J., STOKES, C.R. and BOURNE, F.J. (1985). In *Proceedings of the 3rd International Seminar on Digestive Physiology in the Pig – 1985*, pp. 65–68. Ed. Just, A., Jorgensen, H. and Fernandez, J.A. National Institute of Animal Science, Copenhagen, Denmark

MORRILL, J.L., DAYTON, A.D. and MICKELSEN, R. (1977). *Journal of Dairy Science*, **60**, 1105

MOWAT, A.M. and FERGUSON, A. (1981). *Clinical and Experimental Immunology*, **43**, 574–582

MURALIDHARA, K.S., SHEGGEBY, G.G., ELLIKER, P.R., ENGLAND, D.C. and SANDINE, W.E. (1977). *Journal of Food Protection*, **40**, 288–295

NEWBY, T.J., MILLER, B.G., STOKES, C.R., HAMPSON, D. and BOURNE, F.J. (1985). In *Recent Developments in Pig Nutrition*, pp. 211–221. Ed. Cole, D.J.A. and Haresign, W. Butterworths, London

NEWPORT, M.J. (1980). *British Journal of Nutrition*, **44**, 171–178

NEWPORT, M.J. and KEAL, H.D. (1982). *British Journal of Nutrition*, **48**, 89–95

NEWPORT, M.J. and KEAL, H.D. (1983). *Animal Production*, **37**, 395–400

NISHIMATSU, I. and KUMENO, F. (1966). *Japanese Journal of Zootechnology Science*, **37**, 25–31

NITZAN, Z., VOLCANI, R., GORDIN, S. and HASDAI, A. (1971). *Journal of Dairy Science*, **54**, 1294–1299

NITZAN, Z., VOLCANI, R., HASDAI, A. and GORDIN, S. (1972). *Journal of Dairy Science*, **55**, 811–821

OBARA, T. and WATANABE, Y. (1971). *Cereal Chemistry*, **48**, 523–527

OOMS, L. and DEGRYSE, A. (1986). *Veterinary Research Communications*, **10**, 355–397

PALMER, N.C. and HULLAND, T.J. (1965). *Canadian Veterinary Journal*, **6**, 310–316

PEDERSEN, H.E. (1986). In *Studies of Soyabean Protein Intolerance in the Preruminant Calf*, PhD Thesis, University of Reading, UK

PEDERSEN, H.E., SISSONS, J.W., TURVEY, A. and SONDERGAARD, I. (1984). *Proceedings of the Nutrition Society*, **43**, 114A

PETTIGREW, J.E., HARMON, B.G., CURTIS, S.E., CORNELIUS, S.G., NORTON, H.W. and JENSEN, A.H. (1977). *Journal of Animal Science*, **45**, 261–268

RACKIS, J.J., HONIG, D.H., SASAME, H.A. and STEGGARDCA, F.R. (1970). *Journal of Agriculture and Food Chemistry*, **18**, 977–982

RATCLIFFE, B., SMITH, M.W., MILLER, B.G. and BOURNE, F.J. (1987). *Proceedings of the Nutrition Society*, **46**, 101A

ROITT, I.M. (1974). In *Essential Immunology*. Ed. Roitt, I.M. Blackwell Scientific Publications, London

ROY, J.H.B. (1969). *Proceedings of the Nutrition Society*, **28**, 160–170

ROY, J.H.B. (1984). In *Function and Dysfunction of the Small Intestine – 1984*, pp. 95–132. Ed. Batt, R.M. and Lawrence, T.L.J. Liverpool University Press, UK

ROY, J.H.B., GASTON, H.J., SHILLAM, K.W.G., THOMPSON, S.Y., STOBO, I.J.F. and GREATOREX, J.C. (1964). *British Journal of Nutrition*, **18**, 467–502

ROY, J.H.B., STOBO, I.J.F. and GASTON, H.J. (1970). *British Journal of Nutrition*, **24**, 459–475

RUCKEBUSCH, Y. and BUENO, L. (1976). *British Journal of Nutrition*, **35**, 397–405

SEDGMAN, C.A. (1980). In *Studies of the Digestion, Absorption and Utilization of Single Cell Protein in the Preruminant Calf*. PhD Thesis, University of Reading, UK

SEEGRABER, F.J. and MORRILL, J.L. (1979). *Journal of Dairy Science*, **62**, 972–977

SEVE, B., AUMAITRE, A., MOUNIER, A.M., LAPANOUSE, A., BRUNET, P. and PAUL-URBAIN, G. (1983). *Sciences des Aliments*, **3**, 53–67

SEVE, B., AUMAITRE, A., BOUCHEZ, P., MESSANGER, A., LEBRETON, Y. and LEVREL, R. (1985). *Sciences des Aliments*, **5**, 119–126

SISSONS, J.W. (1981). *Journal of Science Food and Agriculture*, **32**, 105–114

SISSONS, J.W. (1982). *Proceedings of the Nutrition Society*, **41**, 53–61

SISSONS, J.W. (1983). *Journal of Dairy Research*, **50**, 387–395

SISSONS, J.W. and SMITH, R.H. (1976). *British Journal of Nutrition*, **36**, 421–438

SISSONS, J.W. and SMITH, R.H. (1979). *Annales de Recherche Veterinaire*, **10**, 176–178

SISSONS, J.W. and THURSTON, S.M. (1984). *Research in Veterinary Science*, **37**, 242–246

SISSONS, J.W., PEDERSEN, H.E. and WELLS, K. (1984). *Proceedings of the Nutrition Society*, **43**, 113A

SISSONS, J.W., SMITH, R.H. and HEWITT, D. (1979). *British Journal of Nutrition*, **42**, 477–485

SISSONS, J.W., NYRUP, A., KILSHAW, P.J. and SMITH, R.H. (1982a). *Journal of Science Food and Agriculture*, **33**, 706–710

SISSONS, J.W., SMITH, R.H., HEWITT, D. and NYRUP, A. (1982b). *British Journal of Nutrition,* **47**, 311–318

SMITH, H.W. and HALLS, S. (1968). *Journal of Medical Microbiology,* **1**, 45–49

SMITH, R.H. and SISSONS, J.W. (1975). *British Journal of Nutrition,* **33**, 329–349

SRIHARA, P. (1984). In *Processing to Reduce the Antigenicity of Soybean Products for Preruminant Calf Diets,* PhD Thesis, University of Guelph, Canada

STEINER, R.F. (1965). *Biochemistry, Biophysics Acta,* **100**, 111–121

STOBO, I.J.F., GANDERTON, P. and CONNORS, H. (1983). *Animal Production,* **36**, 512–513

STOBO, I.J.F. and ROY, J.H.B. (1977). *Animal Production,* **24**, 143

SUDWEEKS, E.M. and RAMSEY, H.A. (1972). *Journal of Dairy Science,* **55**, 705

TERNOUTH, J.H., ROY, J.H.B., THOMPSON, S.Y., TOOTHILL, J., GILLIES, C.M. and EDWARDS-WEBB, J.D. (1975). *British Journal of Nutrition,* **33**, 181–196

TOULLEC, R., GUILLOTEAU, P. and VILLETTE, Y. (1984). In *Physiologie et Pathologie Perinatales Chez Les Animaux de Ferme – 1984.* Ed. Jarrige, R. INRA, Paris

TOULLEC, R., MATHIEU, C.M. and PION, R. (1974). *Annales de Zootechnie,* **23**, 75–87

VAN DER POEL, A.F.B. and HUISMAN, J. (1988). *Proceedings of the 4th International Seminar on Digestive Physiology in the Pig,* p. 49. Jablonna, Poland

WALKER, W.A. (1987). In *Food Allergy – 1987,* pp. 45–66. Ed. Chandra, R.K. Nutrition Research Education Foundation, St John's, Newfoundland

WILLIAMS, V.J., ROY, J.H.B. and GILLIES, G.M. (1976). *British Journal of Nutrition,* **36**, 317–335

WISEMAN, J. and COLE, D.J.A. (1988). In *Recent Advances in Animal Nutrition – 1988,* pp. 13–37. Ed. Haresign, W. and Cole, D.J.A. Butterworths, London

WOODWARD, J.C. (1972). *Federation Proceedings,* **31**, 695

LIST OF PARTICIPANTS

The twenty-third Feed Manufacturers Conference was organized by the following committee:

Mr J. W. C. Allen (David Patton, Ltd)
Dr C. Brenninkmeijer (Hendrix Voeders, BV)
Dr L. G. Chubb (Private Consultant)
Dr B. Hardy (Dalgety Agriculture, Ltd)
Mr J. J. Holmes (Private Consultant)
Mr J. Lowe (Heygate and Sons, Ltd)
Mr F. G. Perry (BP Nutrition (UK) Ltd)
Mr J. R. Pickford (Tecracon, Ltd)
Dr H. D. Raine (J. Bibby Agriculture, Ltd)
Mr J. S. K. Round (J. Bibby Agriculture, Ltd)
Mr M. H. Stranks (MAAF, Bristol)
Dr A. J. Taylor (BOCM Silcock, Ltd)
Dr K. N. Boorman
Professor P. J. Buttery
Dr D. J. A. Cole (Chairman)
Dr P. C. Garnsworthy   } University of Nottingham
Dr W. Haresign (Secretary)
Professor G. E. Lamming
Dr J. Wiseman

The conference was held at the University of Nottingham School of Agriculture, Sutton Bonington, 4th–6th January 1989 and the committee would like to thank the authors for their valuable contributions. The following persons registered for the meeting:

Adams, Dr C.A.	Kemin Europa NV, Industriezone Wolfstee, 2410 Herentals, Belgium
Alderman, Mr G.	Dept of Agriculture, University of Reading, Earley Gate, Whiteknights, Reading RG6 2AT
Allder, Mr M.	Smith Kline Animal Health Ltd, Cavendish Road, Stevenage, Herts SG1 2EJ
Allen, Mr J.D.	Frank Wright Ltd, Blenheim House, Blenheim Road, Ashbourne, Derbys
Allen, Mr W.	David Patton Ltd, Milltown Mills, Monaghan, Ireland
Anderson, Mr K.R.	Messers Peter Hand (GB) Ltd, 15–19 Church Road, Stanmore, Middlesex HA7 4AR
Ashley, Dr J.H.	Rhone Poulenc Nutrition Animale, 20 Avenue Raymond Aron, 92165 Antony Cedex, France
Aspland, Mr F.P.	Aspland & James Ltd, 118 Bridge Street, Chatteris, Cambs

Atherton, Dr D.	119 Plumstead Road, Norwich NR1 4JT
Atkinson, Mr J.B.	Avonmore Foods Ltd, Avonmore House, Kilkenny, Republic of Ireland
Baker, Professor D.H.	University of Illinois at Urbana-Champaign, Dept of Animal Sciences, 328 Mumford Hall, 1301 West Gregory Drive, Urbana, Illinois 61801, USA
Barber, Dr G.D.	Nutrition & Microbiology Dept, The West of Scotland College for Agriculture, Horticultural and Food Studies, Auchincruive, Ayr KA6 5HW
Barnes, Mr W.J.	BP Nutrition (UK) Ltd, Wincham, Northwich, Cheshire CW9 6DF
Barrie, Mr M.	Elanco Products Limited, Dextra Court, Chapel Hill, Basingstoke, Hants
Barrigan, Mr W.	Unichema Chemicals Ltd, Bebington, Wirral, Cheshire
Bartram, Dr C.G.	Feed Flavours Group, Waterlip, Cranmore, Shepton Mallet, Somerset
Bates, Ms A.	Vitrition Ltd, Ryhall Road, Stamford, Middx PE9 1TZ
Batley, Miss S.	ICI C & P Ltd, PO Box 14, The Heath, Runcorn, Cheshire
Baxter, Mr A.	Smith Kline Animal Health Ltd, Cavendish Road, Stevenage, Herts SG1 2EJ
Beaumont, Mr D.	BP Nutrition (UK) Ltd, Wincham, Northwich, Cheshire CW9 6DF
Beer, Mr J.H.	Messers W & J Pye Ltd, Fleet Square, Lancaster, Lancs LA1 1NA
Beesley, Mr G.	Procter & Gamble, PO Box 9, Hayes Gate HO, Hayes, Middlesex VB4 0JD
Bell, Mr J.G.	Preston Farmers Ltd, Kinross, New Hall Lane, Preston PR1 5JX, Lancs
Bentley, Mr I.	A B M Brewing & Food Group, Poleacre Lane, Woodley, Stockport, Cheshire
Bettle, Mr M.	Alltech UK, 16/17 Abenbury Way, Wrexham Industrial Estate, Wrexham, Clwyd
Boorman, Dr K.N.	University of Nottingham, School of Agriculture, Sutton Bonington, Loughborough, Leics LE12 5RD
Booth, Ms A.M.	Page Feeds Ltd, Darlington Road, Northallerton, N. Yorks
Borgida, Mr L.	COFNA, 25 Rue Du Rempart, 37018 Tours Cedex, France
Bourne, Mr S.J.	Cranswick Mill Ltd, The Airfield, Driffield, N. Humberside
Bowler, Mr M.G.	S.P.A. Ltd, Avenue 3, Station Lane, Witney, Oxon OX8 6BB
Boyd, Dr J.	BOCM Silcock Ltd, Olympia Mills, Barlby Road, Selby, Yorks
Boyd, Mr P.A.	University of Nottingham, School of Agriculture, Sutton Bonington, Loughborough, Leics LE12 5RD
Brennan, Mr O.	Nutec, Greenhills Centre, Greenhills Road, Tallaght, Dublin, Ireland
Brenninkmeijer, Dr C.	Hendrix Voeders BV, Veerstraat 38, 5831 JN Boxmeer, Netherlands
Brett, Dr P.A.	G2 Boughton Hall Drive, Great Boughton, Chester, Cheshire CH3 5QQ
Brewster, Mrs A.S.M.	Messrs E.B. Bradshaw & Sons Ltd, Bell Mills, Skerne Road, Driffield, N. Humberside
Brooking, Ms P.J.	Messrs W.J. Oldacre Ltd, Cleeve Hall, Bishops Cleeve, Cheltenham, Gloucs
Brophy, Mr A.	Alltech (Ireland), Unit 28, Cookstown Industrial Estate, Tallaght, Dublin, Ireland
Brown, Mr G.J.P.	Colborn Dawes Nutrition Ltd, Heanor Gate, Heanor, Derbyshire
Browne, Dr J.W.	ICI plc, Alexander House, Crown Gate, Runcorn, Cheshire WA7 2UP

Brumby, Mr P.	WCF Limited, Burn Lane, Hexham, Northumberland
Bryan-Jones, Dr G.	United Distillers plc, Glenochil Research Station, Menstrie, Clacks FK11 7ES
Burt, Dr A.W.A.	Burt Research Ltd, 23 Stow Road, Kimbolton, Huntingdon PE18 0HU
Bush, Mr T.J.	Colborn Dawes Nutrition Ltd, Heanor Gate, Heanor, Derbyshire
Busto, Dr M.	Piensos Hens SA, Infanta Carlota 129,9,08029 Barcelona, Spain
Buttery, Professor P.J.	University of Nottingham, School of Agriculture, Sutton Bonington, Loughborough, Leics
Buysing Damste, Mr B.	Trouw International BV, Research & Development, PO Box 50,3880 AB Putten, Holland
Cain, Mr S.	Elanco Products Ltd, Dextra Court, Chapel Hill, Basingstoke, Hants
Campbell, Mr B.	Cyanamid UK, Cyanamid House, Fareham Road, Gosport, Hants PO13 0AS
Carter, Mr T.	Kemin UK Ltd, Waddington, Lincoln LN5 9NT
Casterton, Mr R.	Milltime Coop Services, 36 High Street, Chobham, Surrey GU24 8AA
Caygill, Fr J.C.	Ministry of Agriculture Fisheries & Food, Room G9, Nobel House, 17 Smith Square, London SW1P 3HX
Chalmers, Mr D.A.	WCF Ltd, Yorkshire Regional Office, York Road, Pocklington, Yorkshire YO4 2NS
Chamberlain, Dr D.	Hannah Research Institute, Ayr, Scotland KA6 5HL
Chandler, Mr N.J.	Chandler Commodities Ltd, 85 Meols Drive, West Kirby, Merseyside
Chatfield, Mr D.M.	Farmore Farmers Ltd, Farmore Mill, Craven Arms, Shropshire
Chubb, Dr L.G.	39 Station Road, Harston, Cambridge CB2 5PP
Church, Mr I.	Salsbury Laboratories, Solvay House, Flanders Road, Hedge End, Southampton, Hants
Churchman, Mr D.	Dalgety Agriculture Ltd, The Promenade, Clifton, Bristol, Avon BS8 3NJ
Clarke, Mr A.N.	Four-F Nutrition, Darlington Road, Northallerton, N. Yorks DL6 8SL
Clay, Mr J.	Alltech UK, 16/17 Abenbury Way, Wrexham Industrial Estate, Wrexham, Clwyd
Close, Dr W.	AFRC Institute for Grassland and Anim. Prod., Shinfield Research Station, Shinfield, Reading, Berks RG2 9AQ
Cole, Dr D.J.A.	University of Nottingham, School of Agriculture, Sutton Bonington, Loughborough, Leics
Cole, Mr J.	International Additives Ltd, The Flavour Centre, Old Gorsey Lane, Wallasey, Merseyside L44 4AH
Cole, Mr M.A.	Institute of Grassland and Animal Production, Maidenhead, Berks
Colenso, Mr J.	BP Nutrition (UK) Ltd, Wincham, Northwich, Cheshire CW9 6DF
Cooke, Dr B.C.	Dalgety Agriculture Ltd, Dalgety House, The Promenade, Clifton, Bristol BS8 3NJ
Corral Allegue, Miss A.M.	Nutral SA, PO Box 58, Colmenar Viejo, Madrid, Spain
Courtin, Mr B.	EMC Belgium, Square De Meeus 1,1040 Brussels, Belgium
Cox, Mr C.N.	S.C. Associates (Feeding Stuffs) Ltd, Sowerby, Thirsk, N. Yorks
Crawford, Mr J.R.	Carrs Farm Foods Ltd, Old Croft, Stanwix, Carlisle, Cumbria
Crehan, Mr M.	Nutec Ltd, Eastern Avenue, Lichfield, Staffs

Cullin, Mr. A.W.R.	Forum Feeds, Forum House, 41–52 Brighton Road, Redhill, Surrey
Dann, Mr R.	Rodd Dann Marketing, Holme Farm, Cropton, Pickering, N. Yorks
Davies, Mr J.	Insta Pro International, PO Box 61, Cheltenham GL50 1BB, Gloucs
Davies, Miss J.	Vitafoods Ltd, Riverside House, East Street, Birkenhead, Merseyside L41 1BY
De Blas, Dr J.C.	Dpto de Produccion Animal, ETS Ingenieros Agronomos, Ciudad Universitaria, 28003 Madrid, Spain
De Bruyne, Ir K.	EMC Belgium, De Meeus Square 1,1040 Brussels, Belgium
De Heus, Ir J.	Pricor BV, Postbus 51,3420 DB Oudewater, Holland
De Laporte, Mr A.	Amylum NV, Burchstraat 10,B-9300 Aalst, Belgium
De Man, Dr T.J.	Kerkstraat 40,3741 AK Baarn, The Netherlands
De Mol, Mr J.	Monsanto Europe SA, Rue Laid Burniat, 7348 Louvain-La-Neuve, Belgium
Demeersman, Mr M.	Amylum NV, Burchstraat 10,B-9300 Aalst, Belgium
Deverell, Mr P.	BASF United Kingdom Ltd, Cheadle Hulme, Cheshire
Dickins, Mr A.C.	BOCM Silcock Ltd, Basing View, Basingstoke, Hants
Dixon, Mr D.	Brown & Gilmer Ltd, Seville Place, Dublin 1, Ireland
Douglas, Mr A.	Nordos Feed Materials Ltd, Wincham, Northwich, Cheshire CW9 6DF
Drea, Mr E.	Cyanamid UK, Cyanamid House, Fareham Road, Gosport, Hants PO13 0AS
Duran, Mr R.	NANTA, Bista, Allegra, Madrid, Spain
Durrant, Mr C.A.P.	Dove Valley Poultry Ltd, Airfield Industrial Estate, Ashbourne, Derbyshire
Ebbon, Dr G.P.	BP Nutrition, Unit B, Silwood Park, Buckhurst Road, Ascot, Berks SL5 7TB
Edmunds, Dr B.	Intermol, King George Dock, Hedon Rd, Hull, HU9 5PR
Edwards, Mr A.	Elanco Products, Dextra Court, Chapel Hill, Basingstoke, Hants
Edwards, Dr S.A.	North of Scotland College of Agriculture, 581 King Street, Aberdeen AB9 1UK
Ellis, Dr N.	Macfarlan Smith Ltd, Wheatfield Road, Edinburgh EH11 2QA
Evans, Dr P.J.	Unilever Research Ltd, Colworth House, Sharnbrook, Bedfordshire
Ewing, Mr W.N.	Animax Ltd, Gilray Road, Diss, Norfolk
Fairbairn, Dr C.B.	Ministry of Agriculture, Fisheries & Food, ADAS, Government Buildings, Brooklands Avenue, Cambridge CB2 2DR
Fawthrop, Mr G.	Smith Kline Animal Health Ltd, Cavendish Road, Stevenage, Herts SG1 2EJ
Filmer, Mr D.	David Filmer Limited, Wascelyn, Brent Knoll, Somerset
Fishpool, Mr J.N.	R.M. English & Son Ltd, 86 Market Street, Pocklington, Yorks
Fitt, Dr T.J.	Colborn Dawes Nutrition, Heanor Gate, Heanor, Derbyshire
Flack, Mrs H.	BP Nutrition (UK) Ltd, Wincham, Northwich, Cheshire CW9 6DF
Fletcher, Mr C.J.	Aynsome Laboratories, Kentsford Road, Grange over Sands, Cumbria
Fletcher, Mr H.N.	Pixpalm (Oil) Ltd, Fruit Exchange (Room 50), 10–18 Victoria Street, Liverpool L22 6RB
Fontaine, Mr A.	Axis Agriculture, 36 High Street, Eccleshall, Staffs ST21 6BZ

Fordyce, Mr J.	West of England Farmers Ltd, Bradford Road, Melksham, Wilts
Fullarton, Mr P.J.	Forum Chemicals Ltd, Forum House, Brighton Road, Redhill, Surrey
Garland, Mr P.W.	Pauls Agriculture Ltd, 47 Key Street, Ipswich, Suffolk IP4 1BX
Garnsworthy, Dr P.C.	University of Nottingham, School of Agriculture, Sutton Bonington, Loughborough, Leics LE12 5RD
Geddes, Mr N.	Nutec Ltd, Eastern Avenue, Lichfield, Staffs
Geerse, Mr C.	Gist-brocades, Postbox 1,2600 MA Delft, Holland
Givson, Mr J.E.	Parnutt Foods Ltd, High Street Ind. Estate, Heckington, Sleaford, Lincs
Gill, Dr R.D.	BOCM Silcock Ltd, Basin View, Basingstoke, Hants
Gillard, Mr P.S.	Tithebarn Ltd, Weld Road, Southport PR8 2LY, Cheshire
Gilmour, Mr D.I.	ICI, The Heath, Runcorn, Cheshire
Gjefsen, Dr T.	Norske Felleskjop, Lille Grensen 7,0159 Oslo 1, Norway
Goodberry, Mr A.J.	Frank Wright Feeds International Ltd, Hampden House, Hampden Rd, Chalfont St Peter, Bucks
Gordon, Professor F.J.	Agricultural Research Institute of N. Ireland, Hillsborough, Co. Down BT26 6DR, Northern Ireland
Gould, Mrs M.	Volac Limited, Orwell, Royston, Herts
Grant, Mr T.	PVA (UK) Ltd, Cambridge House, Woolpit, Suffolk IP30 9SG
Gray, Mr W.	Boliden UK Ltd, Yorkshire House, East Parade, Leeds, Yorks
Greaves, Mr R.	The Mill Feed Co Ltd, Stow Park, Lincoln LN1 2AN
Greenhalgh, Ms M.	Park Tonks Ltd, 48 North Road, Great Abington, Cambridge CB1 6AS
Grierson, Mr R.	Park Tonks Ltd, 48 North Road, Great Abington, Cambridge CB1 6AS
Haggar, Mr C.W.	Britphos Ltd, Rawdon House, Green Lane, Yeadon, Leeds
Hall, Mr C.	USC (Industrial) Ltd, Sterling House, Heeddon Street, London W1R 8BP
Hall, Mr G.	Kemin UK Ltd, Waddington, Lincoln, LN5 9NT
Hampshire, Mr R.C.	RCH (Oils and Fats) Ltd, Hillview, Old Bridge, Drogneda, Ireland
Hanley, Mr B.	Biocon Ltd, Kilnagleary, Carrigaline, Co. Cork, Ireland
Hannagan, Mr M.J.	West Coates, 11 Durban Park Road, Clevedon, Avon BS21 7EU
Hardy, Dr B.	Dalgety Agriculture Ltd, Dalgety House, The Promenade, Clifton, Bristol BS8 3NJ
Haresign, Dr W.	University of Nottingham, School of Agriculture, Sutton Bonington, Loughborough, Leics LE12 5RD
Harker, Dr A.J.	Carrs Farm Foods Ltd, Old Croft, Stanwix, Carlisle, Cumbria
Harland, Dr J.	British Sugar plc, PO BOx 26, Oundle Road, Peterborough, Northants
Harris, Mr C.	ADAS, Block A, Government Buildings, Coley Park, Reading, Berks RG1 6DT
Harrison, Mrs J.	Messrs Peter Hand (GB) Ltd, Northern Division, Unit 32 Moss Side Industrial Estate, Leyland, Lancs PR5 3QN
Haythornthwaite, Mr A.	Willow Lodge, Church Road, Warton, Preston PR4 1BD Lancs
Hegeman, Ir F.J.M.	Borculo Whey Products, PO Box 76,7270 AA Bocrulo, Netherlands
Henderson, Mr I.R.	Chapman & Frearson Ltd, Victoria Street, Grimsby DN31 1PX
Higginbotham, Dr J.D.	Tate & Lyle Research Technology, PO Box 68, Whiteknights, Reading RE6 2BX

Hirst, Mr J.	John Hirst (Animal Feedstuffs) Ltd, Sworton Heath Farm, Swineyard Lane, High Legh, Knutsford, Cheshire WA16 0RY
Hirst, Mr M.	John Hirst (Animal Feedstuffs) Ltd, Sworton Heath Farm, Swineyard Lane, High Legh, Knutsford, Cheshire WA16 0RY
Hockey, Mr. R.	Smith Kline Animal Health Ltd, Cavendish Road, Stevenage, Herts SG1 2EJ
Hollows, Mr I.W.	Wood Farm, Coppice Lane, Coton, Nr Whitchurch, Shropshire WY13 3LT
Holman, Mr M.	Hoechst UK Ltd, Walton Manor, Walton, Milton Keynes MK7 7AJ
Hooper, Mr G.	R.J. Seaman & Sons Ltd, Bunkers Hill Mill, Walsingham, Norfolk
Horn, Dr J.P.	BOCM Silcock Ltd, Basing View, Basingstoke, Hants
Horwood, Mrs P.M.	Butterworth Scientific Ltd, Westbury House, PO Box 60, Bury Street, Guildford, Surrey GU2 5BH
Houseman, Dr R.	Britphos Ltd, Rawdon House, Green Lane, Yeadon, Leeds, Yorks
Hudson, Mr K.A.	Beecham Animal Health, Gt West Road, Brentford, Middlesex
Huggett, Miss C.D.	University of Nottingham, School of Agriculture, Sutton Bonington, Loughborough, Leics LE12 5RD
Hughes, Dr J.	Carrs Farm Foods Ltd, Old Croft, Stanwix, Carlisle, Cumbria
Hyam, Mr J.M.	Nutral SA, PO Box 58, Colmenar Viejo, Madrid, Spain
I'Anson, Mr C.J.	I'Anson Bros Ltd, The Mill, Thorpe Road, Masham, Ripon, N. Yorks HG4 4JB
Inborr, Mr J.C.	Finnfeeds International Ltd, Forum House, 41–51 Brighton Road, Redhill, Surrey RH1 6YS
Ingham, Mr P.A.	A One Feed Supplements Ltd, North Hill, By RAF Dishforth, Thirsk, North Yorkshire
Jagger, Dr S.	S.C. Associates Ltd, The Limes, Sowerby Road, Sowerby, Thirsk, N. Yorks
Jardine, Mr G.	Unitrition International Ltd, Basing View, Basingstoke, Hants
Jones, Dr E.	BOCM Silcock Ltd, Basing View, Basingstoke RG21 2EQ
Jones, Mr E.J.	Format International Ltd, Owen House, Heathside Crescent, Woking, Surrey
Jones, Mr M.G.S.	Ministry of Agriculture & Fisheries, ADAS, Woodthorne, Wolverhampton WV6 8TQ
Jones, Mr R.	National Agricultural Centre, Sheep Unit, Stoneleigh, Warwickshire
Jones, Mr R.E.	Tuckfeed Ltd, Burston, Diss, Norfolk
Kaartinen, Mr J.	Rauma Repola (UK) Ltd, Tolworth Tower, Ewell Road, Surbiton, Surrey KT6 7IQ
Kaye, Mr A.	ICI Chemicals & Polymers Ltd, PO Box 14, The Heath, Runcorn, Cheshire WA7 4QG
Kennedy, Mr. D.	International Additives Ltd, Old Gorsey Lane, Wallasey, Merseyside
Kennedy, Mr G.	BASF United Kingdom Ltd, Cheadle Holme, Cheshire
Key, Mr D.S.	United Molasses Company, 167 Regent Road, Liverpool L20 8DD
Knight, Mr R.	BP Nutrition (UK) Ltd, Wincham, Northwich, Cheshire CW9 6DF
Knox, Mr G.S.	Dalgety Agriculture Ltd, 102 Corporation Street, Belfast
Lamming, Professor G.E.	University of Nottingham, School of Agriculture, Sutton Bonington, Loughborough, Leics LE12 5RD

Lane, Mr P.F.	Parnutt Foods Ltd, High Street Industrial Estate, Heckington, Sleaford, Lincs
Law, Mr J.R.	Sheldon Jones plc, Priory Mill, West Street, Wells, Somerset
Lee, Mrs H.	J. Bibby Agriculture Ltd, Adderbury, Banbury, Oxon OX15 2LY
Lee, Dr P.A.	AFRC IGAP, Church Farm, Shinfield, Reading, Berks RG2 9AQ
Lima, Mr S.	T. Skretting A/S, Postboks 319,4001 Stavanger, Norway
Low, Dr A.G.	I.G.A.P., Shinfield, Reading, Berks RG2 9AQ
Lowe, Dr J.A.	Messrs Heygate & Sons, Bugbrooke Mills, Northampton
Lowe, Mr R.A.	Frank Wright Ltd, Blenheim House, Blenheim Road, Ashbourne, Derbyshire
Lucey, Mr P.	Ballyclough Co-op Ltd, Mallow, Co. Cork, Ireland
Lyons, Dr P.	Alltech Inc, Biotechnology Center, Nicholasville, Kentucky, USA
Macgregor, Dr R.C.	Cyanamid UK, Cyanamid House, Fareham Road, Gosport, Hampshire PO13 0AS
Mackey, Mr W.S.	G.E. McLarnon & Sons Ltd, 126 Moneynick Road, Randalstown, Antrim, Northern Ireland
Mackie, Mr I.L.	S.C.A.T.S. (Eastern Region), Robertsbridge Mill, Robertsbridge, E. Sussex
Maene, Ir E.L.J.	NV RADAR, Dorpsstraat 4,9800 Deinze, Belgium
Maidstons, Mr T.	R.J. Seaman & Sons Ltd, Bunkers Hill Mill, Walsingham, Norfolk
Makela, Mr S.	Hankkija Group, Feed Industry, PO Box 1, SF-00641 Helsinki, Sweden
Malandra, Dr F.	Siladamin spa, Sostegno di Spessa, 27010 Spessa, Pavia, Italy
Marangos, Dr A.G.	S.J.D. Humphrey Holdings Ltd, Northfield, Twyford, Winchester, Hants FO21 1NZ
Marchment, Dr S.	Pauls Agriculture Ltd, Mill Road, Radstock, Bath BA3 5TT
Marco, Mr E.	PO Box 58, Colmenar Viejo, Madrid, Spain
Marriage, Mr P.	Messrs W & H Marriage & Sons Ltd, Chelmer Mills, Chelmsford, Essex
Marsden, Dr S.	Dalgety Agriculture Ltd, Dalgety House, The Promenade, Clifton, Bristol BS8 3NJ
Martin, Dr P.A.	Hannah Research Institute, Ayr KA6 5HL
Martin, Mr W.S.D.	Curry Morrison & Co Ltd, West Bank Road, Belfast Harbour Estate, Belfast BT3 9JL
Mata, Mr G.	Division of Animal Production, CSIRO, Floreat Park, Perth, Western Australia
Mather, Mr S.	International Additives Ltd, The Flavour Centre, Old Gorsey Lane, Wallasey, Merseyside L44 4AH
Mauger, Mr F.	Rhone Poulenc Chemicals, 27a High Street, Uxbridge, Middlesex UB8 1LQ
McClure, Mr E.J.	Messrs W.J. Oldacre Ltd, Technical Division, Church Road, Bishops Cleeve, Cheltenham, Glos GL52 4RP
McCollum, Mr I.N.	BP Nutrition (NI) Ltd, 8 Governors Place, Carrickferrgus, Co. Antrim, Northern Ireland
McGrain, Mr M.	Alltech (Ireland), Unit 28, Cookstown Industrial Estate, Tallaght, Dublin 24, Ireland
McKibbin, Mr R.F.	I'Anson Bros Ltd, The Mill, Thorpe Road, Masham, Ripon, N. Yorkshire HG4 4JB

McLauglan, Mr G.	International Additives Ltd, The Flavour Centre, Old Gorsey Lane, Wallasey, Merseyside
McLean, Dr A.F.	Volac Ltd, Orwell, Royston, Herts SG8 5QX
McLean, Mr D.R.	R.J. Seaman & Sons Ltd, Bunkers Hill Mill, Walsingham, Norfolk
Mead, Mr S.J.	WCF Ltd (North West), Alanbrooke Road, Rosehill Trading Estate, Carlisle, Cumbria CA1 2UX
Merrin, Mr A.	A One Feed Supplements Ltd, North Hill, by RAF Dishforth, Thirsk, N. Yorkshire
Millard, Mr K.	RMB Animal Health Ltd, Rainham Road South Dagenham, Essex
Miller, Mr C.	Waterford Co-op, Dugarvan, Co. Waterford, Ireland
Miller, Dr E.	University of Cambridge, Dept of Applied Biology, Pembroke Street, Cambridge CB2 3DX
Mills, Mr C.	University of Nottingham, School of Agriculture, Sutton Bonington, Loughborough, Leics LE12 5RD
Moore, Mr D.R.	David Moore (Flavours) Ltd, 29 High Street, Harpenden, Herts AL5 2RU
Moran, Professor E.T.	Poultry Science Department and Alabama Agricultural Experimental Station, Auburn University, Alabama 36849-5416, USA
Morgan, Dr J.T.	Four Gables, The Fosseway, Stow on the Wold, Cheltenham, Glos GL54 1JU
Mounsey, Mr H.	The Feed Compounder, Abney House Baslow, Derbyshire DE4 1RZ
Mounsey, Mr S.	The Feed Compounder, Abney House Baslow, Derbyshire DE4 1RZ
Mullan, Dr B.	AFRC Institute for Grassland and Animal Production, Shinfield Research Station, Shinfield, Reading, Berks RG2 9AQ
Murphy, Mr J.	Grindsted Products Ltd, Northern Way, Bury St Edmunds, Suffolk
Murray, Mr A.G.	WCF Limited, Burn Lane, Hexham, Northumberland
Naylor, Mr P.	International Additives Ltd, Flavour Centre, Old Gorsey Lane, Wallasey, Merseyside L44 4AH
Newcombe, Mrs J.	University of Nottingham, School of Agriculture, Sutton Bonington, Loughborough, Leics LE12 5RD
Niskanen, Ms T.	Hankkija Group, Feed Industry, PO Box 1, SF-00641 Helsinki, Sweden
Noordenbos, Ir H.K.	Orfam BV, Peppelkade 46,3992 AK Houten, Holland
O'Toole, Mr C.	Park Tonks Ltd, 48 North Road, Great Abington, Cambridge CB1 6AS
Openshaw, Mr P.J.	Torberry Farm, Hust, Petersfield, Hampshire GU31 5RG
Palmer, Mr F.G.	ABR Foods Limited, Hunters Road, Corby NN17 1JR
Papasolomontos, Dr S.A.	Dalgety Agriculture Ltd, Dalgety House, The Promenade, Clifton, Bristol BS8 3NJ
Pass, Mr R.T.	Pentlands Scotch Whisky Research Ltd, 84 Slateford Road, Edinburgh EN11 1QU
Patience, Dr J.F.	Prairie Swine Centre, University of Saskatchewan, Saskatoon, Canada S7N OWO
Paton, Dr F.J.	PA Technology, Melbourn, Royston, Herts
Pattinson, Mr F.	Rhone Merieux, c/o RMB Animal Health, Rainham Road South, Dagenham, Essex
Pearson, Mr A.	Hoechst UK Ltd, Walton Manor, Walton, Milton Keynes MK7 7AJ, Bucks

Perry, Mr F.	BP Nutrition (UK) Ltd, Wincham, Northwich, Cheshire CW9 6DF
Peters, Dr A.R.	Hoechst Animal Health, Walton Manor, Milton Keynes, Bucks
Peters, Miss S.	Beaufort Nutrition Pty Ltd, Rosencath Stud, Cobbitty NSW 2570, Australia
Pettigrew, Mr R.	Distillers Co (Cereals) Ltd, Distillers House, 33 Ellersley Road, Edinburgh EG12 6JW
Pfeiffer, Dr K.	F. Hoffman-La Roche & Co AG, 4002 Basle, Switzerland
Phillips, Mr G.	Greenway Farm, Greenway Lane, Charlton Kings, Cheltenham, Gloucs
Pickford, Mr J.R.	Tecracon Ltd, Bocking Hall, Bocking Church Street, Braintree, Essex CM7 5JY
Pickles, Dr R.W.	Elanco Products Ltd, Dextra Court, Chapel Hill, Basingstoke, Hants RG21 2SY
Pike, Dr I.H.	Int. Assoc. Fish Meal Manufacturers, Hoval House, Mutton Lane, Potters Bar, Herts EN6 3AR
Piva, Prof G.	Facolta Agraria UCSC, 29100 Piacenza, Italy
Plowman, Mr G.B.	G.W. Plowman & Son Ltd, Selby House, High Street, Spalding, Lincs PE11 1TW
Poornan, Mr P.K.	Format International Ltd, Owen House, Woking, Surrey GU22 7AG
Portas, Mr G.	Vitas (Micronutrients) Ltd, 17 Central Buildings, Thirsk, North Yorks YO7 1HD
Portsmouth, Mr J.I.	Messrs Peter Hand (GB) Ltd, 15–19 Church Road, Stanmore, Middlesex HA7 4AR
Priest, Mr L.	19 Malvern Close, High Crompton, Shaw, Lancs OL2 7RE
Pringle, Mr R.G.	United Molasses Company, Royal Portbury Dock, Portbury, Bristol BS20 9XW
Putnam, Mr M.E.	Roche Products Ltd, PO Box 8, Welwyn Garden City, Herts AL7 3AY
Quinlivan, Mr T.	Department of Agriculture and Food, Agriculture House, Kildare Street, Dublin 2, Ireland
Rabbetts, Miss S.	J. Bibby Agriculture Ltd, Adderbury, Banbury, Oxon OX15 2LY
Raine, Dr H.D.	J. Bibby Agriculture Ltd, Oxford Road, Adderbury, Banbury, Oxon OX15 2LY
Raper, Mr G.J.	Laboratories Pancosma (UK) Ltd, Anglia Industrial Estate, Saddlebow Road, Kings Lynn, Norfolk PE30 5BN
Rattenbury, Mr D.	Clark & Butcher Ltd, Lion Mills, Soham, Ely, Cambs CB7 5HY
Read, Mr M.	Smith Kline Animal Health Ltd, Cavendish Road, Stevenage, Herts SG1 2EJ
Read, Mr S.	Pauls Agriculture Ltd, 47 Key Street, Ipswich, Suffolk IP4 1BX
Reeve, Dr A.	ICI Fertilizers, Jealotts Hill Research Station, Bracknell, Berks RG12 6EY
Reeve, Mr J.G.	R.S. Feed Blocks Ltd, Orleigh Mill, Bideford, Devon
Richardson, Mr J.S.	BOCM Silcock Ltd, Basing View, Basingstoke, Hants
Rigg, Mr G.	Elanco Products Ltd, Dextra Court, Chapel Hill, Basingstoke, Hants
Roach, Dr D.	PA Technology, Melbourn, Royson, Herts
Roberts, Mr J.C.	Harper Adams Agricultural College, Edgmond, Newport, Shropshire
Robertson, Mr S.	Hannah Research Institute, Ayr KA6 5HL

Robinson, Mrs G.	University of Nottingham, School of Agriculture, Sutton Bonington, Loughborough, Leics LE12 5RD
Rosen, Dr G.D.	66 Bathgate Road, London SW19 5PH
Round, Mr J.S.K.	J. Bibby Agriculture Ltd, Adderbury, Banbury, Oxon OX15 2LY
Ruddle, Mr M.G.	Soyol Trading Ltd, Southbank, Drogneda, Co. Louth, Ireland
Rusticus, Mr D.	C.V. "Sloten", PO Box 474,7400 Al Deventer, Holland
Rutherford, Mr B.	c/o UKASTA, 3 Whitehall Court, London SW1A 2EQ
Ryan, Mr M.	Biocon Limited, Kilnagleary, Carrigaline, Co. Cork, Ireland
Ryan, Mr T.	RMB Animal Health Ltd, Dagenham, Essex RM10 7XS
Santoma, Dr G.	Cyanamid Iberica, SA, Apartado – 471,28080 Madrid, Spain
Schofield, Mr D.	Hoechst UK Ltd, Walton Manor, Walton, Milton Keynes MK7 7AJ
Schofield, Dr S.A.	Volac Limited, Orwell, Royston, Herts SG8 5QX
Shales, Miss N.J.	University of Nottingham, School of Agriculture, Sutton Bonington, Loughborough, Leics LE12 5RD
Sharp, Mr D.J.V.	Frank Wright Ltd, Blenheim House, Blenheim Road, Ashbourne, Derbyshire
Shepperson, Dr N.P.G.	UFAC (UK) Ltd, Waterwitch House, Exeter Road, Newmarket, Suffolk CB8 8LR
Shipston, Mr A.H.	Dalgety Agriculture Ltd, Dalgety House, The Promenade, Clifton, Bristol BS8 3NJ, Avon
Shipton, Mr P.	Messrs Dardis & Dunns Coarse Feeds Ltd, Ashbourne, County Meath, Southern Ireland
Shrimpton, Dr D.H.	Milling, Flour and Feed, 177 Hagden Lane, Watford, Herts WD1 8LN
Silcock, Mr R.	Whitworth Bros Ltd, Agricultural Division, Fletton Mills, Peterborough, Cambs PE2 8AD
Sinclair, Mr L.L.G.	University of Nottingham, School of Agriculture, Sutton Bonington, Loughborough, Leics, LE12 5RD
Singer, Dr M.	Roche Products Ltd, PO Box 8, Welwyn Garden City, Herts AL7 3AY
Singleton, Miss A.M.	Nutrition Trading, 14 Muntz Crescent, Hockley Heath, Solihull, W. Midlands
Sissins, Mr J.	J. Bibby Agriculture Ltd, Adderbury, Banbury, Oxon OX15 2LY
Sissons, Dr J.W.	AFRC Institute for Grassland & Animal Prod., Pig Department, Shinfield Research Station, Reading, Berkshire RG2 9AQ
Sloan, Mr J.	Colborn Dawes Nutrition Ltd, Musgrave Park Industrial Estate, Stockmans Way, Belfast, Northern Ireland
Smith, Mr D.	Biocon (UK) Ltd, Eardiston, Tenbury Wells, Worcester
Smith, Mr G.P.	Salsbury Laboratories, Solvay House, Flanders Road, Hedge End, Southampton, Hants
Smith, Dr H.	Cyanamid UK Ltd, Cyanamid House, Fareham Road, Gosport, Hants
Smith, Mr M.P.	J. Bibby Agriculture Ltd, Adderbury, Banbury, Oxon, OX15 2LY
Speight, Mr D.	Consultant, Newlands, Rainton, Thirsk, N. Yorks YO7 3PX
Spencer, Mr P.G.	Bernard Matthews plc, Great Witchingham Hall, Norwich NE9 5QD
Stainsby, Mr A.K.	BATA Ltd, Railway Steet, Malton, N. Yorks YO17 0NU
Stark, Dr B.	Baydella, Bassetsbury Lane, High Wycome, Bucks
Statham, Mr R.	J. Pyke & Sons, Harvey Lane, Golborne, Nr Warrington, Cheshire

Stevens, Mr C.	Volac Limited, Orwell, Royston, Herts SG8 5QX
Stranks, Mr M.H.	ADAS, Block III, Government Buildings, Burghill Road, Westbury on Trym, Bristol, Avon
Sumner, Mr R.	North Eastern Farmers Ltd, Bannermill, Aberdeen
Talbot, Mrs S.	W. & H.Marriage & Sons Ltd, Chelmer Mills, Chelmsford, Essex
Taylor, Dr A.J.	BOCM Silcock Ltd, Basing View, Basingstoke, Hants
Taylor, Dr S.J.	Volac Limited, Orwell, Royston, Herts
Thomas, Professor P.C.	West of Scotland College, Auchincruive, Ayr KA6 5HW
Thomas, Dr S.	North of Scotland College of Agriculture, 581 King Street, Aberdeen AB9 1UD
Thompson, Mr D.	Right Feeds Ltd, Castlecarde, Cappamore, Co. Limerick, Ireland
Thompson, Dr F.	Rumenco, Stretton House, Derby Road, Stretton, Burton on Trent, Staffs
Thompson, Mr J.	Feed Flavours Group, Waterlip, Cranmore, Shepton Mallet, Somerset
Thompson, Mr R.J.	Preston Farmers Ltd, Kinross, New Hall Lane, Preston PR1 5JX, Lancs
Threlfall, Mrs D.	Vitafoods Ltd, Riverside House, East Street, Birkenhead, Merseyside L41 1BY
Tonks, Mr W.P.	Park Tonks Limited, 48 North Road, Great Abington, Cambridge CB1 6AQ
Toplis, Mr P.	Four F Nutrition, Darlington Road, Northallerton, N. Yorks
Twigge, Mr J.	BP Nutrition (UK) Ltd, Wincham, Northwich, Cheshire CW9 6DF
Underwood, Mr N.P.	Cyanamid UK, Cyanamid House, Fareham Road, Gosport, Hampshire PO13 0AS
Van den Broecke, Ir J.	Eurolysine, 16 Rue Ballu, 75009 Paris, France
Van der Elst, Mr P.	EMC Belgium, Square Du Meeus 1,1040 Brussels, Belgium
Van der Ploeg, Ir H.	Stationsweg 4, Trade Magazin "de Molenaar", Stationsweg 4, 3603 EE Maarssen, Holland
Vernon, Dr B.	Dalgety Agriculture Ltd, Dalgety House, The Promenade, Clifton, Bristol BS8 3NJ
Virkki, Mr M.	Forum House, 41–51 Brighton Road, Redhill, Surrey RH1 6YS
Waddell, Mrs S.	9 Woodnut Road, Wrecclesham, Farnham, Surrey GU10 4QF
Wakelam, Mr J.A.	George A. Palmer Ltd, Oxney Road, Peterborough, PE1 542
Wallace, Mr J.R.	Nutrition Trading, 14 Muntz Crescent, Hockley Heath, Solihull, W. Midlands
Walters, Mr C.I.H.	Messrs Peter Hand (GB) Ltd, 15–19 Church Road, Stanmore, Middlesex HA7 4AR
Ward, Mr J.H.	Favor Parker Ltd, The Hall, Stoke Ferry, Kings Lynn, Norfolk PE33 9SE
Ward, Mr T.	J.N. Miller Ltd, Old Steam Mill, Corn Hill, Wolverhampton, W. Midlands WV10 0DD
Waterworth, Mr D.G.	ICI Biological Products, PO Box 1, Billingham, Cleveland
Waterworth, Mrs K.	BP Nutrition (UK) Ltd, Wincham, Northwich, Cheshire CW9 6DF
Webster, Mrs M.	Format International Ltd, Owen House, Heathside Crescent, Woking Surrey GU22 7AG

Weeks, Mr R.	Pauls Agriculture Ltd, 141 Brierley Road, Walton Summit, Preston PR5 8AH
Welsh, Mr R.	Hoechst UK Ltd, Walton Manor, Walton, Milton Keynes, Bucks
Weston, Miss G.L.	University of Nottingham, School of Agriculture, Sutton Bonington, Loughborough, Leics LE12 5RD
Whitehead, Dr C.	AFRC Institute for Grassland & Animal Production, Roslin, Midlothian EH25 9PS
Whiteoak, Mr R.A.	West Midland Farmers Association Ltd, Llanthony Mill, Merchants Road, Gloucester GL1 5RJ
Wigger, Ing J.J.	NAFAG, Nahr- u. Futtermittel AG,9292 Gossau SG, Switzerland
Wilby, Mr D.T.	Tuckfeed Ltd, The Mills, Burston, Diss, Norfolk
Wilding, Mr C.N.	Vitrition Ltd, Ryhall Road, Stamford, Middlesex PE9 1TZ
Wilkinson, Dr J. M.	13 Highwoods Drive, Marlow Bottom, Marlow, Bucks SL7 3PU
Williams, Mr C.	ABM Brewing & Food Group, Poleacre Lane, Woodley, Stockport, Cheshire
Williams, Dr D.R.	BOCM Silcock Ltd, PO Box 4, Barlby Road, Selby, Yorks YO8 7DT
Wilson, Dr J.	Precision Liquids Ltd, 10 Donegal Square South, Belfast BT1 5JN, Northern Ireland
Wilson, Mr S.	Pauls Agriculture Ltd, 47 Key Street, Ipswich, Suffolk IP4 1BX
Wilson, Dr S.	Prosper De Mulder, Ings Road, Doncaster, S. Yorks
Wiseman, Dr J.	University of Nottingham, School of Agriculture, Sutton Bonington, Loughborough, Leics LE12 5RD
Woodford, Mr R.	Feed Flavours Group, Waterlip, Cranmore, Shepton Mallet, Somerset
Woodgate, Mr S.	Prosper de Mulder, Ings Road, Doncaster, S. Yorks
Woodward, Mr P.	Sun Valley Poultry Ltd, Feed Mill Division, Tram Inn, Allensmore, Hereford
Woolford, Dr M.K.	Agil Ltd, Fishponds Road, Wokingham RG11 2QL, Berks
Wright, Mr I.D.	Microbial Developments Ltd, Spring Lane North, Malvern Link, Worcestershire
Wright, Mrs M.G.	ABM Brewing & Food, Poleacre Lane, Woodley, Stockport, Cheshire SK6 1PQ
Wyatt, Mr D.H.	John Wyatt Ltd, Braithwaite Street, Holbeck Lane, Leeds LS11 9XE, W. Yorks
Youdan, Dr J.	Nutrimix, Boundary Industrial Estate, Boundary Road, Lytham, Lancs FY8 5HU
Young, Dr L.	University of Guelph, Department of Animal Science, Guelph, Canada

INDEX

Acanthosis, 54
Aflatoxins, 5
Amino acid
  bioavailability, 250
  excesses, 257
  limitations, assessment of, 256
  nutrition, pigs and poultry, 245
  requirements, 249
    body composition factors affecting, 249
    environmental temperature, 253
    food composition factors and food intake
      factors, 250
    frequency of feeding, 250
Amylopectin, 88, 90
Angora rabbits, 109
Antidiuretic hormone, 214
Antinutritional factors, 263
  dietary antigens, 266
  enzyme inhibitors, 264
  lectins, 266
Arginine, 130, 199
Ataxia, 37

Betaine, 62
Bicarbonate, 211
Biotin
  bioavailability, 65
  deficiency, 54, 61
    fatty liver and kidney syndrome (FLKS), 55
    hypoglycaemia, 55
    parrot beak, 54
  function, 54, 61
  requirements for poultry, 55
Bovine somatotrophin, 10, 15, 23

Caecotrophy, 119
Calcium, 211
  requirements for lambs, 204
Calves, pre-ruminant, diarrhoea, 261
Chloride, 217, 222
Cholecalciferol, 38

Choline, 70
  bioavailability, 65
  deficiency, 58, 61
    fatty livers, 58
  functions, 58, 61
  relationship with methionine, 59
  requirements for poultry, 59
Cooperatives, 30
Cobalamin
  bioavailability, 65
  deficiency, 56
  functions, 56, 61
  requirements for poultry, 56
Coccidiosis, 204
Copper toxicity, lambs, 202
Cysteine, 251
Cystine, 129

Dermatitis
  nicotinic acid deficiency, 48
  riboflavin deficiency, 47
Diarrhoea
  aetiology of, in pigs and pre-ruminant calves,
    261
  dietary approaches to control, 272
  effect on performance, 272
  electrolyte disturbance, 217
  pathogenesis, 261
  post-weaning, 272
  probiotics, 276
  nicotinic acid deficiency, 48
  riboflavin deficiency, 47
Dietary electrolyte balance, 221
Dietary undetermined anion, 221
Drugs
  anabolic steroids, 22
  analysis, 19
  anthelmintics, 13, 14
  antimicrobial, 13, 14
  antiprotozals, 13, 14
  coccidiostats, 13, 14

Drugs (*cont.*)
  ectoparasiticides, 13, 14
  hormones, 13, 14
  licensing, 15
  residues, 13, 18, 22
  safety, 13

Early lambing, 195
Egg production
  pyroxidine deficiency, 53
  vitamin D
    deficiency, 39
    requirements for poultry, 39
Egg shell quality
  vitamin D requirements, 39
Electrolyte balance
  tibial dischondroplasia, 224
Electrolyte disturbance, 217
  diagnosis of, 220
  diarrhoea, 217
  starving and refeeding, 219
  stress, 219, 224
  vomiting, 217
Electrolytes, 211
Escherichia coli, 262
European Economic Community, 3
  legislation, 27
  trade barriers, 27
European Monetary System, 27

Fatty liver and kidney syndrome (FLKS), 55, 78
Feed additives, 9
Feed manufacturing
  declaration of ingredients, 3
  legislation, 3
    health and safety, 10
  marketing, 3
  registration, 8
Feed
  aflatoxin levels, 6
  drug levels, 8
  medicated, 6
  pesticide levels, 6
  storage life, 4
Fibre
  digestion, 177
Folic acid
  bioavailability, 65
  deficiency, 57, 61
  function, 57, 61
  requirements for poultry, 57

GAFTA, 31

Hatchability
  pyroxidine deficiency, 53
  vitamin D deficiency, 39
  vitamin K deficiency, 44
  vitamin supplementation, 73
Histidine, 199, 256
Hyperkeratosis, 37, 54
Hypoglycaemia, 55

Ionic composition of body fluids, 212

Lactating sow
  factorial estimation of nutrient requirements,
    230
  nutrient partition, 235
  prediction of
    body composition, 237–239
    nutrient responses, 229
  reproduction, 239
  voluntary food intake, 235, 236
Lambs
  carcass composition, 200
  energy nutrition, 196
  energy requirements, 196
  intensive rearing, 195, 207
  mineral disorders, 202
  protein requirements, 196
Litter size, 236
Lysine, 5, 129, 199, 245, 246, 249
Lysine-arginine antagonism, 223

Magnesium 211
  requirements for lambs, 204
Menaquinone, 43
Metabolizable energy, 5
  supply and utilization from silage-based diets,
    187
  for milk production, 187
Methionine, 5, 129, 199, 245, 249, 251
Mineral disorders, lambs, 202
Molybdenum, 202
Monetary compensatory amounts, 31

Niacin, *see* Nicotinic acid
Nicotinic acid, 48, 61
  bioavailability, 64
  black tongue, 48
  dementia, 48
  dermatitis, 48
  diarrhoea, 48
  requirements for poultry, 49

Osteomalacia, 39

Pantothenic acid
  bioavailability, 64
  deficiency, 51, 61
  function, 50, 61
  requirements for poultry, 51
Pellet binders, 87, 95–101, 105
  betonite, 97
  carboxymethylcellulose, 97
  fat, 98–100
  fibre, 96
  gelatin, 98
  hemicellulose, 96
  lignosulphonates, 95
  starch, 98, 99
Pellets
  caramelization, 94

Pellets (*cont.*)
  conditioning, 88
  extrusion, 87
  quality, 87
  stability, 87, 88, 105
    bird performance, 101–104
    fibre, 91
    protein, 90
    starch, 88–89, 91
Pharmacokinetic principles, 14
Phosphate, 211
Phosphorus
  requirements for lambs, 204
Phylloquinone, 43
Piglet, suckling
  nutritional requirements, 232
Pigs
  amino acid nutrition, 245
  diarrhoea, 261
Potassium, 211, 215
Poultry
  amino acid nutrition, 245
Pyroxidone
  bioavailability, 64
  deficiency, 52, 61
    egg production, 53
    hatchability, 53
  function, 52, 61

Rabbits, 109
  amino acid requirements, 124–126, 130
  caecotrophy, 119
  digestion
    fat, 121
    fibre, 115–118
    protein, 118–120
    starch, 120
  digestive physiology, 113–115
  energy requirements, 124–126
  energy utilization, 121–124
  fibre requirements, 127
  meat production, 109–113, 132
  milk production, 123, 125, 128
  nutritive value of feedstuffs, 131
  protein requirements, 124–126
Retinol, 36
Riboflavin
  bioavailability, 64
  deficiency, 47, 61
    clubbed down, 47
    dermatitis, 47
    diarrhoea, 47
  functions, 47, 61
  requirements for poultry, 48
  supplementation, 75
Rickets, 39

Salmonella, 262
Silage additives, 159
Silage
  amino acid supplementation, 181

Silage (*cont.*)
  amino acid supply, 180
  compound feeds, to complement, 175
    voluntary intake, 176
    carbohydrate, 176
    crude fibre, 144
    lipid, 177
    starch, 176
  fermentation, 159
    clostridia, 159
    coliforms, 159
  fermentation inhibitors, 160
    formaldehyde, 160–164
    formic acid, 160–164
    sulphuric acid, 162–164
  fermentation stimulants, 165
    enzymes, 167
    inoculants, 166
  harvesting, 170
  intake and milk energy output, 167
  metabolizable energy, 142, 145, 147
    prediction of efficiency of utilization, 149
  lipid supplementation, 181, 190
  net energy, 146, 149
  prediction of
    energy, 143, 151
    intake, 143
    nutritive value, 141
    organic matter in the dry matter, 150
  protein supplementation, 181
  sodium bicarbonate supplementation, 181
  sugar supplementation, 181
  wilting, 159, 167
    growing cattle, 167
    lactating cattle, 169
Sodium, 211, 215
Sulphonamides, 21, 205
Sulphur, 202

Thiamin, 70
  bioavailability, 64
  deficiency, 46, 60
  functions, 45, 60
  requirement, 46
Threonine, 249
Tibial dischondroplasia
  electrolyte balance, 224
Trenbolone, 20, 22
Tryptophan, 249
Turkeys
  vitamin supplementation, 78

Undegradable crude protein, 181, 199
Urolithiasis, 203

Value added tax, 28, 31
Vitamin
  bioavailability, 63
  content in feeds, 63, 66–69

Vitamin (*cont.*)
  supplementation, 70, 76–81
    free range production, 81
    interactions, 73
    reproduction, 73
    stress, 81
    turkeys, 78
  requirements for poultry, 35
Vitamin A, 70
  bioavailability, 63, 71–72
  deficiency, 36, 60
  function, 36, 60
  requirements for poultry, 37
    factors affecting, 37
  storage, 36
  supplementation, 74
  toxicity, 38
Vitamin B, rabbits, 120
Vitamin B$_1$, *see* Thiamin
Vitamin B$_2$, *see* Riboflavin
Vitamin B$_6$, *see* Pyroxidine
Vitamin B$_{12}$, *see* Cobalamin, 56
Vitamin C
  supplementation, 75

Vitamin D, 82
  deficiency, 39, 60
  functions, 38, 60
  requirements for poultry, 39
  supplementation, 74
Vitamin E
  bioavailability, 64
  deficiency, 42, 60
  functions, 41, 60
  requirements for poultry, 41
  supplementation, 75
Vitamin K
  anticoagulants, 45
  deficiency, 44, 60
    hyperplasia, 44
  functions, 43, 60
  rabbits, 120
  requirements, 44
    chickens, 62
    turkeys, 63

Zeranol, 17, 20, 22